CAMBRIDGE TRACTS IN MATHEMATICS

General Editors

229 Large Deviations for Markov Chains

CAMBRIDGE TRACTS IN MATHEMATICS

A complete list of books in the series can be found at www.cambridge.org/mathematics.
Recent titles include the following:

193. Distribution Modulo One and Diophantine Approximation. By Y. Bugeaud
194. Mathematics of Two-Dimensional Turbulence. By S. Kuksin and A. Shirikyan
195. A Universal Construction for Groups Acting Freely on Real Trees. By I. Chiswell and T. Müller
196. The Theory of Hardy's Z-Function. By A. Ivić
197. Induced Representations of Locally Compact Groups. By E. Kaniuth and K. F. Taylor
198. Topics in Critical Point Theory. By K. Perera and M. Schechter
199. Combinatorics of Minuscule Representations. By R. M. Green
200. Singularities of the Minimal Model Program. By J. Kollár
201. Coherence in Three-Dimensional Category Theory. By N. Gurski
202. Canonical Ramsey Theory on Polish Spaces. By V. Kanovei, M. Sabok, and J. Zapletal
203. A Primer on the Dirichlet Space. By O. El-Fallah, K. Kellay, J. Mashreghi, and T. Ransford
204. Group Cohomology and Algebraic Cycles. By B. Totaro
205. Ridge Functions. By A. Pinkus
206. Probability on Real Lie Algebras. By U. Franz and N. Privault
207. Auxiliary Polynomials in Number Theory. By D. Masser
208. Representations of Elementary Abelian p-Groups and Vector Bundles. By D. J. Benson
209. Non-homogeneous Random Walks. By M. Menshikov, S. Popov, and A. Wade
210. Fourier Integrals in Classical Analysis (Second Edition). By C. D. Sogge
211. Eigenvalues, Multiplicities and Graphs. By C. R. Johnson and C. M. Saiago
212. Applications of Diophantine Approximation to Integral Points and Transcendence. By P. Corvaja and U. Zannier
214. The Mathieu Groups. By A. A. Ivanov
215. Slenderness I: Foundations. By R. Dimitric
216. Justification Logic. By S. Artemov and M. Fitting
217. Defocusing Nonlinear Schrödinger Equations. By B. Dodson
218. The Random Matrix Theory of the Classical Compact Groups. By E. S. Meckes
219. Operator Analysis. By J. Agler, J. E. McCarthy, and N. J. Young
220. Lectures on Contact 3-Manifolds, Holomorphic Curves and Intersection Theory. By C. Wendl
221. Matrix Positivity. By C. R. Johnson, R. L. Smith, and M. J. Tsatsomeros
222. Assouad Dimension and Fractal Geometry. By J. M. Fraser
223. Coarse Geometry of Topological Groups. By C. Rosendal
224. Attractors of Hamiltonian Nonlinear Partial Differential Equations. By A. Komech and E. Kopylova
225. Noncommutative Function-Theoretic Operator Function and Applications. By J. A. Ball and V. Bolotnikov
226. The Mordell Conjecture. By A. Moriwaki, H. Ikoma, and S. Kawaguchi
227. Transcendence and Linear Relations of 1-Periods. By A. Huber and G. Wüstholz
228. Point-Counting and the Zilber–Pink Conjecture. By J. Pila

Large Deviations for Markov Chains

ALEJANDRO D. DE ACOSTA
Case Western Reserve University

CAMBRIDGE
UNIVERSITY PRESS

CAMBRIDGE
UNIVERSITY PRESS

University Printing House, Cambridge CB2 8BS, United Kingdom

One Liberty Plaza, 20th Floor, New York, NY 10006, USA

477 Williamstown Road, Port Melbourne, VIC 3207, Australia

314–321, 3rd Floor, Plot 3, Splendor Forum, Jasola District Centre, New Delhi – 110025, India

103 Penang Road, #05–06/07, Visioncrest Commercial, Singapore 238467

Cambridge University Press is part of the University of Cambridge.

It furthers the University's mission by disseminating knowledge in the pursuit of education, learning, and research at the highest international levels of excellence.

www.cambridge.org
Information on this title: www.cambridge.org/9781316511893
DOI: 10.1017/9781009053129

First published 2022

A catalogue record for this publication is available from the British Library.

Library of Congress Cataloging-in-Publication Data
Names: Acosta, Alejandro D. de, 1941– author.
Title: Large deviations for Markov chains / Alejandro D. de Acosta, Case Western Reserve University.
Description: Cambridge, United Kingdom ; New York, NY : Cambridge University Press, 2022. | Includes bibliographical references and index.
Identifiers: LCCN 2021062728 | ISBN 9781316511893 (hardback) | ISBN 9781009053129 (ebook)
Subjects: LCSH: Large deviations. | Markov processes. | BISAC: MATHEMATICS / Probability & Statistics / General
Classification: LCC QA273.67 .A265 2022 | DDC 519.2/33–dc23/eng20220521
LC record available at https://lccn.loc.gov/2021062728

ISBN 978-1-316-51189-3 Hardback

To Martha, Alejandro, and Diego

Contents

Preface

The purpose of this book is to study the large deviations for empirical measures and vector-valued additive functionals of Markov chains with general state space.

Under suitable recurrence conditions, the ergodic theorem for additive functionals of a Markov chain asserts the almost sure convergence of the averages of a real or vector-valued function of the chain to the mean of the function with respect to the invariant measure. In the case of empirical measures, the ergodic theorem states the almost sure convergence in a suitable sense to the invariant measure. The large deviation theorems provide precise asymptotic estimates at logarithmic level of the probabilities of deviating from the preponderant behavior asserted by the ergodic theorems.

In the ergodic theorems, the state space S of the Markov chain is a measurable space, and no topology on S is involved. This setup appears to be the natural one, at least initially, for the study of large deviations, and it is the setup we adopt. In the case of empirical measures, the space of probability measures on S is endowed with the topology induced by a vector subspace of the space of bounded measurable functions, subject to certain restrictions.

Both the perspective and many results in the book have not previously appeared in the literature.

The prerequisites for the study of the book are: standard graduate-level measure and integration theory; basic graduate-level probability theory, including light exposure to large deviations; basic notions of general topology and functional analysis. Some familiarity with general Markov chains is desirable, but we have included several appendices which essentially cover the relevant aspects of the subject.

Acknowledgement I thank Xia Chen, Peter Ney and Sandy Zabell for stimulating exchanges on various aspects of the book.

Alejandro D. de Acosta
Shaker Heights, Ohio
October 2021

1

Introduction

We will present the outlook and some of the main contents of the book.

Let P be a Markov kernel and ν a probability measure on a measurable space $(S, \mathcal{S})$, with $\mathcal{S}$ countably generated. Our basic framework will be the canonical Markov chain with transition kernel P and initial distribution ν, given by $(\Omega = S^{\mathbb{N}_0}, \mathcal{S}^{\mathbb{N}_0}, \mathbb{P}_\nu, \{X_j\}_{j\geq 0})$, where $\mathbb{N}_0$ is the set of nonnegative integers, $\{X_j\}_{j\geq 0}$ are the coordinate functions on $S^{\mathbb{N}_0}$, and $\mathbb{P}_\nu$ is the unique probability measure on $(S^{\mathbb{N}_0}, \mathcal{S}^{\mathbb{N}_0})$ such that $\{X_j\}_{j\geq 0}$ is a Markov chain with transition kernel P and initial distribution ν.

Most of this introduction will be devoted to the empirical measure case. Let $\mathcal{P}(S)$ be the space of probability measures on $(S, \mathcal{S})$, and let $B(S)$ be the space of real-valued bounded $\mathcal{S}$-measurable functions on S. For a set $\mathcal{F} \subset B(S)$, the $\sigma(\mathcal{P}(S), \mathcal{F})$ topology on $\mathcal{P}(S)$, or simply the $\mathcal{F}$ topology, is the topology induced on $\mathcal{P}(S)$ by $\mathcal{F}$, that is, by the action $\mu \mapsto \int g\, d\mu$, $g \in \mathcal{F}$: a net $\{\mu_\alpha\} \subset \mathcal{P}(S)$ converges in this topology to $\mu \in \mathcal{P}(S)$ if and only if for all $f \in \mathcal{F}$,

$$\lim_\alpha \int f\, d\mu_\alpha = \int f\, d\mu.$$

Let $L_n \colon \Omega \to \mathcal{P}(S)$, $n \geq 1$, be the nth empirical measure associated with $\{X_j\}$:

$$L_n = n^{-1} \sum_{j=0}^{n-1} \delta_{X_j},$$

where for $x \in S$, δ_x is point mass at x.

Proposition 1.1 *Assume that P is positive Harris recurrent (Section B.3) and let π be its unique invariant probability measure.*

1. *Assume that $\mathcal{F} \subset B(S)$ is separable for the uniform norm on $B(S)$. Then for any $\nu \in \mathcal{P}(S)$, $\{L_n\}$ converges $\sigma(\mathcal{P}(S), \mathcal{F})$ to π, $\mathbb{P}_\nu$-a.s..*

2. *Let $(S, \mathcal{S})$ be a separable metric space with its Borel σ-algebra, and let $C_b(S)$ be the space of real-valued bounded continuous functions on S. Then for any $\nu \in \mathcal{P}(S)$, $\{L_n\}$ converges $\sigma(\mathcal{P}(S), C_b(S))$ to π, $\mathbb{P}_\nu$-a.s..*

This result is proved in Appendix A.

One of the main objectives of the present work is to study the large deviations associated with this result; in particular, to determine when the probabilities $\{\mathbb{P}_\nu[L_n \notin B]\}$ decay exponentially and at what rate, where $B \subset \mathcal{P}(S)$ is measurable in a suitable sense and π is an interior point of B in a suitable topology.

We will state now the definition of the large deviation principle in the present context. Let V be a vector space $\subset B(S)$. For a set $B \subset \mathcal{P}(S)$, its V-closure is denoted $\text{cl}_V(B)$ and its V-interior $\text{int}_V(B)$. The σ-algebra $\mathcal{B}(\mathcal{P}(S), B(S))$ is the smallest σ-algebra on $\mathcal{P}(S)$ for which each map $\mu \mapsto \int g \, d\mu$, $g \in B(S)$, is measurable. In the rest of this text, $B \subset \mathcal{P}(S)$ being measurable will mean $B \in \mathcal{B}(\mathcal{P}(S), B(S))$. Note that the map $L_n : \Omega \to \mathcal{P}(S)$ is measurable.

Definition 1.2 For $\nu \in \mathcal{P}(S)$, $\{\mathbb{P}_\nu[L_n \in \cdot]\}$ *satisfies the large deviation principle in the V topology with rate function J* if

1. $J : \mathcal{P}(S) \to \overline{\mathbb{R}^+}$ is V-lower semicontinuous.
2. For every measurable set $B \subset \mathcal{P}(S)$,

$$\overline{\lim_n}\, n^{-1} \log \mathbb{P}_\nu[L_n \in B] \le -\inf\{J(\mu) : \mu \in \text{cl}_V(B)\}, \tag{1.1}$$

$$\underline{\lim_n}\, n^{-1} \log \mathbb{P}_\nu[L_n \in B] \ge -\inf\{J(\mu) : \mu \in \text{int}_V(B)\}. \tag{1.2}$$

Inequality (1.1) (and, respectively, (1.2)) will be referred to as the upper (respectively, lower) bound.

We will say that J is *V-tight* if for each $a \ge 0$,

$$L_a \triangleq \{\mu \in \mathcal{P}(S) : J(\mu) \le a\}$$

is V-compact (we have adopted here the terminology of Rassoul-Agha and Seppäläinen, 2015).

A set $M \subset \mathcal{P}(S)$ is a *uniformity set* for the upper (lower) bound in the V topology with rate function J if for every measurable set $B \subset \mathcal{P}(S)$,

$$\overline{\lim_n}\, n^{-1} \log \sup_{\nu \in M} \mathbb{P}_\nu[L_n \in B] \le -\inf\{J(\mu) : \mu \in \text{cl}_V(B)\} \tag{1.3}$$

and, respectively,

$$\underline{\lim_n}\, n^{-1} \log \inf_{\nu \in M} \mathbb{P}_\nu[L_n \in B] \ge -\inf\{J(\mu) : \mu \in \text{int}_V(B)\}. \tag{1.4}$$

In the literature on large deviations for empirical measures of Markov chains, the cases of $(S, \mathcal{S})$ and the vector space $V \subset B(S)$ defining the topology on $\mathcal{P}(S)$ that have been considered are:

I. S is a Polish space (it is understood that $\mathcal{S}$ is its Borel σ-algebra) and $V = C_b(S)$. In this case, the $C_b(S)$ topology on $\mathcal{P}(S)$ is usually called the weak topology.
II. $(S, \mathcal{S})$ is a measurable space (with $\mathcal{S}$ countably generated) and $V = B(S)$. In this case, the $B(S)$ topology on $\mathcal{P}(S)$ is usually called the τ topology. This includes in particular the case when S is a countable set, $\mathcal{S}$ is its power set, and $V = B(S)$ is the set of all real-valued bounded functions on S.

In order to avoid repetitions and to capture the common features of cases I and II above that are relevant to the study of large deviations for empirical measures of Markov chains, we will introduce in Chapters 3 and 4 certain conditions on V which cover both cases.

We turn now to the statement of some of the main results. Throughout the book, it will be understood that $(S, \mathcal{S})$, P, ν, $\mathbb{P}_\nu$ are as in the basic framework and $\mathcal{P}(S)$ is endowed with the V topology, where V is a vector subspace of $B(S)$; further assumptions on V will be stated as needed. The assumption that S is Polish will be specified in some results.

For simplicity, in this introduction we will limit ourselves to the statement of large deviations results in case II above, omitting results on uniformity sets. But we must first introduce several functions from $\mathcal{P}(S)$ to $\overline{\mathbb{R}^+}$ which will play the role of rate functions in suitable contexts.

For $g \in B(S)$, we define

$$\phi(g) = \overline{\lim_n}\, n^{-1} \log \sup_{x \in S} \mathbb{E}_x \left(\exp S_n(g)\right),$$

where $S_n(g) = \sum_{j=0}^{n-1} g(X_j)$. It will sometimes be convenient to consider $T_n(g) = \sum_{j=1}^{n} g(X_j)$. Since

$$\mathbb{E}_x \left(\exp S_n(g)\right) = e^{g(x)} \mathbb{E}_x \left(\exp T_{n-1}(g)\right),$$

we have

$$\phi(g) = \overline{\lim_n}\, n^{-1} \log \sup_{x \in S} \mathbb{E}_x \left(\exp T_n(g)\right).$$

The *transform kernel* associated with the Markov kernel P and $g \in B(S)$ is defined to be

$$K_g(x, A) = \int_A e^{g(y)} P(x, dy), \qquad x \in S,\ A \in \mathcal{S}.$$

Since for all $x \in S$, $\mathbb{E}_x(\exp T_n(g)) = K_g^n 1(x)$, we have

$$\phi(g) = \overline{\lim_n}\, n^{-1} \log \sup_{x \in S} K_g^n 1(x),$$

and if K_g is regarded as a bounded linear operator on the space $B(S)$ with the supremum norm, then

$$\phi(g) = \log r(K_g),$$

where $r(K_g)$ is the spectral radius of K_g. For, by the spectral radius formula,

$$\begin{aligned} \log r(K_g) &= \lim_n n^{-1} \log \|K_g^n\| \\ &= \lim_n n^{-1} \log \|K_g^n 1\| = \phi(g). \end{aligned}$$

For $\mu \in \mathcal{P}(S)$, we define

$$\phi^*(\mu) = \sup\left\{ \int g\, d\mu - \phi(g) : g \in B(S) \right\}.$$

Since $\phi(0) = 0$, we have $\phi^*(\mu) \geq 0$. For $\nu \in \mathcal{P}(S)$, $g \in B(S)$, we define

$$\phi_\nu(g) = \overline{\lim_n}\, n^{-1} \log \mathbb{E}_\nu (\exp S_n(g));$$

it is easily shown that

$$\phi_\nu(g) = \overline{\lim_n}\, n^{-1} \log \nu K_g^n 1.$$

If $x \in S$ and $\nu = \delta_x$, we write

$$\begin{aligned} \phi_x(g) &= \overline{\lim_n}\, n^{-1} \log \mathbb{E}_x (\exp S_n(g)) \\ &= \overline{\lim_n}\, n^{-1} \log K_g^n 1(x). \end{aligned}$$

For $\mu \in \mathcal{P}(S)$, we define

$$\phi_\nu^*(\mu) = \sup\left\{ \int g\, d\mu - \phi_\nu(g) : g \in B(S) \right\}.$$

Since $\phi_\nu(0) = 0$, we have $\phi_\nu^*(\mu) \geq 0$.

The function $I : \mathcal{P}(S) \to \overline{\mathbb{R}^+}$ is defined as follows:

$$I(\mu) = \sup\left\{ \int \log\left(\frac{e^g}{Pe^g} \right) d\mu : g \in B(S) \right\}, \qquad \mu \in \mathcal{P}(S).$$

Taking $g = 0$ in this expression, we have $I(\mu) \geq 0$. We also define for $\lambda \in \mathcal{P}(S)$, $\mu \in \mathcal{P}(S)$,

$$I_\lambda(\mu) = \begin{cases} I(\mu) & \text{if } \mu \ll \lambda, \\ \infty & \text{otherwise.} \end{cases}$$

Then $I_\lambda = I$ if and only if $\operatorname{dom} I \subset \mathcal{P}(S, \lambda)$, where $\mathcal{P}(S, \lambda) = \{\mu \in \mathcal{P}(S) : \mu \ll \lambda\}$ and for a function $J : \mathcal{P}(S) \to \overline{\mathbb{R}^+}$,

$$\operatorname{dom} J = \{\mu \in \mathcal{P}(S) : J(\mu) < \infty\}.$$

In the next paragraph we will introduce certain functions that involve the irreducibility of P, a condition that will play an essential role in this work (see Appendix B). As will be seen in Chapter 5, irreducibility is necessary for a central formulation of the large deviation principle to hold. The most important case of I_λ is the case when P is irreducible and $\lambda = \psi$, a P-maximal irreducibility probability measure (Appendix B). The symbol ψ will have this meaning throughout the text (except for a slight departure from this convention in Appendices B and C).

If P is irreducible, then so is K_g and its convergence parameter $R(K_g)$ exists (Appendix C). We define for $g \in B(S)$

$$\Lambda(g) = -\log R(K_g).$$

If (s, ν) is a small pair (Appendix B) then by (C.1),

$$\Lambda(g) = \overline{\lim_n}\, n^{-1} \log \nu K_g^n s;$$

one can show that

$$\begin{aligned}\Lambda(g) &= \overline{\lim_n}\, n^{-1} \log \mathbb{E}_\nu \left[(\exp S_n(g))\, s(X_{n-1})\right] \\ &= \overline{\lim_n}\, n^{-1} \log \mathbb{E}_\nu \left[(\exp T_{n-1}(g))\, s(X_{n-1})\right].\end{aligned}$$

If $\mu \in \mathcal{P}(S)$, we define

$$\Lambda^*(\mu) = \sup\left\{\int g\, d\mu - \Lambda(g) : g \in B(S)\right\}.$$

Since $\Lambda(0) \le 0$, we have $\Lambda^*(\mu) \ge 0$.

All functions defined above are convex.

As we will see in Chapter 3, ϕ_ν^* is a rate function for the upper bound for $\{\mathbb{P}_\nu[L_n \in \cdot]\}$ in the τ topology: under suitable conditions, (1.1) holds with $V = B(S)$ and $J = \phi_\nu^*$. The rate function ϕ_ν^* is a convex-analytic construct which emerges as a result of an argument involving compactness and an exponential Markov inequality.

In Chapter 2 it is proved that if P is irreducible, then Λ^* is a rate function for the lower bound for $\{\mathbb{P}_\nu[L_n \in \cdot]\}$ in the τ topology: (1.2) holds with $V = B(S)$ and $J = \Lambda^*$. The rate function Λ^* is a convex-analytic construct the emergence of which is more subtle and requires an argument based on the fundamental minorization property of irreducible kernels, as well as subadditivity and identification arguments.

For P irreducible, it is always the case that $\phi_\nu^* \leq \Lambda^*$. Moreover, as will be seen in Chapter 5, if $\{\mathbb{P}_\nu[L_n \in \cdot]\}$ satisfies the large deviation principle in the τ topology with rate function J, then $\phi_\nu^* \leq J \leq \Lambda^*$. A central issue will be to establish conditions ensuring the equality $\phi_\nu^* = \Lambda^*$.

It is very important to relate these rate functions to the functions I and I_ψ, which are of primary significance and analytically more tractable. In this direction, we have

Proposition 1.3 (part of Theorems 4.1 and 4.2)

1. $\phi^* = I$ *and for all* $\nu \in \mathcal{P}(S)$, $\phi_\nu^* \geq I$.
2. *Let* P *be irreducible. Then* $\Lambda^* = I_\psi$ *and for all* $\nu \in \mathcal{P}(S)$, $\phi_\nu^* \leq I_\psi$.

We will now state some of the large deviation results. Under the sole assumption of irreducibility, for any $\nu \in \mathcal{P}(S)$ the lower bound for $\{\mathbb{P}_\nu[L_n \in \cdot]\}$ in the τ topology with rate function $\Lambda^* = I_\psi$ always holds:

Proposition 1.4 (part of Theorem 2.10) *Let* P *be irreducible and* $\nu \in \mathcal{P}(S)$. *Then for any measurable set* $B \subset \mathcal{P}(S)$,

$$\underline{\lim}_n\, n^{-1} \log \mathbb{P}_\nu[L_n \in B] \geq -\inf\{\Lambda^*(\mu) : \mu \in \operatorname{int}_\tau(B)\}$$
$$= -\inf\{I_\psi(\mu) : \mu \in \operatorname{int}_\tau(B)\}.$$

The following result gives a necessary and sufficient analytic condition for the upper bound for $\{\mathbb{P}_\nu[L_n \in \cdot]\}$ in the τ topology with τ-tight rate function ϕ_ν^*.

Proposition 1.5 (part of Theorem 3.2) *Let* $\nu \in \mathcal{P}(S)$.

1. *For every measurable set* $B \subset \mathcal{P}(S)$ *with compact* τ*-closure,*

$$\overline{\lim}_n\, n^{-1} \log \mathbb{P}_\nu[L_n \in B] \leq -\inf\{\phi_\nu^*(\mu) : \mu \in \operatorname{cl}_\tau(B)\}. \tag{1.5}$$

2. *The following conditions are equivalent:*

   (i) *If* $0 \leq g_k \in B(S)$ *and* $g_k \downarrow 0$ *pointwise, then* $\phi_\nu(g_k) \to 0$.

   (ii) *For every measurable set* $B \subset \mathcal{P}(S)$, (1.5) *holds and* ϕ_ν^* *is* τ*-tight.*

In Chapter 6 we will present several analytic conditions equivalent to 2(i) and in Chapter 7 sufficient conditions for 2(i), hence for 2(ii). We state one such sufficient condition. A set $C \in \mathcal{S}$ is *$(P-\tau)$-tight* if there exists $m \in \mathbb{N}$ such that $\{P^m(x, \cdot) : x \in C\}$ is τ-relatively compact. Let $\tau_C = \inf\{n \geq 1 : X_n \in C\}$.

Proposition 1.6 (part of Theorem 7.8) *Let $\nu \in \mathcal{P}(S)$. Assume that for every $b > 0$, there exists a $(P - \tau)$-tight set C such that*

$$\sup_{x \in C} \mathbb{E}_x e^{b\tau} < \infty, \quad \textit{where } \tau = \tau_C,$$
$$\mathbb{E}_\nu e^{b\tau} < \infty.$$

Then 2(i), *hence also* 2(ii), *of Proposition 1.5 holds.*

In order to obtain a large deviation principle for $\{\mathbb{P}_\nu[L_n \in \cdot]\}$ from Propositions 1.4 and 1.5, we must reconcile the lower bound rate function I_ψ and the upper bound rate function ϕ_ν^*. Under the assumption that P is irreducible, Proposition 1.7 provides necessary and sufficient conditions for the large deviation principle for $\{\mathbb{P}_\nu[L_n \in \cdot]\}$ in the τ topology with τ-tight rate function $\phi_\nu^* = I_\psi$. In particular, it states that this large deviation principle is equivalent to the assertion: $\{\mathbb{P}_\nu[L_n \in \cdot]\}$ satisfies the upper bound in the τ topology with a τ-tight rate function J such that $\operatorname{dom} J \subset \mathcal{P}(S, \psi)$.

Proposition 1.7 (part of Theorem 8.1) *Let P be irreducible and $\nu \in \mathcal{P}(S)$.*

1. *The conditions* (1.6)–(1.9) *are equivalent:*

$$\textit{If } 0 \le g_k \in B(S) \textit{ and } g_k \downarrow 0 \textit{ pointwise, then } \phi_\nu(g_k) \to 0. \tag{1.6a}$$
$$\textit{If } 0 \le g \in B(S) \textit{ and } \textstyle\int g\, d\psi = 0, \textit{ then } \phi_\nu(g) = 0. \tag{1.6b}$$
$$\phi_\nu = \Lambda. \tag{1.7a}$$
$$I_\psi \textit{ is } \tau\textit{-tight.} \tag{1.7b}$$
$$\{\mathbb{P}_\nu[L_n \in \cdot]\} \textit{ satisfies the large deviation principle in the } \tau \textit{ topology with } \tau\textit{-tight rate function } \phi_\nu^* = I_\psi. \tag{1.8}$$
$$\{\mathbb{P}_\nu[L_n \in \cdot]\} \textit{ satisfies the upper bound in the } \tau \textit{ topology with a } \tau\textit{-tight rate function } J \textit{ such that } \operatorname{dom} J \subset \mathcal{P}(S, \psi). \tag{1.9}$$

2. *If any, hence all, of conditions* (1.6)–(1.9) *is satisfied, then P has a unique invariant probability measure π, $\pi \equiv \psi$, and therefore $I_\psi = I_\pi$.*

In the context of Proposition 1.7, suppose that $B \subset \mathcal{P}(S)$ is measurable and $\pi \in \operatorname{int}_\tau(B)$. Since $I_\psi(\mu) = 0$ implies that $\mu = \pi$ (to be shown in Lemma 5.5) and I_ψ is τ-tight, we have

$$a = \inf\{I_\psi(\mu) : \mu \in \operatorname{cl}_\tau(B^c)\} > 0.$$

Therefore the probabilities $\{\mathbb{P}_\nu[L_n \notin B]\}$ indeed decay exponentially at a specified rate:

$$\overline{\lim_n}\, n^{-1} \log \mathbb{P}_\nu[L_n \notin B] \le -a.$$

This form of convergence of $\{L_n\}$ to π is sometimes called *exponential convergence*.

We turn now to the situation when S is countable and P is *matrix irreducible*: for all $x, y \in S$, $\sum_{n=1}^{\infty} P^n(x, y) > 0$. Equivalently, P is irreducible and counting measure is a P-maximal irreducibility measure. In this situation, a significant simplification occurs: $I_\psi = I$ and, by Proposition 1.3, for all $\nu \in \mathcal{P}(S)$, $\phi_\nu^* = I$.

Proposition 1.8 (part of Theorem 9.3 and Remark 9.4) *Let S be countable and assume that P is matrix irreducible.*

1. *The following conditions are equivalent:*

$$\text{For all } b > 0, \text{ there exists } F \text{ finite, } F \subset S, \text{ such that}$$
$$\text{for all } y \in F,\ \mathbb{E}_y e^{b\tau} < \infty, \text{ where } \tau = \tau_F. \tag{1.10}$$
$$\text{If } 0 \le g_k \in B(S) \text{ and } g_k \downarrow 0 \text{ pointwise, then}$$
$$\text{for all } x \in S,\ \phi_x(g_k) \to 0. \tag{1.11}$$
$$\text{For all } x \in S,\ \{\mathbb{P}_x[L_n \in \cdot]\} \text{ satisfies the large deviation principle}$$
$$\text{in the } \tau \text{ topology with } \tau\text{-tight rate function } I. \tag{1.12}$$

2. *If any, hence all, of conditions* (1.10)–(1.12) *is satisfied, then P has unique invariant probability measure π and $\pi(x) > 0$ for $x \in S$.*

Finally, we turn to the case of large deviations for vector-valued additive functionals. Let E be a separable Banach space, $f: S \to E$ a measurable function, $S_n(f) = \sum_{j=0}^{n-1} f(X_j)$. Parallel to Proposition 1.1, we have:

Proposition 1.9 *Assume that P and π are as in Proposition 1.1 and that $\int \|f\|\, d\pi < \infty$. Then if $\pi(f) = \int f\, d\pi$, for any $\nu \in \mathcal{P}(S)$,*

$$\lim_n n^{-1} S_n(f) = \pi(f), \quad \mathbb{P}_\nu\text{-a.s.}.$$

Proposition 1.9 is proved in Appendix A.

The objective of Chapter 11 is to study the large deviations associated with this result; in particular, to determine when the probabilities $\{\mathbb{P}_\nu[n^{-1} S_n(f) \notin G]\}$ decay exponentially and at what rate, where $G \subset E$ is open and $\pi(f) \in G$.

To describe the results on vector-valued functionals, we start with the lower bound. Let $F(S)$ be the space of measurable functions $g: S \to \mathbb{R}$ and, as in the case $g \in B(S)$, for $g \in F(S)$ let

$$K_g(x, A) = \int_A e^{g(y)} P(x, dy), \qquad x \in S,\ A \in \mathcal{S}.$$

If P is irreducible, so is K_g; we denote its convergence parameter by $R(K_g)$ (Appendix C).

If $f\colon S \to E$ is a measurable function and $\xi \in E^*$, the dual space of E, we write

$$K_{f,\xi}(x, A) = K_{\langle f,\xi\rangle}(x, A) = \int_A e^{\langle f(y),\xi\rangle} P(x, dy)$$

and define

$$\Lambda_f(\xi) = -\log R(K_{f,\xi}),$$
$$\Lambda_f^*(u) = \sup\{\langle u, \xi\rangle - \Lambda_f(\xi)\colon \xi \in E^*\}, \qquad u \in E.$$

Since $\Lambda_f(0) \le 0$, we have $\Lambda_f^*(u) \ge 0$.

Under the sole assumption of irreducibility, we have:

Proposition 1.10 (part of Theorem 11.1) *Let P be irreducible and let $f\colon S \to E$ be measurable. Then for every $\mu \in \mathcal{P}(S)$ and every open set $G \subset E$,*

$$\underline{\lim}_n\, n^{-1} \log \mathbb{P}_\mu \left[n^{-1} S_n(f) \in G\right] \ge -\inf_{u\in G} \Lambda_f^*(u).$$

In Theorem 11.13 the rate function Λ_f^* will be identified in terms of I_ψ.

To state the upper bound result, as in the case $g \in B(S)$ we define for $g \in F(S)$, $\mu \in \mathcal{P}(S)$,

$$\phi_\mu(g) = \overline{\lim}_n\, n^{-1} \log \mathbb{E}_\mu (\exp S_n(g)),$$

and for $f\colon S \to E$ measurable, $\xi \in E^*$,

$$\begin{aligned}\phi_{f,\mu}(\xi) &= \phi_\mu(\langle f, \xi\rangle)\\ &= \overline{\lim}_n\, n^{-1} \log \mathbb{E}_\mu (\exp \langle S_n(f), \xi\rangle)\end{aligned}$$

and

$$\phi_{f,\mu}^*(u) = \sup\{\langle u, \xi\rangle - \phi_{f,\mu}(\xi)\colon \xi \in E^*\}, \qquad u \in E.$$

Since $\phi_{f,\mu}(0) = 0$, we have $\phi_{f,\mu}^*(u) \ge 0$.

Proposition 1.11 (part of Theorem 11.16) *Let $f\colon S \to E$ be measurable, and let $\mu \in \mathcal{P}(S)$. Then:*

1. *For every $\sigma(E, E^*)$-compact set $F \subset E$,*

$$\overline{\lim}_n\, n^{-1} \log \mathbb{P}_\mu \left[n^{-1} S_n(f) \in F\right] \le -\inf_{u\in F} \phi_{f,\mu}^*(u). \tag{1.13}$$

2. *Assume:*

   (i) *For some $m \in \mathbb{N}$, $\{P^m(x, \cdot) \circ f^{-1} : x \in S\}$ is tight.*

(ii) *For all $r > 0$,*

$$\int \exp(r\|f(y)\|)\,\mu(dy) < \infty, \tag{1.14a}$$

$$\sup_{x \in S} \int \exp(r\|f(y)\|)\,P(x, dy) < \infty. \tag{1.14b}$$

Then (a) (1.13) *holds for every closed set $F \subset E$ and (b) $\phi^*_{f,\mu}$ is tight.*

The assumptions can be weakened if E is finite dimensional; see Remark 11.18.

We state now a necessary and sufficient condition for the large deviation principle for $\{\mathbb{P}_\mu[n^{-1}S_n(f) \in \cdot]\}$ (for the definition of the large deviation principle relevant here, see, e.g., Rassoul-Agha and Seppäläinen 2015, p. 21).

Proposition 1.12 (part of Theorem 11.21) *Let $\mu \in \mathcal{P}(S)$. Assume:*

1. *P is irreducible.*
2. *P, f, and μ satisfy the assumptions of Proposition 1.11.*

Then the following conditions are equivalent:

$\phi_{f,\mu} = \Lambda_f$.

$\{\mathbb{P}_\mu[n^{-1}S_n(f) \in \cdot]\}$ *satisfies the large deviation principle with tight rate function Λ^*_f.*

In the context of Proposition 1.12, and under certain conditions given in Theorem 11.22 which imply that $\Lambda^*_f(u) = 0$ if and only if $u = \pi(f)$, if $G \subset E$ is open and $\pi(f) \in G$, then

$$b = \inf\left\{\Lambda^*_f(u) : u \in G^c\right\} > 0.$$

Therefore the probabilities $\{\mathbb{P}_\mu[n^{-1}S_n(f) \notin G]\}$ do decay exponentially at a specified rate:

$$\overline{\lim_n}\, n^{-1} \log \mathbb{P}_\mu\left[n^{-1}S_n(f) \notin G\right] \le -b.$$

In Theorem 11.27 we give a sufficient condition for the large deviation principle for $\{\mathbb{P}_\nu[n^{-1}S_n(f) \in \cdot]\}$: under a certain assumption on P, and if P and f satisfy 2(i) and (1.14b) of Proposition 1.11, the large deviation principle holds for all petite ν (Appendix B) and all $\nu = \delta_x$, where x belongs to a set in $\mathcal{S}$ of full ψ measure.

In Theorem 11.29 we discuss the relationship between the large deviations for empirical measures and the large deviations for additive functionals.

1.1 Outline of the Book

We will now give an outline of some of the main contents of each chapter.

Chapter 2 We prove a lower bound for $\{\mathbb{P}_\nu[n^{-1}S_n(f) \in \cdot]\}$, where $f\colon S \to E$ is a bounded measurable function and E is a separable Banach space, and use it to obtain the lower bound for $\{\mathbb{P}_\nu[L_n \in \cdot]\}$ in the V topology with rate function $(\Lambda \mid V)^*$, hence with rate function $\Lambda^* = I_\psi$. We also present a class of uniformity sets $M \subset \mathcal{P}(S)$ for the lower bound; the class includes all petite sets (Appendix B). We prove that the function $\Lambda\colon B(S) \to \mathbb{R}$ is $\sigma(B(S), \mathcal{P}(S,\psi))$-lower semicontinuous. This result will play a significant role in Chapters 8 and 11.

Chapter 3 We introduce the assumptions V.1–V.3 for a vector space $V \subset B(S)$ and obtain upper bounds for random probability measures. We give a necessary and sufficient analytic condition for the upper bound for $\{\mathbb{P}_\nu[L_n \in \cdot]\}$ in the V topology with rate function $(\phi_\nu \mid V)^*$. We discuss the connection between the analytic condition and exponential tightness.

Chapter 4 We introduce the assumptions V.1′–V.4 for a vector space $V \subset B(S)$ and prove that $(\phi \mid V)^* = I$ and $\Lambda^* = I_\psi$. We establish conditions under which $(\phi_\nu \mid V)^* = I_\psi$ and $I = I_\psi$. We also study the relationship between $(\Lambda \mid V)^*$ and Λ^* and conditions for the equality $\phi_M^* = I_\lambda$ in certain cases in which P may not be irreducible.

Chapter 5 We assume that $\{\mathbb{P}_\nu[L_n \in \cdot]\}$ satisfies the large deviation principle in the V topology with a rate function J which is not known a priori and derive several consequences, including bounds on J. The existence and uniqueness of invariant measures is discussed. It is proved that if $\{\mathbb{P}_x[L_n \in \cdot]\}$ satisfies the large deviation principle in the τ topology for every $x \in S$ with a common a priori unknown τ-tight rate function J, then P must be irreducible and $J = I$.

Chapter 6 We study in a more abstract setting the analytic condition introduced in Chapter 3 and refine the results found there. We obtain necessary conditions for the uniformity of a set of initial distributions for the upper bound.

Chapter 7 We obtain different sufficient conditions for the upper bound for $\{\mathbb{P}_\nu[L_n \in \cdot]\}$ in the V topology.

Chapter 8 We present several formulations for the large deviation principle in the V topology. These include the large deviation principle for $\{\mathbb{P}_\nu[L_n \in \cdot]\}$ for an arbitrary $\nu \in \mathcal{P}(S)$, conditions for a set $M \subset \mathcal{P}(S)$ to be a uniformity set for both the upper and lower bounds, and the case when the large deviation principle holds for $\{\mathbb{P}_x[L_n \in \cdot]\}$ for all $x \in S$.

Chapter 9 We study the case when S is countable and P is matrix irreducible.

Chapter 10 We present several examples which show boundaries of the general results obtained in the previous chapters. In particular, it is shown that even under the assumption of irreducibility, it is possible for the large deviation principle to hold for $\{\mathbb{P}_x[L_n \in \cdot]\}$ for every $x \in S$ with different rate functions for each x.

Chapter 11 We study large deviations for $\{\mathbb{P}_\nu[n^{-1}S_n(f) \in \cdot]\}$, where f is a measurable function on S taking values in a separable Banach space. The identification of the rate function Λ_f^* is discussed, as well as its zero set. We study the relationship between the large deviation principle for empirical measures and the large deviation principle for additive functionals.

Appendix A–Appendix K The appendices have a two-fold purpose. We present some well-known analytical or probabilistic results in a form suitable for application in the main text; in certain cases, we discuss some useful consequences or variants. In particular, Appendices B and C contain definitions and results related to general Markov chains and irreducible kernels. We also prove some auxiliary results for which we have no ready reference. Some of the results proved in the appendices may be new and possibly of independent interest (e.g., Propositions C.3 and I.1).

1.2 Notes

Large deviations for empirical measures (occupation times) of Markov chains were first studied by Donsker and Varadhan (1975) in the case where S is a compact metric space under very strong conditions on P and later in greater generality in Donsker and Varadhan (1976). The basic rate function I was introduced in Donsker and Varadhan (1975). For an entropy representation of I see, e.g., Rassoul-Agha and Seppäläinen (2015, Theorem 13.2). We will comment on the relation of the results in the present book to Donsker and Varadhan (1976) in later chapters; see the notes to Chapters 2 and 7.

Closely related is Gärtner's work, Gärtner (1977).

Other papers that studied large deviations for empirical measures of Markov chains under very strong conditions are Bolthausen (1987), where the τ topology was introduced in this setting, and Ellis (1988). Chapters on the subject may be found in Stroock (1984), Deuschel and Stroock (1989), Dembo and Zeitouni (1998) (largely along the lines of Deuschel and Stroock), Dupuis and Ellis (1997), den Hollander (2000), Rassoul-Agha and Seppäläinen (2015), and Feng and Kurtz (2006).

The present work has its origins in de Acosta (1985, 1988, 1990, 1994a,b) and de Acosta and Ney (1998, 2014). The papers by Ney and Nummelin (1987), Dinwoodie (1993), Dinwoodie and Ney (1995), and Nummelin's book Nummelin (1984) have been highly influential in the development of our outlook. A previous paper related to Ney and Nummelin (1987) is Iscoe et al. (1985). A feature of our approach which is significantly different from the sources cited in the previous paragraphs is the derivation of the lower bound on a general state space under the sole assumption of irreducibility, including the construction of the rate function from the convergence parameters of the irreducible transform kernels. We also focus on initial distributions and uniformity sets and on the existence and uniqueness of invariant measures and their relation to rate functions.

In several chapters we incorporate important contributions of Wu (2000a,b), particularly in connection to upper bounds.

Our presentation of large deviations for vector-valued additive functionals of a Markov chain is based on and extends de Acosta (1985, 1988) and de Acosta and Ney (1998, 2014).

For some results when the assumption of irreducibility is relaxed, see Wu (2000a,b) and Jiang and Wu (2005). For process-level large deviations, a topic not covered here, see Donsker and Varadhan (1983), Deuschel and Stroock (1989), and Wu (2000a,b).

2

Lower Bounds and a Property of Λ

For a vector space $V \subset B(S)$, the V topology on $\mathcal{P}(S)$ was defined in Chapter 1 and the definition of the large deviation principle for $\{\mathbb{P}_\nu[L_n \in \cdot]\}$ was stated in terms of it. Let P be irreducible, and for $\mu \in \mathcal{P}(S)$ let

$$(\Lambda \mid V)^*(\mu) = \sup\left\{\int g\, d\mu - \Lambda(g) \colon g \in V\right\},$$

where Λ is as in Chapter 1.

The main objective in this chapter is to prove the lower bound for $\{\mathbb{P}_\nu[L_n \in \cdot]\}$ in the V topology with rate function $(\Lambda \mid V)^*$ (hence, with rate function Λ^*). In Chapter 4 we will identify Λ^* in a more explicit analytic form: $\Lambda^* = I_\psi$. Actually, Theorem 2.10 has a broader scope: we present a class of sets $M \subset \mathcal{P}(S)$ which are uniformity sets for the lower bound. This class contains all petite sets (for the definition of a petite set, see Appendix B), which constitute a broad family; for example, there exists an increasing sequence $\{C_j\}_{j \in \mathbb{N}}$ of petite sets such that $C_j \uparrow S$ (Proposition B.5).

In order to prove the lower bound for $\{\mathbb{P}_\nu[L_n \in \cdot]\}$, we first obtain in Theorem 2.1 a lower bound for $\{\mathbb{P}_\nu[n^{-1}S_n(f) \in \cdot]\}$, where $f \colon S \to E$, E is a separable Banach space and f is a bounded measurable function, and then use it to prove Theorem 2.10. The case of an arbitrary measurable function $f \colon S \to E$ will be studied in Chapter 11.

In Theorem 2.13 we prove a special analytic property of $\Lambda \colon B(S) \to \mathbb{R}$, not shared in general by the other real-valued functions defined on $B(S)$ which we later consider: Λ is $\sigma(B(S), \mathcal{P}(S, \psi))$-lower semicontinuous. This result will play a significant role in both Chapters 8 and 11.

2.1 Lower Bounds for Vector-Valued Additive Functionals: The Bounded Case

Our first task is to prove Theorem 2.1, in which a lower bound is obtained for $\{\inf_{\nu\in M}\mathbb{P}_\nu[n^{-1}S_n(f)\in\cdot]\}$, where $f\colon S\to E$, E is a separable Banach space, f is bounded and measurable, $S_n(f)=\sum_{j=0}^{n-1}f(X_j)$, and M is a suitable subset of $\mathcal{P}(S)$. As the key step towards obtaining lower bounds for empirical measures, it is sufficient to consider the case $E=\mathbb{R}^d$; however, essentially no additional technical work is needed for the general case of a separable Banach space.

We define for $f\colon S\to E$ as above,

$$\Lambda_f(\xi)=\Lambda(\langle f,\xi\rangle),\qquad \xi\in E^*,$$

where E^* is the dual space of E,

$$\Lambda_f^*(u)=\sup_{\xi\in E^*}\left[\langle u,\xi\rangle-\Lambda_f(\xi)\right],\qquad u\in E.$$

We introduce the following condition for a set $M\subset\mathcal{P}(S)$: there exist a petite set D and $h\in\mathbb{N}_0$ such that

$$\inf\left\{\sum_{i=0}^{h}\mu P^i(D)\colon \mu\in M\right\}>0. \tag{2.1}$$

Note that if $M\subset S$ and M is petite, then (2.1) holds with $h=0$ and $D=M$. (Of course, in this context, $x\in S$ is identified with $\delta_x\in\mathcal{P}(S)$.) If $\mu\in\mathcal{P}(S)$ and $M=\{\mu\}$, then (2.1) is satisfied: for example, this follows at once from Proposition B.5.

Theorem 2.1 *Let P be irreducible and let $f\colon S\to E$ be a bounded measurable function. Then for every set $M\subset\mathcal{P}(S)$ satisfying (2.1) and every open set $G\subset E$,*

$$\varliminf_n n^{-1}\log\left\{\inf_{\mu\in M}\mathbb{P}_\mu[n^{-1}S_n(f)\in G]\right\}\geq-\inf_{u\in G}\Lambda_f^*(u). \tag{2.2}$$

In particular, the conclusion holds if $M\subset S$ and M is petite.

The proof has the following steps:

1. It is assumed first that P satisfies the minorization condition (2.3). Under this condition, it is possible to carry out a subadditivity argument, leading to a rate function for a weaker form of the lower bound (2.2).
2. The rate function is then identified as Λ_f^*.
3. A general irreducible Markov kernel P is suitably approximated by Markov kernels satisfying (2.3), and then the lower bound obtained in step 1 is shown to imply (2.2).

We will need several lemmas. In Lemmas 2.2–2.6, we will assume that the following minorization condition holds: for some $\alpha > 0$, $C \in \mathcal{S}^+$ (Section B.2), $\nu \in \mathcal{P}(S)$ with $\nu(C) > 0$,

$$P \geq \alpha(\mathbf{1}_C \otimes \nu), \tag{2.3}$$

where $(\mathbf{1}_C \otimes \nu)(x, A) = \mathbf{1}_C(x)\nu(A)$ for $x \in S$, $A \in \mathcal{S}$. For $n \in \mathbb{N}$, $B \in \mathcal{E}$ (the Borel σ-algebra of E), let $\ell(n, B) = \mathbb{P}_\nu[n^{-1}S_n \in B, X_{n-1} \in C]$; here $S_n = S_n(f)$.

Lemma 2.2 *Assume that* (2.3) *holds. Let* $p, q \in \mathbb{N}$, $A_1, A_2, A \in \mathcal{E}$ *be such that*

$$\frac{p}{r}A_1 + \frac{q}{r}A_2 \subset A, \tag{2.4}$$

where $r = p + q$. *Then for all* $n \in \mathbb{N}$,

$$\ell(n(p+q), A) \geq \alpha\ell(np, A_1)\ell(nq, A_2).$$

Proof. It follows from (2.3) that for all measurable functions $\Phi\colon S^{\mathbb{N}_0} \to \mathbb{R}^+$, $x \in S$,

$$\mathbb{E}_x[\Phi \circ \Theta] \geq \alpha\mathbf{1}_C(x)\mathbb{E}_\nu\Phi, \tag{2.5}$$

where $\Theta\colon S^{\mathbb{N}_0} \to S^{\mathbb{N}_0}$ is the shift: $\Theta((x_n)_{n\geq 0}) = (x_{n+1})_{n\geq 0}$. For, let $g(y) = \mathbb{E}_y\Phi$, $y \in S$. Then by the Markov property,

$$\begin{aligned}\mathbb{E}_x[\Phi \circ \Theta] &= \mathbb{E}_x\mathbb{E}_x[\Phi \circ \Theta \mid \mathcal{F}_1] = \mathbb{E}_x g(X_1)\\ &= \int P(x, dy)g(y) \geq \alpha\mathbf{1}_C(x)\int g(y)\nu(dy)\\ &= \alpha\mathbf{1}_C(x)\mathbb{E}_\nu\Phi,\end{aligned}$$

proving (2.5). We have by (2.4),

$$\begin{aligned}\ell(n(p+q), A) &= \mathbb{P}_\nu\left[\left(\frac{S_{nr}}{nr}\right) \in A, X_{nr-1} \in C\right]\\ &\geq \mathbb{P}_\nu\left[\left(\frac{S_{np}}{np}\right) \in A_1, X_{np-1} \in C, \frac{1}{nq}\sum_{j=np}^{nr-1} f(X_j) \in A_2, X_{nr-1} \in C\right].\end{aligned}$$

By the Markov property, the last expression equals

$$\mathbb{E}_\nu\left[\mathbf{1}_{A_1}\left(\frac{S_{np}}{np}\right)\mathbf{1}_C(X_{np-1})\mathbb{E}_{X_{np-1}}(\Phi \circ \Theta)\right],$$

where

$$\Phi\big((x_j)_{j\geq 0}\big) = \mathbf{1}_{A_2}\left(\frac{1}{nq}\sum_{j=0}^{nq-1} f(x_j)\right)\mathbf{1}_C(x_{nq-1}).$$

By (2.5), the previous expectation is minorized by

$$\mathbb{E}_\nu\left[\mathbf{1}_{A_1}\left(\frac{S_{np}}{np}\right)\mathbf{1}_C(X_{np-1})\right]\alpha\mathbb{E}_\nu\left[\mathbf{1}_{A_2}\left(\frac{S_{nq}}{nq}\right)\mathbf{1}_C(X_{nq-1})\right]$$
$$= \alpha\ell(np, A_1)\ell(nq, A_2). \quad \square$$

Lemma 2.3 *Assume that* (2.3) *holds. Then:*

1. *For every open convex set $B \subset E$,*

$$\ell_f(B) \stackrel{\Delta}{=} \lim_n n^{-1} \log \ell(n, B) \tag{2.6}$$

   exists in $[-\infty, 0]$.
2. *Let $J_f \colon E \to [0, \infty]$ be defined by*

$$J_f(u) = \sup\left\{-\ell_f(B) \colon B \text{ open convex, } u \in B\right\}.$$

   Then J_f is convex and lower semicontinuous.

Proof.

1. By the convexity of B, by Lemma 2.2 for every $p, q \in \mathbb{N}$,

$$\ell(p + q, B) \geq \alpha\ell(p, B)\ell(q, B) \tag{2.7}$$

   and therefore the sequence $\{-\log a_p \colon p \in \mathbb{N}\}$ is subadditive, where $a_p = \alpha\ell(p, B)$. By Fekete's lemma (Rassoul-Agha and Seppäläinen, 2015, p. 63), statement (2.6) will follow provided we prove the following claim:

   If $a_p > 0$ for some $p \in \mathbb{N}$, then $a_n > 0$ for all sufficiently large n. (2.8)

   To prove (2.8), assume $a_p > 0$ and let $\epsilon > 0$ be such that $\ell(p, B_\epsilon) > 0$, where $B_\epsilon = \{u \in B \colon d(u, B^c) > \epsilon\}$ and $d(u, A) = \inf\{\|u - w\| \colon w \in A\}$, $A \subset E$. Let $n_0 \in \mathbb{N}$ be such that $2p\|f\|/n_0 < \epsilon$. For $n \geq n_0$, write $n = mp + r$, $m \in \mathbb{N}$, $0 \leq r < p$. Then

$$\left\|\frac{S_n}{n} - \frac{S_{mp}}{mp}\right\| \leq \frac{2p\|f\|}{n} < \epsilon,$$

   and therefore by (2.5) and the Markov property

$$\ell(n, B) \geq \mathbb{P}_\nu\left[\left(\frac{S_{mp}}{mp}\right) \in B_\epsilon, X_{mp-1} \in C, X_{n-1} \in C\right]$$
$$\geq \begin{cases} \ell(mp, B_\epsilon) & \text{if } r = 0, \\ \ell(mp, B_\epsilon)\alpha\mathbb{E}_\nu \mathbf{1}_C(X_{r-1}) & \text{if } r \geq 1. \end{cases}$$

By (2.7), applied to the open convex set B_ϵ, we have

$$\alpha\ell(mp, B_\epsilon) \geq (\alpha\ell(p, B_\epsilon))^m > 0.$$

Also for $r \geq 1$, it is easily proved from (2.3) by induction that

$$\mathbb{E}_\nu \mathbf{1}_C(X_{r-1}) = \nu P^{r-1}\mathbf{1}_C \geq \alpha^{r-1}(\nu(C))^r > 0.$$

Therefore $a_n = \alpha\ell(n, B) > 0$ for $n \geq n_0$, proving (2.8), hence (2.6).

2. Let $g_B = (-\ell_f(B))\mathbf{1}_B$ for B open convex. Then g_B is lower semicontinuous and therefore so is $J_f = \sup\{g_B \colon B \text{ open convex}\}$. To prove the convexity of J_f, let $\beta \in (0,1) \cap \mathbb{Q}$, say $\beta = p/r$ with $p, r \in \mathbb{N}$, and let $q = r - p$. Given $u_1, u_2 \in E$ and an open convex set B such that $\beta u_1 + (1-\beta)u_2 \in B$, there exist open convex sets $B_1 \ni u_1$, $B_2 \ni u_2$ such that

$$\beta B_1 + (1-\beta)B_2 \subset B.$$

By Lemma 2.2, for all $n \in \mathbb{N}$,

$$\alpha\ell(nr, B) \geq (\alpha\ell(np, B_1))(\alpha\ell(nq, B_2)),$$
$$\frac{1}{nr}\log(\alpha\ell(nr, B)) \geq \frac{p}{r}\cdot\frac{1}{np}\log(\alpha\ell(np, B_1)) + \frac{q}{r}\cdot\frac{1}{nq}\log(\alpha\ell(nq, B_2)),$$

and by (2.6) and the definitions of ℓ_f and J_f,

$$\ell_f(B) \geq \frac{p}{r}\ell_f(B_1) + \frac{q}{r}\ell_f(B_2),$$
$$-\ell_f(B) \leq \frac{p}{r}J_f(u_1) + \frac{q}{r}J_f(u_2),$$

which implies: for every $\beta \in (0,1) \cap \mathbb{Q}$,

$$J_f(\beta u_1 + (1-\beta)u_2) \leq \beta J_f(u_1) + (1-\beta)J_f(u_2). \tag{2.9}$$

Finally, (2.9) and the lower semicontinuity of J_f imply

$$J_f(\gamma u_1 + (1-\gamma)u_2) \leq \gamma J_f(u_1) + (1-\gamma)J_f(u_2),$$

for all $\gamma \in [0,1]$. □

Lemma 2.4 *Assume that* (2.3) *holds. Then:*

1. *For every open set $G \subset E$,*

$$\varliminf_n n^{-1}\log\ell(n, G) \geq -\inf_{u\in G} J_f(u). \tag{2.10}$$

2. *For every compact set $K \subset E$,*

$$\overline{\lim_n}\, n^{-1} \log \ell(n, K) \leq - \inf_{u \in K} J_f(u). \tag{2.11}$$

3. *For every closed convex set $F \subset E$,*

$$\alpha\ell(n, F) \leq \exp\left[-n \inf_{u \in F} J_f(u)\right], \tag{2.12a}$$

$$\overline{\lim_n}\, n^{-1} \log \ell(n, F) \leq - \inf_{u \in F} J_f(u). \tag{2.12b}$$

Proof.

1. Let $u \in G$, and let B be an open convex set such that $u \in B \subset G$. Then

$$\begin{aligned} \underline{\lim_n}\, n^{-1} \log \ell(n, G) &\geq \underline{\lim_n}\, n^{-1} \log \ell(n, B) \\ &= \ell_f(B) \geq -J_f(u), \end{aligned}$$

proving (2.10).

2. Let $a > -\inf_{u \in K} J_f(u)$. Then for every $w \in K$ we have $a > -J_f(w)$ and there exists an open convex set G_w such that $w \in G_w$, $a > \ell(G_w)$. Let $\{G_{w_j} : j = 1, \ldots, k\}$ be a finite subcover of K. Then

$$\begin{aligned} \overline{\lim_n}\, n^{-1} \log \ell(n, K) &\leq \overline{\lim_n}\, n^{-1} \log \ell\left(n, \bigcup_{j=1}^{k} G_{w_j}\right) \\ &\leq \overline{\lim_n}\, n^{-1} \log \left(k \max_{1 \leq j \leq k} \ell(n, G_{w_j})\right) \\ &= \max_{1 \leq j \leq k} \ell_f(G_{w_j}) < a. \end{aligned}$$

Now let $a \to -\inf_{u \in K} J_f(u)$, proving (2.11).

3. Let K be a compact convex set, $K \subset F$. By Lemma 2.2, for $n, p \in \mathbb{N}$,

$$\begin{aligned} (\alpha\ell(n, K))^p &\leq \alpha\ell(np, K), \\ \frac{1}{n} \log \alpha\ell(n, K) &\leq \frac{1}{np} \log \alpha\ell(np, K). \end{aligned}$$

Letting $p \to \infty$, we have by (2.11),

$$\begin{aligned} \frac{1}{n} \log \alpha\ell(n, K) &\leq - \inf_{u \in K} J_f(u), \\ \alpha\ell(n, K) &\leq \exp\left[-n \inf_{u \in F} J_f(u)\right]. \end{aligned} \tag{2.13}$$

Let $\ell(n, \cdot) = \mu$. By the regularity of finite measures on E,

$$\mu(F) = \sup\{\mu(K) : K \text{ compact},\ K \subset F\}.$$

But if $\hat{K}$ is the closed convex hull of a compact set $K \subset F$, then $\hat{K}$ is compact by Mazur's theorem (Conway, 1985, p. 180) and $\hat{K} \subset F$. Therefore

$$\mu(F) = \sup\{\mu(K) \colon K \text{ compact convex}, \ K \subset F\}.$$

The first statement (2.12a) follows from this fact and (2.13). The second statement (2.12b) is then immediate. □

Remark 2.5 Let $c = \sup_{x \in S} \|f(x)\|$, and let $B(c) = \{u \in E \colon \|u\| \le c\}$. Then $J_f(u) = \infty$ for $u \notin B(c)$. This is clear from the fact that $\ell_f(B) = -\infty$ if B is an open convex set disjoint from $B(c)$ and the definition of J_f.

We proceed to identify J_f analytically in terms of the convergence parameters of the transform kernels

$$K_{f,\xi}(x, A) = \int_A e^{\langle f(y), \xi \rangle} P(x, dy), \qquad \xi \in E^*.$$

Recall that $\Lambda_f(\xi) = \Lambda(\langle f, \xi \rangle) = -\log R(K_{f,\xi})$.

Lemma 2.6

$$J_f = \Lambda_f^*.$$

Proof. We will prove

$$\Lambda_f = J_f^*. \tag{2.14}$$

By Lemma 2.3, J_f is convex and lower semicontinuous. Therefore it is $\sigma(E, E^*)$-lower semicontinuous. Also, J_f is not identically equal to ∞. For, if $J_f \equiv \infty$, then $J_f^* \equiv -\infty$; but by (2.14), since $(1_C, \nu)$ is a small pair (Section B.2),

$$\begin{aligned} J_f^*(0) = \Lambda_f(0) &= \overline{\lim_n}\, n^{-1} \log \nu P^n 1_C \\ &\ge \log(\alpha \nu(C)), \end{aligned}$$

since $\nu P^n 1_C \ge \alpha^n (\nu(C))^{n+1}$, as in the proof of Lemma 2.3. It follows now from (2.14) and Proposition F.1 that $\Lambda_f^* = J_f^{**} = J_f$.

To prove (2.14), let $\rho_n(A) = \ell(n, A)$ for $A \in \mathcal{E} \cap B(c)$. By Lemma 2.4, the sequence of subprobability measures $\{\rho_n\}$ on $(B(c), \mathcal{E} \cap B(c))$ satisfies:

1. For every open set $G \subset B(c)$,

$$\underline{\lim_n}\, n^{-1} \log \rho_n(G) \ge -\inf_{u \in G} J_f(u).$$

2. For every closed convex set $F \subset B(c)$,

$$\overline{\lim_n}\, n^{-1} \log \rho_n(F) \leq -\inf_{u \in F} J_f(u).$$

For $\xi \in E^*$ and $u \in B(c)$, let $\varphi_\xi(u) = \langle u, \xi \rangle$. Then $\varphi_\xi \colon B(c) \to \mathbb{R}$ is continuous, bounded and moreover,

$$\varphi_\xi^{-1}(H) = B(c) \cap \{u \in E \colon \langle u, \xi \rangle \in H\}$$

is a closed convex set in $B(c)$ for every closed interval H. It follows from Remark E.2 that Proposition E.1 applies to $\{\rho_n\}$ and φ_ξ and therefore for $\xi \in E^*$,

$$\begin{aligned} \lim_n\, n^{-1} \log \int \exp(n\varphi_\xi)\, d\rho_n &= \sup_{u \in B(c)} \left[\varphi_\xi(u) - J_f(u)\right] \\ &= \sup_{u \in E} \left[\langle u, \xi \rangle - J_f(u)\right] = J_f^*(\xi) \end{aligned} \tag{2.15}$$

by Remark 2.5. But

$$\begin{aligned} \int \exp(n\varphi_\xi)\, d\rho_n &= \mathbb{E}_\nu \left\{ \left[\exp\left(\sum_{j=0}^{n-1} \langle f(X_j), \xi \rangle \right) \right] \mathbf{1}_C(X_{n-1}) \right\} \\ &= \nu_\xi K_{f,\xi}^{n-1} \mathbf{1}_C, \end{aligned}$$

where $d\nu_\xi = e^{\langle f, \xi \rangle} d\nu$. Since by (2.3) $(\mathbf{1}_C, \nu_\xi)$ is a $K_{f,\xi}$-small pair for every $\xi \in E^*$, we have by the definition of $\Lambda_f(\xi)$ and (C.1),

$$\Lambda_f(\xi) = \overline{\lim_n}\, n^{-1} \log \nu_\xi K_{f,\xi}^n \mathbf{1}_C. \tag{2.16}$$

Now (2.15) and (2.16) imply (2.14). □

By Lemmas 2.4 and 2.6, we have: if P is a Markov kernel satisfying (2.3), then for every open set $G \subset E$,

$$\underline{\lim_n}\, n^{-1} \log \mathbb{P}_\nu \left[n^{-1} S_n \in G\right] \geq -\inf_{u \in G} \Lambda_f^*(u). \tag{2.17}$$

Let P_t, $\mathbb{P}_\nu^{(t)}$, Λ_t be as in Appendix D. Then by (2.17) and (D.2) we have: for every open set $G \subset E$,

$$\underline{\lim_n}\, n^{-1} \log \mathbb{P}_\nu^{(t)} \left[n^{-1} S_n \in G\right] \geq -\inf_{u \in G} \left(\Lambda_f^{(t)}\right)^* (u), \tag{2.18}$$

where $\Lambda_f^{(t)}(\xi) = \Lambda_t(\langle f, \xi \rangle)$.

Let $\sigma_t(n)$ be as in Appendix D.

Lemma 2.7 *Let $c = \sup_{x \in S} \|f(x)\|$. For G open in E, $\epsilon > 0$, let $G_\epsilon = \{u \in E \colon d(u, G^c) > \epsilon\}$. Then for every $\mu \in \mathcal{P}(S)$,*

$$\mathbb{P}_\mu^{(t)} \left[n^{-1} S_n(f) \in G_\epsilon\right] \leq \mathbb{P}_\mu \left[n^{-1} S_n(f) \in G\right] + r_n(t, \epsilon),$$

where

$$r_n(t,\epsilon) = \mathbb{P}\left[\frac{\sigma_t(n)}{n} \geq 1 + \frac{\epsilon}{2c}\right].$$

Proof. In Lemma D.1, let

$$F((x_j)_{j\geq 0}) = \mathbf{1}_{G_\epsilon}\left(n^{-1}\sum_{j=0}^{n-1} f(x_j)\right).$$

Then by Lemma D.1,

$$\begin{aligned}\mathbb{P}_\mu^{(t)}\left[n^{-1}S_n(f) \in G_\epsilon\right] &= \mathbb{E}_\mu^{(t)}F = \mathbb{E}\,\mathbb{E}_\mu(F \circ \Phi_t) \\ &= \mathbb{E}\,\mathbb{E}_\mu \mathbf{1}_{G_\epsilon}\left(n^{-1}\sum_{j=0}^{n-1} f\left(X_{\sigma_t(j)}\right)\right).\end{aligned} \tag{2.19}$$

Let $H = \{\sigma_t(0), \ldots, \sigma_t(n-1)\}$, $K = \{0, \ldots, n-1\}$. Then

$$\begin{aligned}\left\|\sum_{j=0}^{n-1} f\left(X_{\sigma_t(j)}\right) - \sum_{j=0}^{n-1} f(X_j)\right\| &\leq c\,[\mathrm{card}(H \setminus K) + \mathrm{card}(K \setminus H)] \\ &= 2c\,\mathrm{card}(H \setminus K) \\ &\leq 2c\,(\sigma_t(n-1) - n + 1) \leq 2c(\sigma_t(n) - n),\end{aligned} \tag{2.20}$$

since $\mathrm{card}(H) = \mathrm{card}(K)$ and $H \setminus K \subset [n, \sigma_t(n-1)] \cap \mathbb{N}$.

Next,

$$\begin{aligned}\mathbf{1}_{G_\epsilon}\left(n^{-1}\sum_{j=0}^{n-1} f\left(X_{\sigma_t(j)}\right)\right) &\leq \mathbf{1}_G\left(n^{-1}\sum_{j=0}^{n-1} f(X_j)\right) \\ &\quad + \mathbf{1}_{[\epsilon,\infty)}\left(n^{-1}\left\|\sum_{j=0}^{n-1}\left(f\left(X_{\sigma_t(j)}\right) - f(X_j)\right)\right\|\right),\end{aligned}$$

and by (2.19)–(2.20),

$$\mathbb{P}_\mu^{(t)}\left[n^{-1}S_n(f) \in G_\epsilon\right] \leq \mathbb{P}_\mu\left[n^{-1}S_n(f) \in G\right] + \mathbb{P}\left[\frac{2c}{n}(\sigma_t(n) - n) \geq \epsilon\right]. \qquad \square$$

The following lemma is elementary.

Lemma 2.8 *For* $b > 1$,

$$\lim_{t\to 0}\overline{\lim_{n}}\, n^{-1}\log\mathbb{P}\left[\frac{\sigma_t(n)}{n} \geq b\right] = -\infty.$$

Proof. As in (D.7),

$$\mathbb{E}\exp(\lambda\sigma_t(n)) = \left(\frac{(1-t)e^{\lambda}}{1-te^{\lambda}}\right)^n \qquad \text{for } 0 < \lambda < -\log t,$$

$$\mathbb{P}\left[\frac{\sigma_t(n)}{n} \geq b\right] \leq e^{-n\lambda b}\left(\frac{(1-t)e^{\lambda}}{1-te^{\lambda}}\right)^n.$$

Therefore for $0 < \lambda < -\log t$,

$$\overline{\lim_n}\, n^{-1}\log\mathbb{P}\left[\frac{\sigma_t(n)}{n} \geq b\right] \leq -\lambda(b-1) + \log\left(\frac{1-t}{1-te^{\lambda}}\right),$$

and it follows that for all $\lambda > 0$,

$$\overline{\lim_{t\to 0}}\,\overline{\lim_n}\, n^{-1}\log\mathbb{P}\left[\frac{\sigma_t(n)}{n} \geq b\right] \leq -\lambda(b-1).$$

Now let $\lambda \to \infty$. □

Proof of Theorem 2.1

1. For $0 < t < 1$, G open $\subset E$,

$$\underline{\lim_n}\, n^{-1}\log\inf_{\mu\in M}\mathbb{P}^{(t)}_{\mu}\left[n^{-1}S_n \in G\right] \geq -\inf_{u\in G}\left(\Lambda^{(t)}_f\right)^*(u). \tag{2.21}$$

For, let a petite set D and $h \in \mathbb{N}_0$ be such that

$$\delta = \inf\left\{\sum_{i=0}^{h}\mu P^i(D) : \mu \in M\right\} > 0.$$

Let $p \in \mathbb{N}$, $\alpha > 0$, $\lambda \in \mathcal{P}(S)$ be such that $\sum_{j=1}^{p} P^j \geq \alpha(\mathbf{1}_D \otimes \lambda)$. Then

$$\sum_{i=0}^{h} P^i \sum_{j=1}^{p} P^j \geq \alpha\left(\sum_{i=0}^{h} P^i\mathbf{1}_D\right)\otimes\lambda,$$

which implies that for some $\alpha' > 0$,

$$\sum_{j=1}^{h+p} P^j \geq \alpha'\left(\sum_{i=0}^{h} P^i\mathbf{1}_D\right)\otimes\lambda. \tag{2.22}$$

By irreducibility, there exists $k \in \mathbb{N}$ such that $\gamma = \lambda P^k\mathbf{1}_C > 0$. Then by (2.22) and (D.1),

$$Q \triangleq \left(\sum_{j=1}^{h+p} P^j\right)(P^{k+m}) \geq \alpha'\left[\left(\sum_{i=0}^{h} P^i\mathbf{1}_D\right)\otimes\lambda\right]P^k\beta(\mathbf{1}_C\otimes\nu)$$

$$= \alpha'\beta\gamma\left(\sum_{i=0}^{h} P^i\mathbf{1}_D\right)\otimes\nu.$$

Next, for some $\rho_t > 0$,

$$P_t \geq \rho_t Q \geq \rho_t \alpha' \beta \gamma \left(\sum_{i=0}^{h} P^i \mathbf{1}_D \right) \otimes \nu.$$

For $\mu \in M$,

$$\mu P_t \geq \rho_t \alpha' \beta \gamma \left(\sum_{i=0}^{h} \mu P^i \mathbf{1}_D \right) \nu \geq \eta_t \nu, \tag{2.23}$$

where $\eta_t = \rho_t \alpha' \beta \gamma \delta$.

Let $u \in G$, and let $\epsilon > 0$ be such that $u \in G_\epsilon$. Let $n_0 \in \mathbb{N}$ be such that $2\|f\| < \epsilon n_0$. Then for $n \geq n_0$,

$$\left\| n^{-1} S_n - n^{-1} T_n \right\| \leq n^{-1} 2\|f\| < \epsilon,$$

where $T_n = \sum_{j=1}^{n} f(X_j)$, and therefore

$$\left[n^{-1} T_n \in G_\epsilon \right] \subset \left[n^{-1} S_n \in G \right].$$

By (2.23), setting $\mu_t = \mu P_t$, for $n \geq n_0$,

$$\begin{aligned} \inf_{\mu \in M} \mathbb{P}_\mu^{(t)} \left[n^{-1} S_n \in G \right] &\geq \inf_{\mu \in M} \mathbb{P}_\mu^{(t)} \left[n^{-1} T_n \in G_\epsilon \right] \\ &= \inf_{\mu \in M} \mathbb{P}_{\mu_t}^{(t)} \left[n^{-1} S_n \in G_\epsilon \right] \\ &\geq \eta_t \mathbb{P}_\nu^{(t)} \left[n^{-1} S_n \in G_\epsilon \right]. \end{aligned} \tag{2.24}$$

Statement (2.21) follows from (2.18) and (2.24).

2. For $u \in E$,

$$\left(\Lambda_f^{(t)} \right)^* (u) \leq \Lambda_f^*(u) - \log(1 - t). \tag{2.25}$$

This follows from (D.5) in Proposition D.2 with $g = \langle f, \xi, \rangle$.

3. Again, given $u \in G$, let $\epsilon > 0$ be such that $u \in G_\epsilon$. By Lemma 2.7, (2.21) and (2.25), for $0 < t < 1$,

$$\begin{aligned} -\Lambda_f^*(u) + \log(1 - t) &\leq \underline{\lim}_n \, n^{-1} \log \inf_{\mu \in M} \mathbb{P}_\mu^{(t)} \left[n^{-1} S_n \in G_\epsilon \right] \\ &\leq \max \left\{ \underline{\lim}_n \, n^{-1} \log \inf_{\mu \in M} \mathbb{P}_\mu \left[n^{-1} S_n \in G \right], \right. \\ &\qquad \left. \overline{\lim}_n \, n^{-1} \log r_n(t, \epsilon) \right\}. \end{aligned}$$

Letting $t \to 0$, the conclusion in Theorem 2.1 follows from Lemma 2.8. □

Remark 2.9 If P is uniformly recurrent, then the set M in Theorem 2.1 can be taken to be S. For, in this case S is petite; see Proposition B.9.

2.2 Lower Bounds for Empirical Measures

We turn now to large deviation lower bounds for empirical measures. Let V be a vector subspace of $B(S)$.

Theorem 2.10 *Let P be irreducible. Then for every set $M \subset \mathcal{P}(S)$ satisfying* (2.1) *and every measurable set $B \subset \mathcal{P}(S)$,*

$$\underline{\lim}_n \, n^{-1} \log \inf_{\mu \in M} \mathbb{P}_\mu[L_n \in B] \geq -\inf\{(\Lambda \mid V)^*(\lambda) \colon \lambda \in \operatorname{int}_V(B)\}$$
$$\geq -\inf\{\Lambda^*(\lambda) \colon \lambda \in \operatorname{int}_V(B)\}.$$

In particular, the conclusion holds if $M \subset S$ and M is petite.

Proof. Let $\lambda \in \operatorname{int}_V(B)$. By the definition of the V topology on $\mathcal{P}(S)$, there exist $f_1, \ldots, f_d \in V$ and U open in $\mathbb{R}^d$ such that

$$\lambda \in \Phi_f^{-1}(U) \subset B, \tag{2.26}$$

where $f = (f_1, \ldots, f_d)$ and $\Phi_f \colon \mathcal{P}(S) \to \mathbb{R}^d$ is defined by

$$\Phi_f(\rho) = \int f \, d\rho, \qquad \rho \in \mathcal{P}(S).$$

Now $\Phi_f(L_n) = n^{-1} S_n(f)$. Therefore by (2.26) and Theorem 2.1,

$$\begin{aligned} \underline{\lim}_n \, n^{-1} \log \inf_{\mu \in M} \mathbb{P}_\mu[L_n \in B] &\geq \underline{\lim}_n \, n^{-1} \log \inf_{\mu \in M} \mathbb{P}_\mu\left[L_n \in \Phi_f^{-1}(U)\right] \\ &= \underline{\lim}_n \, n^{-1} \log \inf_{\mu \in M} \mathbb{P}_\mu\left[n^{-1} S_n(f) \in U\right] \\ &\geq -\Lambda_f^*(\Phi_f(\lambda)). \end{aligned} \tag{2.27}$$

But

$$\begin{aligned} \Lambda_f^*(\Phi_f(\lambda)) &= \sup_{\xi \in \mathbb{R}^d} \left[\langle \Phi_f(\lambda), \xi \rangle - \Lambda_f(\xi)\right] \\ &= \sup_{\xi \in \mathbb{R}^d} \left[\int \langle f, \xi \rangle \, d\lambda - \Lambda(\langle f, \xi \rangle)\right] \\ &\leq \sup_{g \in V} \left[\int g \, d\lambda - \Lambda(g)\right] \\ &= (\Lambda \mid V)^*(\lambda), \end{aligned} \tag{2.28}$$

since $\langle f, \xi \rangle = \sum_{i=1}^d \xi_i f_i \in V$, where $\xi = (\xi_1, \ldots, \xi_d)$. The first inequality in Theorem 2.10 follows from (2.27)–(2.28). For the second inequality, note that $(\Lambda \mid V)^* \leq \Lambda^*$. □

Remark 2.11

1. In the context of Theorem 2.10, if P is uniformly recurrent then $M = S$ is a uniformity set because, by Proposition B.9, S is petite.
2. In general, the lower bound in terms of $(\Lambda \mid V)^*$ could be sharper than that in terms of Λ^*. However, if Λ^* is V-tight, then under suitable conditions on V (V.1′–V.4 in Chapter 4) we have $(\Lambda \mid V)^* = \Lambda^*$; see Proposition 4.6.

In view of Theorem 2.10, it is natural to ask what conditions are necessary for $(\Lambda \mid V)^*$ to be also a rate function for the upper bound for $\{\mathbb{P}_\mu[L_n \in \cdot]\}$ in the V topology. Theorem 2.12 addresses this question, which will be studied in a broader context in Chapter 8.

For $M \subset \mathcal{P}(S)$, $g \in B(S)$, let

$$\phi_M(g) = \overline{\lim_n}\, n^{-1} \log \sup_{\mu \in M} \mathbb{E}_\mu (\exp S_n(g)).$$

Theorem 2.12 *Assume that P is irreducible, $M \subset \mathcal{P}(S)$, and for every measurable set $B \subset \mathcal{P}(S)$,*

$$\overline{\lim_n}\, n^{-1} \log \sup_{\mu \in M} \mathbb{P}_\mu[L_n \in B] \leq -\inf\{(\Lambda \mid V)^*(\lambda) : \lambda \in \mathrm{cl}_V(B)\}.$$

Then $\phi_M \mid V = \Lambda \mid V$.

Proof. $\phi_M \mid V \leq \Lambda \mid V$: Let $g \in V$. We will apply Corollary E.4 with $\gamma_{\mu,n} = \mathbb{P}_\mu[L_n \in \cdot]$, $J = (\Lambda \mid V)^*$. By that result,

$$\phi_M(g) \leq \sup\left\{\int g\, d\lambda - (\Lambda \mid V)^*(\lambda) : \lambda \in \mathcal{P}(S)\right\} \leq \Lambda(g).$$

The inequality $\phi_M(g) \geq \Lambda(g)$ follows from (2.8) in de Acosta and Ney (2014) by taking any $\mu \in M$ (or, see part 2 of the proof of Theorem 4.2). □

We will see in Chapter 3 that under certain conditions $(\phi_M \mid V)^*$ is a rate function for the upper bound for $\{\sup_{\mu \in M} \mathbb{P}_\mu[L_n \in \cdot]\}$ in the V topology.

2.3 The $\sigma(B(S), \mathcal{P}(S, \psi))$ Lower Semicontinuity of Λ

Our main goal in this section is to prove the stated property of Λ, which will play an important role in Chapter 8. However, for other applications in Chapter 11, it will be convenient to prove a somewhat more general result.

Let $F : S \to \mathbb{R}$ be a measurable function such that $\inf F > 0$, and let $\Lambda_F : B(S) \to \mathbb{R}$ be defined by $\Lambda_F(g) = \Lambda(gF)$, $g \in B(S)$; if $h : S \to \mathbb{R}$ is measurable and P is irreducible, $R(K_h)$ is defined in Appendix C and $\Lambda(h) = -\log R(K_h)$.

Theorem 2.13 *Let P be irreducible. Then Λ_F is $\sigma(B(S), \mathcal{P}(S,\psi))$-lower semicontinuous. That is: if D is a directed set and $\{g_\delta\}_{\delta\in D} \subset B(S)$, $g \in B(S)$, and for all $\mu \in \mathcal{P}(S,\psi)$, $\lim_\delta \int g_\delta \, d\mu = \int g \, d\mu$, then*

$$\underline{\lim}_{\delta} \Lambda_F(g_\delta) \geq \Lambda_F(g).$$

We will need two lemmas.

Lemma 2.14 *Let $g, h \in F(S)$, and assume $g = h[\psi]$. Then $\Lambda(g) = \Lambda(h)$.*

Proof. Let (s, ν) be a $K_{g\wedge h}$-small pair. Then (s, ν) is a K_g (respectively, K_h)-small pair. By the definition of the convergence parameter (Appendix C), it suffices to show: for all $n \in \mathbb{N}$,

$$\nu K_g^n s = \nu K_h^n s.$$

We will prove by induction the stronger statement: for all $\rho \in \mathcal{P}(S,\psi)$, all $n \in \mathbb{N}$, $\rho K_g^n = \rho K_h^n$.

1. $n = 1$. For $A \in \mathcal{S}$,

$$\rho K_g(A) = \rho \left[P(e^g \mathbf{1}_A)\right] = (\rho P)(e^g \mathbf{1}_A).$$

   But $\rho P \ll \psi P \ll \psi$ (Proposition B.2). Since $g = h[\psi]$,

$$(\rho P)(e^g \mathbf{1}_A) = (\rho P)(e^h \mathbf{1}_A) = \rho K_h(A).$$

2. Assume that the statement holds for n. Then for $A \in \mathcal{S}$,

$$\rho K_g^{n+1}(A) = \rho K_g^n \left[P(e^g \mathbf{1}_A)\right] = \rho K_h^n \left[P(e^g \mathbf{1}_A)\right] = (\rho K_h^n P)(e^g \mathbf{1}_A).$$

   But $\rho K_h^n P \ll \psi K_h^n P \ll \psi$ (Proposition B.2). Therefore

$$\rho K_h^n P(e^g \mathbf{1}_A) = \rho K_h^n P(e^h \mathbf{1}_A) = \rho K_h^{n+1}(A). \qquad \square$$

Lemma 2.15 *Let Q be an irreducible sub-Markov kernel on $(S, \mathcal{S})$ such that ψ is a Q-maximal irreducibility probability measure. For $h \in F(S)$, let*

$$H_h(x, A) = \int_A e^{h(y)} Q(x, dy), \qquad x \in S, \ A \in \mathcal{S}.$$

Let $g_j \in B(S)$, $g_j \geq 0$ for $j \geq 1$, $g_0 \in B(S)$, $g_0 \geq 0$, and assume that $g_j \to g_0$ in the $\sigma(B(S), \mathcal{P}(S,\psi))$ topology. Then if λ is a Q-small measure and $f \in B(S)$, $f \geq 0$, for all $k \in \mathbb{N}$,

$$\underline{\lim}_{j} \lambda H_{g_j F}^k f \geq \lambda H_{g_0 F}^k f.$$

Proof. For $b \in \mathbb{N}$, let $F_b = F \wedge b$, and let $h_j = g_j F_b$, for $j \in \mathbb{N}_0$. For $A \in \mathcal{S}^{k+1}$, let

$$\mu(A) = \int \lambda(dx_0) \int Q(x_0, dx_1) \int \cdots \int Q(x_{k-1}, dx_k) \mathbf{1}_A(x_0, \ldots, x_k).$$

Since $e^t \geq 1 + t$ ($t \in \mathbb{R}$), we have, setting $\bar{x} = (x_0, \ldots, x_k)$,

$$\begin{aligned} \lambda H^k_{g_j F} f &\geq \lambda H^k_{h_j} f \\ &= \int \left(\exp \sum_{i=1}^{k} h_j(x_i) \right) f(x_k)\, d\mu(\bar{x}) \\ &\geq \int \left(\exp \sum_{i=1}^{k} h_0(x_i) \right) f(x_k)\, d\mu(\bar{x}) \\ &\quad + \int \left(\sum_{i=1}^{k} (h_j - h_0)(x_i) \right) \left(\exp \sum_{i=1}^{k} h_0(x_i) \right) f(x_k)\, d\mu(\bar{x}). \end{aligned} \tag{2.29}$$

Next, for $i = 1, \ldots, k$,

$$\begin{aligned} &\int (h_j - h_0)(x_i) \left(\exp \sum_{l=1}^{k} h_0(x_l) \right) f(x_k)\, d\mu(\bar{x}) \\ &\quad = \int \lambda(dx_0) \int Q(x_0, dx_1) \\ &\qquad \int \cdots \int Q(x_{i-1}, dx_i) \left(\exp \sum_{l=1}^{i} h_0(x_l) \right) (h_j - h_0)(x_i) \\ &\qquad \times \int Q(x_i, dx_{i+1}) \int \cdots \int Q(x_{k-1}, dx_k) \left(\exp \sum_{l=i+1}^{k} h_0(x_l) \right) f(x_k) \\ &\quad = \int \lambda H^i_{h_0}(dx) \left(H^{k-i}_{h_0} f(x) \right) (h_j - h_0)(x). \end{aligned} \tag{2.30}$$

By (2.29) and (2.30),

$$\lambda H^k_{g_j F} f \geq \lambda H^k_{h_0} f + \int (g_j - g_0)\, d\mu_k, \tag{2.31}$$

where

$$d\mu_k = F_b \sum_{i=1}^{k} \left(H^{k-i}_{h_0} f \right) d\left(\lambda H^i_{h_0} \right).$$

But $\lambda H^i_{h_0}$ is an H_{h_0}-small measure, hence an irreducibility measure, hence $\lambda H^i_{h_0} \ll \psi$ (Proposition B.2). Therefore μ_k is a finite measure such that $\mu_k \ll \psi$, and it follows that

$$\lim_j \int (g_j - g_0)\, d\mu_k = 0. \tag{2.32}$$

From (2.31) and (2.32) we obtain

$$\underline{\lim}_j\, \lambda H^k_{g_j F} f \geq \lambda H^k_{g_0 F_b} f.$$

By the monotone convergence theorem, as $b \to \infty$ we have

$$\lambda H^k_{g_0 F_b} f \uparrow \lambda H^k_{g_0 F} f.$$

This completes the proof. □

Proof of Theorem 2.13. We must prove: for all $a \in \mathbb{R}$,

$$L_a = \{g \in B(S) \colon \Lambda_F(g) \leq a\} \text{ is } \sigma(B(S), \mathcal{P}(S, \psi))\text{-closed.} \tag{2.33}$$

1. Let $T \colon B(S) \to L^\infty(\psi)$ be the canonical map assigning to each $g \in B(S)$ its equivalence class $\tilde{g} = T(g)$ in $L^\infty(\psi)$. Then T is $\sigma(B(S), \mathcal{P}(S, \psi))/\sigma(L^\infty(\psi), L^1(\psi))$-continuous. For $h \in L^\infty(\psi)$, let $\widetilde{\Lambda_F}(h) = \Lambda_F(g)$, where $h = \tilde{g}$. Then $\widetilde{\Lambda_F}$ is well defined by Lemma 2.14 and if

$$\tilde{L}_a = \left\{h \in L^\infty(\psi) \colon \widetilde{\Lambda_F}(h) \leq a\right\},$$

then $L_a = T^{-1}(\tilde{L}_a)$. Therefore to prove (2.33) it suffices to prove:

$$\tilde{L}_a \text{ is } \sigma(L^\infty(\psi), L^1(\psi))\text{-closed.} \tag{2.34}$$

2. Let $C_r = \{h \in L^\infty(\psi) \colon \|h\| \leq r\}$. Since $\tilde{L}_a$ is convex, by the Krein–Smulian theorem (Conway, 1985, p. 159), (2.34) will follow if we can prove: for all $r \geq 0$, $a \in \mathbb{R}^+$,

$$\tilde{L}_a \cap C_r \text{ is } \sigma(L^\infty(\psi), L^1(\psi))\text{-closed.} \tag{2.35}$$

3. Since the σ-algebra $\mathcal{S}$ is countably generated, the Banach space $L^1(\psi)$ is separable, and therefore each ball C_r of its dual space $L^\infty(\psi)$ is a compact metrizable space for the $\sigma(L^\infty(\psi), L^1(\psi))$ topology by the Banach–Alaoglu theorem (Conway, 1985, pp. 130, 134). Therefore in order to establish (2.35) it suffices to prove: if $\{h_j\}_{j\in\mathbb{N}}$ is a sequence in C_r and $h_j \to h \in C_r$ in the $\sigma(L^\infty(\psi), L^1(\psi))$ topology, then

$$\underline{\lim}_j\, \widetilde{\Lambda_F}(h_j) \geq \widetilde{\Lambda_F}(h). \tag{2.36}$$

4. In order to prove (2.36), it suffices to show: if $g_j, g_0 \in B_r$, where $B_r = \{g \in B(S) : \|g\| \le r\}$, and $g_j \to g_0$ in the $\sigma(B(S), \mathcal{P}(S,\psi))$ topology, then

$$\underline{\lim}_{j} \Lambda_F(g_j) \ge \Lambda_F(g_0). \tag{2.37}$$

For, suppose $\{h_j\}_{j\in\mathbb{N}}$ is a sequence in C_r and $h_j \to h \in C_r$ in the $\sigma(L^\infty(\psi), L^1(\psi))$ topology. For each $j \in \mathbb{N}$, let $g_j \in B(S)$ be such that $\tilde{g}_j = h_j, g_j \in B_r$; also, let $g_0 \in B(S)$ be such that $\tilde{g}_0 = h, g_0 \in B_r$. Then $g_j \to g_0$ in the $\sigma(B(S), \mathcal{P}(S,\psi))$ topology and by (2.37),

$$\underline{\lim}_{j} \widetilde{\Lambda_F}(h_j) = \underline{\lim}_{j} \Lambda_F(g_j) \ge \Lambda_F(g_0) = \widetilde{\Lambda_F}(h).$$

5. We will prove (2.37). Let $Q(x,A) = \int_A e^{q(y)}P(x,dy)$, where $q = -rF$. Let $m \in \mathbb{N}$, and let (t,λ) be a Q-small pair such that $\int t\,d\lambda > 0$ and $Q^m \ge t \otimes \lambda$.
Let H_h be as in Lemma 2.15. If $h \ge 0$, then

$$H_h^m \ge Q^m \ge t \otimes \lambda. \tag{2.38}$$

Assume $g_j, j \ge 1, g_0 \in B_r$, and $g_j \to g_0$ in the $\sigma(B(S), \mathcal{P}(S,\psi))$ topology. Let $h_j = (g_j+r)F$ for $j \in \mathbb{N}_0$. By (2.38) and Proposition C.3, since $g_j+r \ge 0$, for $n \ge 1$,

$$\begin{aligned}\Lambda_F(g_j) = \Lambda\left((g_j + r)F - rF\right) &= -\log R(K_{h_j+q}) \\ &= -\log R(H_{h_j}) \ge (nm)^{-1} \log \lambda H_{h_j}^{(n-1)m} t.\end{aligned}$$

Therefore for $n \ge 1$, by Lemma 2.15,

$$\begin{aligned}\underline{\lim}_{j} \Lambda_F(g_j) &\ge (nm)^{-1} \log \underline{\lim}_{j} \lambda H_{h_j}^{(n-1)m} t \\ &\ge (nm)^{-1} \log \lambda H_{h_0}^{(n-1)m} t,\end{aligned}$$

so again by (2.38) and Proposition C.3,

$$\begin{aligned}\underline{\lim}_{j} \Lambda_F(g_j) &\ge -\log R(H_{h_0}) \\ &= -\log R(K_{h_0+q}) = \Lambda(g_0F) = \Lambda_F(g_0).\end{aligned}$$

This completes the proof of Theorem 2.13. □

Remark 2.16 If P is an irreducible sub-Markov kernel – that is, $P(x,S) \le 1$ for all $x \in S$ – and Λ is defined as in the Markov case, so

$$\Lambda(g) = \overline{\lim_{n}}\, n^{-1} \log \nu K_g^n s,$$

then Theorem 2.13 still holds, with the same proof.

The following corollary of Theorem 2.13 provides more detailed analytic information about the functions Λ_f and Λ_f^* in Theorem 2.1. In Theorem 2.17, statements 2 and 3, Λ_f^* is identified in terms of Λ^*; see also Remark 2.18.

Theorem 2.17 *Let P be an irreducible sub-Markov kernel and let $f: S \to E$ be a bounded measurable function, where E is a separable Banach space. Then:*

1. *$\Lambda_f: E^* \to \mathbb{R}$ is $\sigma(E^*, E)$-lower semicontinuous.*
2. *For $u \in E$, let*

$$\widetilde{\Lambda_f}(u) = \begin{cases} \inf\left\{\Lambda^*(\mu) : \mu \in \Phi_f^{-1}(u)\right\} & \text{if } u \in \Phi_f(\mathcal{P}(S)), \\ \infty & \text{otherwise,} \end{cases}$$

   where $\Phi_f: \mathcal{P}(S) \to E$ is defined by $\Phi_f(\mu) = \int f\, d\mu$. Then $\Lambda_f^* = \widetilde{\Lambda_f}^{**}$.
3. *If V is a vector subspace of $B(S)$, $\langle f, \xi\rangle \in V$ for all $\xi \in E^*$, and Λ^* is V-tight, then $\Lambda_f^* = \widetilde{\Lambda_f}$.*

Proof. For $\xi \in E^*$,

$$\begin{aligned} \widetilde{\Lambda_f}^*(\xi) &= \sup_{u \in E}\left\{\langle u, \xi\rangle - \widetilde{\Lambda_f}(u)\right\} \\ &= \sup\left\{\langle u, \xi\rangle - \inf\left[\Lambda^*(\mu) : \mu \in \Phi_f^{-1}(u)\right] : u \in \Phi_f(\mathcal{P}(S))\right\} \\ &= \sup\left\{\sup\left[\left\langle \textstyle\int f\, d\mu, \xi\right\rangle - \Lambda^*(\mu) : \mu \in \Phi_f^{-1}(u)\right] : u \in \Phi_f(\mathcal{P}(S))\right\} \\ &= \sup\left\{\int \langle f, \xi\rangle\, d\mu - \Lambda^*(\mu) : \mu \in \mathcal{P}(S)\right\} = \Lambda(\langle f, \xi\rangle) = \Lambda_f(\xi), \end{aligned} \tag{2.39}$$

by Theorem 2.13, Remark 2.16, and Proposition F.3. Since $\widetilde{\Lambda_f}^*$ is $\sigma(E^*, E)$-lower semicontinuous, this proves statement 1.

It follows from (2.39) that $\Lambda_f^* = \widetilde{\Lambda_f}^{**}$, proving statement 2.

To prove that $\Lambda_f^* = \widetilde{\Lambda_f}$, we must show that $\widetilde{\Lambda_f}^{**} = \widetilde{\Lambda_f}$. To prove this, by Proposition F.1 we must show that $\widetilde{\Lambda_f}: E \to \mathbb{R}$ is convex, proper, and $\sigma(E, E^*)$-lower semicontinuous. From its definition, it is easily seen that $\widetilde{\Lambda_f}$ is convex, and by (2.39) it is proper. Let $a \geq 0$,

$$\begin{aligned} L_a &= \{\mu \in \mathcal{P}(S) : \Lambda^*(\mu) \leq a\}, \\ M_a &= \left\{u \in E : \widetilde{\Lambda_f}(u) \leq a\right\}. \end{aligned}$$

We will prove that the map Φ_f is $\sigma(\mathcal{P}(S), V)/\sigma(E, E^*)$-continuous and $\Phi_f(L_a) = M_a$. Since L_a is V-compact, it will follow that M_a is $\sigma(E, E^*)$-compact and therefore $\widetilde{\Lambda_f}$ is $\sigma(E, E^*)$-lower semicontinuous.

To prove the first assertion, assume that $\{\mu_\alpha\}_{\alpha\in D}$ is a net in $\mathcal{P}(S)$, $\mu \in \mathcal{P}(S)$, and $\int g\,d\mu_\alpha \to \int g\,d\mu$ for all $g \in V$. Then for all $\xi \in E^*$,

$$\left\langle \Phi_f(\mu_\alpha), \xi \right\rangle = \int \langle f, \xi \rangle\, d\mu_\alpha \longrightarrow \int \langle f, \xi \rangle\, d\mu = \left\langle \Phi_f(\mu), \xi \right\rangle,$$

proving the continuity of Φ_f.

To prove the second assertion: clearly $\Phi_f(L_a) \subset M_a$. On the other hand, if $u \in M_a$ then for all $\epsilon > 0$, $\Phi_f^{-1}(u) \cap L_{a+\epsilon}$ is a nonempty V-compact set. Therefore

$$\Phi_f^{-1}(u) \cap L_a \neq \emptyset, \quad \text{or } u \in \Phi_f(L_a).$$

The proof of statement 3 is complete. □

Remark 2.18 By Theorem 4.2, $\Lambda^* = I_\psi$, so one can write

$$\widetilde{\Lambda_f}(u) = \begin{cases} \inf\left\{ I_\psi(\mu) \colon \mu \in \Phi_f^{-1}(u) \right\} & \text{if } u \in \Phi_f(\mathcal{P}(S)), \\ \infty & \text{otherwise.} \end{cases}$$

Theorem 2.17.3 is consistent with the following considerations: if $\{\mathbb{P}_\nu[L_n \in \cdot]\}$ satisfied the large deviation principle in the V topology with V-tight rate function $\Lambda^* = I_\psi$ then, since Φ_f is $\sigma(\mathcal{P}(S), V)/\sigma(E, E^*)$-continuous, by the contraction principle (Rassoul-Agha and Seppäläinen, 2015, Chapter 3)

$$\left\{ \mathbb{P}_\nu \left[n^{-1} S_n(f) \in \cdot \right] \right\} = \left\{ \mathbb{P}_\nu \left[L_n \in \cdot \right] \circ \Phi_f^{-1} \right\}$$

would satisfy the large deviation principle in the $\sigma(E, E^*)$ topology with $\sigma(E, E^*)$-tight rate function $\widetilde{\Lambda_f}$.

In the following result we obtain some further analytic properties of Λ, mainly as a consequence of Theorem 2.13.

Corollary 2.19 *Assume that P is irreducible.*

1. *Let $g_n, g \in B(S)$, and assume $g_n \uparrow g$ pointwise. Then $\Lambda(g_n) \uparrow \Lambda(g)$.*
2. *The following conditions are equivalent:*
   (i) *If $0 \le g_n \in B(S)$ and $g_n \downarrow 0$ pointwise, then* $\lim_n \Lambda(g_n) \le 0$.
   (ii) *Λ^* is τ-tight.*
3. *If either* 2(i) *or* 2(ii) *holds and $g_n, g \in B(S)$, $g_n \downarrow g$ pointwise, then $\Lambda(g_n) \downarrow \Lambda(g)$.*

Proof.

1. Since $g_n \le g$, we have $\lim_n \Lambda(g_n) \le \Lambda(g)$. On the other hand, $g_n \uparrow g$ pointwise implies by monotone convergence that for all $\mu \in \mathcal{P}(S)$,

$$\int g_n\, d\mu \uparrow \int g\, d\mu,$$

so $g_n \to g$ in the $\sigma(B(S), \mathcal{P}(S))$ topology. By Theorem 2.13, $\lim_n \Lambda(g_n) \geq \Lambda(g)$.

2. Condition 2(i) $\implies$ 2(ii): By Proposition H.1, it suffices to prove: for all $a \geq 0$, $0 \leq g_n \in B(S)$, $g_n \downarrow 0$ pointwise imply

$$\sup_{\mu \in L_a} \int g_n \, d\mu \to 0,$$

where $L_a = \{\mu \in \mathcal{P}(S) : \Lambda^*(\mu) \leq a\}$. For $t > 0$, $\mu \in L_a$,

$$\int (tg_n) \, d\mu \leq \Lambda(tg_n) + \Lambda^*(\mu) \leq \Lambda(tg_n) + a.$$

Therefore

$$\sup_{\mu \in L_a} \int g_n \, d\mu \leq t^{-1} \Lambda(tg_n) + t^{-1} a.$$

By Condition 2(i), $\lim_n \Lambda(tg_n) \leq 0$, and therefore

$$\limsup_n \sup_{\mu \in L_a} \int g_n \, d\mu \leq t^{-1} a.$$

Now let $t \to \infty$.

3. Condition 2(ii) $\implies$ 2(i): By Theorem 2.13 and Proposition F.3 with $V = B(S)$, $\Gamma = \Lambda$, for all $g \in B(S)$,

$$\Lambda(g) = \sup \left\{ \int g \, d\mu - \Lambda^*(\mu) : \mu \in \mathcal{P}(S) \right\}. \tag{2.40}$$

Let $0 \leq g_n \in B(S)$, $g_n \downarrow 0$ pointwise. By 2(ii) and Proposition H.1, for all $a \geq 0$,

$$\lim_n \sup_{\mu \in L_a} \int g_n \, d\mu = 0. \tag{2.41}$$

By (2.40) and (2.41),

$$\lim_n \Lambda(g_n) \leq \lim_n \max \left\{ \sup_{\mu \in L_a} \int g_n \, d\mu, c - a \right\},$$

where $c = \sup g_1$,

$$= \max\{0, c - a\}.$$

But a is arbitrary. This proves condition 2(i).

4. Let $p > 0$, $q > 0$, $p + q = 1$. By the convexity of Λ,

$$\begin{aligned} \Lambda(g_n) &= \Lambda\left(p(p^{-1}g) + q(q^{-1}(g_n - g))\right) \\ &\leq p\Lambda(p^{-1}g) + q\Lambda(q^{-1}(g_n - g)). \end{aligned}$$

Since $q^{-1}(g_n - g) \downarrow 0$ pointwise, by 2(i), $\Lambda(q^{-1}(g_n - g)) \to 0$. Therefore

$$\lim_n \Lambda(g_n) \leq p\Lambda(p^{-1}g). \tag{2.42}$$

Since the function $t \mapsto \Lambda(tg)$, $t \in \mathbb{R}$, is continuous, being a finite convex function on $\mathbb{R}$, letting $p \to 1$ in (2.42) we get

$$\lim_n \Lambda(g_n) \leq \Lambda(g).$$

On the other hand, since $g_n \geq g$, we have

$$\lim_n \Lambda(g_n) \geq \Lambda(g).$$ □

Remark 2.20 Corollary 2.19(1) is generalized in Corollary C.5.

2.4 Notes

Theorem 2.1 for $M = \{\mu\}$ was proved in de Acosta (1988) and later in de Acosta and Ney (1998) with a much simpler proof. In the present work, this proof is further simplified; also, background material is presented in detail. The proof in de Acosta and Ney (1998) is based on ideas first developed in Dinwoodie and Ney (1995) for the study of lower bounds for empirical measures when S is Polish and $V = C_b(S)$. In Dinwoodie and Ney (1995) the basic minorization property of irreducible kernels is used to carry out a subadditivity/identification/convex analysis argument yielding lower bounds for empirical measures with rate function Λ^*.

The subadditivity/identification/convex analysis approach to large deviations had been previously developed in the i.i.d. case in Bahadur and Zabell (1979), and is presented in most of the books mentioned in Section 1.2. This approach was extended to the Markov case in Stroock (1984) and Deuschel and Stroock (1989); more on this follows below.

Theorem 2.10 for the τ topology was proved in de Acosta (1990) on the basis of work done in de Acosta (1988), with a more restrictive condition on M. The present, much simpler proof, based on Theorem 2.1, was given in de Acosta (1994b) in the case $M = \{\mu\}$.

We will now discuss several lower bound results in the literature, obtained by different methods. All of these results follow from Theorem 2.10 and actually hold in the τ topology. The assumptions in these results imply that $I = I_\psi = \Lambda^*$.

In Donsker and Varadhan (1976), the lower bound for $\{\mathbb{P}_x[L_n \in \cdot]\}$, $x \in S$, in the weak topology with rate function I is proved when S is Polish, P is irreducible and satisfies $P(x, \cdot) \ll \psi$ for all $x \in S$. As we will see in Chapter 4,

Remark 4.14(1), under these conditions $I = I_\psi = \Lambda^*$. It is also shown that if P is strong Feller (Appendix B), then every compact set in S is a uniformity set for the lower bound. But under the strong Feller condition every compact set is petite (Proposition B.10). Therefore the uniformity result in Donsker and Varadhan (1976) follows from Theorem 2.10.

In the same context, in Jain (1990) P is assumed to be irreducible but the absolute continuity assumption in Donsker and Varadhan (1976) is relaxed, replacing it by the weaker condition $\operatorname{dom} I \subset \mathcal{P}(S, \psi)$, which is equivalent to $I = I_\psi$. The condition given in Jain (1990) implying that every compact set is a uniformity set for the lower bound again implies that every compact set is petite (Proposition B.10), and the result in Jain (1990) follows from Theorem 2.10.

The first condition given in Dupuis and Ellis (1997) for the lower bound in the weak or τ topology for S Polish implies that P is irreducible and there exists $\ell \in \mathbb{N}$ such that $P^\ell(x, \cdot) \ll \psi$ all $x \in S$. By Remark 4.14(1), this implies $I = I_\psi = \Lambda^*$. In order to prove that every compact set in S is a uniformity set for the lower bound, the following condition is imposed in Dupuis and Ellis (1997): every point in S has a petite neighborhood. But since the union of petite sets is petite (Proposition B.5), it easily follows from this condition that every compact set is petite. Therefore the uniformity result in Dupuis and Ellis (1997) in the weak or τ topology follows from Theorem 2.10.

For S Polish and under very strong assumptions on P, a subadditivity/identification/convex analysis approach is developed in Stroock (1984) and Deuschel and Stroock (1989), proving that S is a uniformity set for the lower bound in the weak topology with rate function I. As we will see in Remark 4.14(2), the condition on P in Deuschel and Stroock (1989) implies that P is irreducible, $I = I_\psi = \Lambda^*$, and S is petite. Therefore the result follows from Theorem 2.10.

Lemma 3.4 of de Acosta and Ney (1998) – the proof of which showed the relevance of the Krein–Smulian theorem – is recovered here in Theorem 2.17.1.

3

Upper Bounds I

In this chapter we will give a necessary and sufficient analytic condition for the upper bound in the V topology for

$$\left\{\sup_{\nu \in M} \mathbb{P}_\nu[L_n \in \cdot]\right\}$$

with V-tight rate function $(\phi_M \mid V)^*$; here $M \subset \mathcal{P}(S)$ and, extending some definitions in Chapter 1, for $g \in B(S)$,

$$\phi_M(g) = \overline{\lim_n}\, n^{-1} \log \sup_{\nu \in M} \mathbb{E}_\nu\left(\exp S_n(g)\right)$$

and for $\mu \in \mathcal{P}(S)$,

$$(\phi_M \mid V)^*(\mu) = \sup\left\{\int g\, d\mu - \phi_M(g) \colon g \in V\right\}.$$

Since $\phi_M(0) = 0$, we have $(\phi_M \mid V)^*(\mu) \geq 0$. We will also discuss the connection between the condition for the upper bound presented here and the condition of exponential tightness.

3.1 Upper Bounds for Random Probability Measures and Empirical Measures

Although the upper bound is sharp in many situations, the Markov framework and the specific form of L_n are not used in the proof. We are thus led to formulate and prove an upper bound for more general random measures. The setup is as follows.

Let $(\Omega', \mathcal{A})$ be a measurable space, $\{\mathbb{P}_\alpha\}_{\alpha\in A}$ a family of probability measures on $(\Omega', \mathcal{A})$. Let $M_n\colon \Omega' \to \mathcal{P}(S)$, $n \in \mathbb{N}$, be a $\mathcal{A}/\mathcal{B}(\mathcal{P}(S), B(S))$-measurable map. For $g \in B(S)$, we define

$$\Gamma_A(g) = \overline{\lim_n}\, n^{-1} \log \sup_{\alpha\in A} \mathbb{E}_\alpha \left[\exp\left(n \int g \, dM_n\right)\right],$$

and for $\mu \in \mathcal{P}(S)$,

$$(\Gamma_A \mid V)^*(\mu) = \sup\left\{\int g \, d\mu - \Gamma_A(g)\colon g \in V\right\}. \tag{3.1}$$

In the case of lower bounds, in Theorem 2.10 we took V to be any vector subspace of $B(S)$. However, an arbitrary vector subspace of $B(S)$ in general lacks enough structure to provide a basis for the proof of the upper bound with a V-tight rate function, covering the cases i and ii described in Chapter 1. The following restrictions on V capture what is needed for the proof of the upper bound with a V-tight rate function, specially for the use of Daniell's theorem (Proposition G.1) in Theorem 3.1.

V is assumed to satisfy the following conditions:

V.1 V is a vector lattice $\subset B(S)$; that is, V is a vector subspace of $B(S)$ and for $f, g \in V$, $f \vee g$, $f \wedge g \in V$.

V.2 $1 \in V$.

V.3 The σ-algebra generated by V on S is $\mathcal{S}$.

Particular cases are $V = B(S)$ and, when S is a Polish space, $V = C_b(S)$.

Let V^* be the dual space of the normed vector space $(V, \|\cdot\|)$, where $\|\cdot\|$ is the supremum norm. For $\mu \in \mathcal{P}(S)$, let $\ell_\mu\colon V \to \mathbb{R}$ be defined by $\ell_\mu(f) = \int f \, d\mu$, $f \in V$. Then $\ell_\mu \in V^*$ and $\|\ell_\mu\| = 1$. By Lemma G.2, the map $\mu \mapsto \ell_\mu$ is injective; consequently, we regard $\mathcal{P}(S)$ as a subset of the unit ball U of V^*. Note that the V topology on $\mathcal{P}(S)$ is the relativized $\sigma(V^*, V)$ topology.

For use in Theorem 3.1, we extend (3.1) to V^*: for $\ell \in V^*$,

$$(\Gamma_A \mid V)^*(\ell) = \sup\{\ell(g) - \Gamma_A(g)\colon g \in V\}.$$

Theorem 3.1

1. *Let V be a vector space $\subset B(S)$. For every measurable set $B \subset \mathcal{P}(S)$ such that $\mathrm{cl}_V(B)$ is compact,*

$$\overline{\lim_n}\, n^{-1} \log \sup_{\alpha\in A} \mathbb{P}_\alpha[M_n \in B] \le -\inf\{(\Gamma_A \mid V)^*(\mu)\colon \mu \in \mathrm{cl}_V(B)\}. \tag{3.2}$$

2. *Assume that V satisfies V.1–V.3. Then the following conditions are equivalent:*

(i) *If* $0 \le g_k \in V$ *and* $g_k \downarrow 0$ *pointwise, then* $\Gamma_A(g_k) \to 0$.
(ii) *Inequality* (3.2) *holds for every measurable set* $B \subset \mathcal{P}(S)$ *and* $(\Gamma_A \mid V)^*$ *is* V*-tight.*

Proof.

1. Set $\Gamma_A = \Gamma$. Let $a = \inf\{(\Gamma \mid V)^*(\mu) \colon \mu \in \mathrm{cl}_V(B)\}$, and assume that $a < \infty$. Let $\epsilon > 0$, and for $g \in V$ let

$$C(g) = \left\{\mu \in \mathcal{P}(S) \colon \int g \, d\mu - \Gamma(g) > a - \epsilon\right\}.$$

Then

$$\mathrm{cl}_V(B) \subset \{\mu \in \mathcal{P}(S) \colon (\Gamma \mid V)^*(\mu) > a - \epsilon\} = \bigcup_{g \in V} C(g).$$

Since $\mathrm{cl}_V(B)$ is compact and $C(g)$ is V-open, there exist $g_1, \ldots, g_k \in V$ such that

$$\mathrm{cl}_V(B) \subset \bigcup_{i=1}^{k} C(g_i).$$

We have

$$\begin{aligned}
\mathbb{P}_\alpha[M_n \in B] &\le \sum_{i=1}^{k} \mathbb{P}_\alpha[M_n \in C(g_i)] \\
&= \sum_{i=1}^{k} \mathbb{P}_\alpha\left(\int g_i \, dM_n > \Gamma(g_i) + a - \epsilon\right) \\
&\le \sum_{i=1}^{k} e^{-n(\Gamma(g_i)+a-\epsilon)} \mathbb{E}_\alpha\left[\exp\left(n \int g_i \, dM_n\right)\right].
\end{aligned}$$

Let

$$h_n(g_i) = n^{-1} \log \sup_{\alpha \in A} \mathbb{E}_\alpha\left[\exp\left(n \int g_i \, dM_n\right)\right] - \Gamma(g_i).$$

Then

$$\sup_{\alpha \in A} \mathbb{P}_\alpha[M_n \in B] \le k e^{(-a+\epsilon)n} \exp\left(n \sup_i h_n(g_i)\right)$$

and therefore

$$\overline{\lim_n}\, n^{-1} \log \sup_{\alpha \in A} \mathbb{P}_\alpha[M_n \in B] \le -a + \epsilon.$$

But ϵ is arbitrary. This proves the statement for $a < \infty$. If $a = \infty$, one proceeds in a similar way.

2. Assumption 2(i) $\Rightarrow$ 2(ii): Let $B \subset \mathcal{P}(S)$ be a measurable set, and let $\overline{B}$ be its $\sigma(V^*, V)$-closure. Since $B \subset U$ and U is compact in the $\sigma(V^*, V)$ topology by the Banach–Alaoglu theorem (Conway, 1985, p. 130), $\overline{B}$ is $\sigma(V^*, V)$-compact.

   We prove first: setting $(\Gamma \mid V)^*(\ell) = \theta(\ell)$ for $\ell \in V^*$,

$$\overline{\lim_n}\, n^{-1} \log \sup_{\alpha \in A} \mathbb{P}_\alpha[M_n \in B] \le -\inf\left\{\theta(\ell) \colon \ell \in \overline{B}\right\}. \tag{3.3}$$

   The proof is similar to that of step 1 above. We will indicate the initial steps. Let $b = \inf\{\theta(\ell) \colon \ell \in \overline{B}\}$, and assume $b < \infty$. Let $\epsilon > 0$, and for $g \in V$ let

$$D(g) = \{\ell \in U \colon \ell(g) - \Gamma(g) > b - \epsilon\}.$$

   Then

$$\overline{B} \subset \{\ell \in U \colon \theta(\ell) > b - \epsilon\} = \bigcup_{g \in V} D(g).$$

   Exploiting the compactness of $\overline{B}$ and proceeding as in step 1, we arrive at (3.3). Again, the case $b = \infty$ is similar.

   Next, in order to prove (3.2) we must show

$$\inf\left\{\theta(\ell) \colon \ell \in \overline{B}\right\} = \inf\{(\Gamma \mid V)^*(\mu) \colon \mu \in \mathrm{cl}_V(B)\}. \tag{3.4}$$

   This will follow from the subsequent fact:

$$F \triangleq \{\ell \in V^* \colon \theta(\ell) < \infty\} \subset \mathcal{P}(S). \tag{3.5}$$

   For, assuming (3.5), since $\mathrm{cl}_V(B) = \overline{B} \cap \mathcal{P}(S)$, we have

$$\begin{aligned} \inf\left\{\theta(\ell) \colon \ell \in \overline{B}\right\} &= \inf\left\{\theta(\ell) \colon \ell \in \overline{B} \cap F\right\} \ge \inf\left\{\theta(\ell) \colon \ell \in \overline{B} \cap \mathcal{P}(S)\right\} \\ &= \inf\{(\Gamma \mid V)^*(\mu) \colon \mu \in \mathrm{cl}_V(B)\}, \end{aligned}$$

   proving (3.4), hence (3.2).

   In view of properties V.1–V.3, by Proposition G.1, to prove (3.5) it suffices to show: if $\ell \in V^*$ and $\theta(\ell) < \infty$, then

$$g \in V,\ g \ge 0 \text{ imply } \ell(g) \ge 0; \tag{3.6}$$

$$\ell(1) = 1; \tag{3.7}$$

$$\text{if } 0 \le g_k \in V \text{ and } g_k \downarrow 0 \text{ pointwise, then } \ell(g_k) \to 0. \tag{3.8}$$

   (a) Assume $\ell(g) < 0$ for some $g \in V$, $g \ge 0$. Then for $t < 0$ we have

$$\Gamma(tg) \le \Gamma(0),$$
$$t\ell(g) = \ell(tg) \le \Gamma(tg) + c,$$

   where $c = \theta(\ell)$. Letting $t \to -\infty$, we obtain a contradiction, proving (3.6).

(b) For all $t \in \mathbb{R}$,

$$t\ell(1) = \ell(t1) \leq \Gamma(t1) + c = t + c.$$

Therefore $t(\ell(1) - 1) \leq c$, which implies $\ell(1) = 1$. This is (3.7).

(c) Suppose $0 \leq g_k \in V$, $g_k \downarrow 0$ pointwise. Then for $t \in \mathbb{R}$,

$$t\ell(g_k) = \ell(tg_k) \leq \Gamma(tg_k) + c,$$
$$t\overline{\lim_k}\, \ell(g_k) \leq \overline{\lim_k}\, \Gamma(tg_k) + c = c,$$

by assumption 2(i), and it follows that $\ell(g_k) \to 0$. This is (3.8).

This completes the proof of (3.2) for any measurable set $B \subset \mathcal{P}(S)$. We turn now to the second assertion in 2(ii). For $a \geq 0$, let

$$L_a = \{\mu \in \mathcal{P}(S) \colon (\Gamma \mid V)^*(\mu) \leq a\}.$$

We will give two proofs of the V-compactness of L_a.

(a) Since $L_a \subset U$, its $\sigma(V^*, V)$-closure $\overline{L_a}$ is $\sigma(V^*, V)$-compact. Let $\ell \in \overline{L_a}$. By the $\sigma(V^*, V)$ lower semicontinuity of θ, we have $\theta(\ell) \leq a$. But, as shown above, this implies that $\ell \in \mathcal{P}(S)$, hence $\ell \in L_a$. Therefore $\overline{L_a} = L_a$, proving that L_a is V-compact.

(b) We will apply Proposition H.1. Clearly L_a is V-closed. Assume that $0 \leq g_k \in V$, $g_k \downarrow 0$ pointwise. Since for all $t > 0$,

$$t \int g_k \, d\mu \leq \Gamma(tg_k) + (\Gamma \mid V)^*(\mu),$$

we have by assumption 2(i),

$$t \lim_k \sup_{\mu \in L_a} \int g_k \, d\mu \leq a,$$

which implies

$$\lim_k \sup_{\mu \in L_a} \int g_k \, d\mu = 0.$$

By Proposition H.1, L_a is V-compact.

3. Assumption 2(ii) $\Rightarrow$ 2(i): By Corollary E.4, for $g \in V$,

$$\Gamma(g) \leq \sup\left\{\int g_k \, d\mu - (\Gamma \mid V)^*(\mu) \colon \mu \in \mathcal{P}(S)\right\}. \tag{3.9}$$

Assume that $0 \leq g_k \in V$ and $g_k \downarrow 0$ pointwise. By Proposition H.1, since L_a is V-compact,

$$\lim_k \sup_{\mu \in L_a} \int g_k \, d\mu = 0. \tag{3.10}$$

By (3.9) and (3.10),

$$\Gamma(g_k) \leq \max\left\{\sup_{\mu \in L_a} \int g_k \, d\mu, c - a\right\},$$

where $c = \sup g_1$,

$$\overline{\lim_k}\, \Gamma(g_k) \leq \max\{0, c - a\}.$$

But a is arbitrary. This proves assumption 2(i). □

Taking into account that $S_n(g) = n \int g \, dL_n$, so

$$\phi_M(g) = \overline{\lim_n}\, n^{-1} \log \sup_{\nu \in M} \mathbb{E}_\nu \left[\exp\left(n \int g \, dL_n\right)\right],$$

we record the corollary for $\{L_n\}$.

Theorem 3.2

1. *For every set $M \subset \mathcal{P}(S)$ and every measurable set $B \subset \mathcal{P}(S)$ such that $\mathrm{cl}_V(B)$ is compact,*

$$\overline{\lim_n}\, n^{-1} \log \sup_{\nu \in M} \mathbb{P}_\nu[L_n \in B] \leq -\inf\{(\phi_M \mid V)^*(\mu) \colon \mu \in \mathrm{cl}_V(B)\}. \qquad (3.11)$$

2. *Assume that V satisfies V.1–V.3. Then the following conditions are equivalent:*
   (i) *If $0 \leq g_k \in V$ and $g_k \downarrow 0$ pointwise, then $\phi_M(g_k) \to 0$.*
   (ii) *Inequality (3.11) holds for every measurable set $B \subset \mathcal{P}(S)$ and $(\phi_M \mid V)^*$ is V-tight.*

In Chapter 6 we will present several conditions equivalent to assumption 2(i) of Theorems 3.1 and 3.2 and will give a different proof of 2(i) ⇒ 2(ii) in Theorem 3.1. In Chapter 7 we will present different sufficient conditions for 2(i), hence for the upper bound.

In view of Theorems 2.10 and 3.2, two problems arise:

1. To find analytically more tractable expressions for the rate functions in the two cases, or at least to gain some more explicit analytic information on them.
2. To find conditions under which the two rate functions coincide, so that a large deviation principle holds.

These problems will be addressed in Chapter 4.

3.2 On the Condition of Exponential Tightness

It is often the case in the large deviations literature that the extension of the upper bound from the class of sets with compact closure to general measurable sets is proved by verifying the condition of exponential tightness (see, e.g., Rassoul-Agha and Seppäläinen 2015, p. 23).

In view of Theorem 3.1, it is natural to ask: what is the relation between a (suitably formulated) exponential tightness condition on $\{M_n\}$ in the V topology and assumption 2(i) of Theorem 3.1? We will show

i. For a vector space $V \subset B(S)$, an exponential tightness condition on $\{M_n\}$ in the V topology implies assumption 2(i) of Theorem 3.1.
ii. If $V = B(S)$, then the exponential tightness condition is too restrictive and often not satisfied, and therefore is not an adequate assumption for upper bounds.
iii. If S is Polish and $V = C_b(S)$, then assumption 2(i) of Theorem 3.1 implies exponential tightness, at least if $\{\mathbb{P}_\alpha\}_{\alpha\in A}$ is a finite family.

Proposition 3.3 *Let V be a vector space $\subset B(S)$, and let $\{\mathbb{P}_\alpha\}_{\alpha\in A}$, $\{M_n\}$, Γ_A be as above. Assume: for every $a > 0$, there exists a measurable set $C_a \subset \mathcal{P}(S)$ with V-compact closure, such that*

$$\overline{\lim_n}\, n^{-1} \log \sup_{\alpha\in A} \mathbb{P}_\alpha \left[M_n \in C_a^c\right] \le -a.$$

Then $0 \le g_k \in V$, $g_k \downarrow 0$ pointwise imply $\Gamma_A(g_k) \to 0$.

Proof. By the assumption, given $a > 0$ there exist $n_0 \in \mathbb{N}$, C_a measurable with V-compact closure K_a, such that for $n \ge n_0$,

$$\sup_{\alpha\in A} \mathbb{P}_\alpha \left[M_n \in C_a^c\right] \le e^{-an}.$$

For any $0 \le g \in V$, $\alpha \in A$, $n \ge n_0$,

$$\begin{aligned}
\mathbb{E}_\alpha\left[\exp\left(n\int g\, dM_n\right)\right] &= \mathbb{E}_\alpha\left[\exp\left(n\int g\, dM_n\right)\right]\mathbf{1}_{\{M_n\in C_a\}} \\
&\quad + \mathbb{E}_\alpha\left[\exp\left(n\int g\, dM_n\right)\right]\mathbf{1}_{\{M_n\in C_a^c\}} \\
&\le \exp\left(n\sup_{\mu\in K_a}\int g\, d\mu\right) + e^{nc}\mathbb{P}_\alpha\left[M_n \in C_a^c\right], \text{ where } c = \sup g, \\
&\le 2\max\left\{\exp\left(n\sup_{\mu\in K_a}\int g\, d\mu\right), e^{n(c-a)}\right\}. \qquad (3.12)
\end{aligned}$$

Assume now that $0 \le g_k \in V$, $g_k \downarrow 0$ pointwise. Then by Proposition H.1 (taking into account Remark H.2),

$$\lim_k \sup_{\mu \in K_a} \int g_k \, d\mu = 0. \tag{3.13}$$

Therefore by the definition of $\Gamma_A(g_k)$, we have by (3.12) and (3.13),

$$\lim_k \Gamma_A(g_k) \le \max\{0, c_1 - a\},$$

where $c_1 = \sup g_1$. But a is arbitrary. □

The following example shows that exponential tightness in the τ topology fails to hold under broad conditions.

Example 3.4 Suppose that $\mathcal{S}$ contains all singletons and P is a Markov kernel such that $P(x, C) = 0$ for $x \in S$, C countable. Let B be a measurable set $\subset \mathcal{P}(S)$ such that $\mathrm{cl}_\tau(B)$ is compact. Then $\mathbb{P}_\nu[L_n \in B] = 0$ for all $\nu \in \mathcal{P}(S)$.

For, by Dembo and Zeitouni (1998, Ex. 6.2.24), there exists a probability measure ρ on $\mathcal{S}$ such that for all $\mu \in B$, $\mu \ll \rho$. For $\lambda \in \mathcal{P}(S)$, let $A(\lambda)$ be the set of atoms of λ, and for $M \subset \mathcal{P}(S)$, $A(M) = \bigcup\{A(\lambda) \colon \lambda \in M\}$. Clearly $A(B) \subset A(\rho)$, a countable set. Since $[L_n \in B] \subset [X_1 \in A(B)]$, we have

$$\mathbb{P}_\nu[L_n \in B] \le \mathbb{P}_\nu[X_1 \in A(\rho)] = \int \nu(dx) P(x, A(\rho)) = 0.$$

Proposition 3.5 *Let S be a Polish space, and let $\{P_\alpha\}_{\alpha \in A}$ be finite. Assume: if $0 \le g_k \in C_b(S)$ and $g_k \downarrow 0$ pointwise, then $\Gamma_A(g_k) \to 0$. Then for every $a > 0$ there exists a $\sigma(\mathcal{P}(S), C_b(S))$-compact set $C_a \subset \mathcal{P}(S)$ such that*

$$\overline{\lim_n}\, n^{-1} \log \sup_{\alpha \in A} \mathbb{P}_\alpha [M_n \in C_a^c] \le -a.$$

Proof. For a set $E \subset S$, $\delta > 0$, we will denote

$$E^\delta = \{x \in S \colon d(x, E) < \delta\}, \qquad E^{[\delta} = \{x \in S \colon d(x, E) \le \delta\}.$$

For $T \subset S$, $\delta > 0$, $\epsilon > 0$, let

$$A(T, \delta, \epsilon) = \left\{\mu \in \mathcal{P}(S) \colon \mu\left(\left(T^{[\delta}\right)^c\right) > \epsilon\right\}.$$

We claim: for every $\delta > 0$, $\epsilon > 0$, $c > 0$, there exists $T \subset S$, T finite such that for all $n \in \mathbb{N}$,

$$\sup_{\alpha \in A} \mathbb{P}_\alpha [M_n \in A(T, \delta, \epsilon)] \le e^{-nc}. \tag{3.14}$$

For: let $\{S_j\}_{j \in \mathbb{N}}$ be an increasing sequence of finite sets such that $\bigcup_j S_j$ is dense in S,

$$U_j = \left(S_j^{[\delta}\right)^c, \quad F_j = \left(S_j^{\delta}\right)^c, \quad G_j = \left(S_j^{[\delta/2}\right)^c.$$

Then U_j and G_j are open, F_j is closed, and

$$U_j \subset F_j \subset G_j.$$

Let $g_j \in C_b(S)$ be defined by

$$g_j(x) = \frac{d(x, G_j^c)}{d(x, G_j^c) + d(x, F_j)}, \qquad x \in S.$$

Then $\mathbf{1}_{F_j} \le g_j \le \mathbf{1}_{G_j}$ and since $G_j \downarrow \emptyset$, we have $g_j \downarrow 0$ pointwise. For $b > 0$, $n \in \mathbb{N}$, $j \in \mathbb{N}$, $\alpha \in A$,

$$\begin{aligned}\mathbb{P}_\alpha\left[M_n \in A(S_j, \delta, \epsilon)\right] &\le \mathbb{P}_\alpha\left[M_n(F_j) > \epsilon\right] \\ &\le \mathbb{P}_\alpha\left[\int g_j \, dM_n > \epsilon\right] \\ &\le e^{-nb\epsilon}\mathbb{E}_\alpha\left[\exp\left(nb\int g_j \, dM_n\right)\right],\end{aligned}$$

and it follows that

$$\overline{\lim_n}\, n^{-1} \log \sup_{\alpha \in A} \mathbb{P}_\alpha\left[M_n \in A(S_j, \delta, \epsilon)\right] \le -b\epsilon + \Gamma_A(bg_j).$$

Therefore by the assumption there exist $j_0 \in \mathbb{N}$, $n_0 \in \mathbb{N}$, such that for $n \ge n_0$,

$$\sup_{\alpha \in A} \mathbb{P}_\alpha\left[M_n \in A(S_{j_0}, \delta, \epsilon)\right] \le e^{-nc}. \tag{3.15}$$

Next, since $U_j \downarrow \emptyset$, we have $A(S_j, \delta, \epsilon) \downarrow \emptyset$ and therefore there exists $j_1 \ge j_0$ such that

$$\sup_{\alpha \in A} \mathbb{P}_\alpha\left[M_n \in A(S_{j_1}, \delta, \epsilon)\right] \le e^{-nc} \tag{3.16}$$

for $n = 1, \ldots, n_0 - 1$. By (3.15) and (3.16), (3.14) holds with $T = S_{j_1}$.

Let $\epsilon_i > 0$, $\sum_{i=1}^\infty \epsilon_i < \infty$, $\delta_i \downarrow 0$. Given $a > 0$, let $c_i > a$ be such that $\sum_{i=1}^\infty e^{-(c_i - a)} \le 1$. In (3.14) let $\delta = \delta_i$, $\epsilon = \epsilon_i$, $c = c_i$, let T_i be the corresponding finite set, and let

$$K = \bigcap_{i=1}^\infty \left\{\mu \in \mathcal{P}(S) \colon \mu\left(\left(T_i^{[\delta_i]}\right)^c\right) \le \epsilon_i\right\}.$$

We claim that K is $\sigma(\mathcal{P}(S), C_b(S))$-compact. For:

1. K is $\sigma(\mathcal{P}(S), C_b(S))$-closed. For, since $(T_i^{[\delta_i]})^c$ is open, each set in the intersection defining K is closed.

2. K is $\sigma(\mathcal{P}(S), C_b(S))$-relatively compact. To prove this, we apply Proposition H.1. Let $0 \le f_j \in C_b(S)$, $f_j \downarrow 0$ pointwise, and for $h \in \mathbb{N}$ let

$$C_h = \bigcap_{i=h}^{\infty} T_i^{[\delta_i]}.$$

For $\mu \in K$, $c = \sup f_1$,

$$\int f_j \, d\mu = \int_{C_h} f_j \, d\mu + \int_{C_h^c} f_j \, d\mu \le \sup_{x \in C_h} f_j(x) + c\mu\left(C_h^c\right). \tag{3.17}$$

We have

$$\mu\left(C_h^c\right) \le \sum_{i=h}^{\infty} \mu\left(\left(T_i^{[\delta_i]}\right)^c\right) \le \sum_{i=h}^{\infty} \epsilon_i. \tag{3.18}$$

Since C_h is compact, by Dini's theorem,

$$\sup_{x \in C_h} f_j(x) \to 0.$$

Therefore from (3.17) and (3.18), for each $h \in \mathbb{N}$,

$$\limsup_j \sup_{\mu \in K} \int f_j \, d\mu \le c \sum_{i=h}^{\infty} \epsilon_i.$$

Letting $h \to \infty$, statement 2 follows from Proposition H.1.

Finally, we claim that the conclusion of the theorem holds with $C_a = K$. For: by (3.14), for all $n \in \mathbb{N}$,

$$\begin{aligned}\sup_{\alpha \in A} \mathbb{P}_\alpha[M_n \in K^c] &\le \sum_{i=1}^{\infty} \sup_{\alpha \in A} \mathbb{P}_\alpha\left\{M_n\left(\left(T_i^{[\delta_i]}\right)^c\right) > \epsilon_i\right\} \\ &\le \sum_{i=1}^{\infty} e^{-nc_i} \le e^{-na},\end{aligned}$$

and then

$$\overline{\lim_n}\, n^{-1} \log \sup_{\alpha \in A} \mathbb{P}_\alpha[M_n \in K^c] \le -a. \qquad \square$$

3.3 Notes

The proof of Theorem 3.1.1 (and, in part, that of Theorem 3.1.2) follows de Acosta (1985). An argument of this type previously appeared in Gärtner (1977). Condition (3.5) and the use of Daniell's theorem appeared in a different context in Dawson and Gärtner (1987) and then in the Markov setting in de Acosta (1990). Assumption 2(i) of Theorem 3.1 is implicit in Dawson and Gärtner (1987) and de Acosta (1990). The immersion $\mathcal{P}(S) \subset V^*$ and the explicit statement of 2(i) in the case when S is a metric space and $V = C_b(S)$ appear in Léonard (1992), where (3.2) for (weakly) closed sets is proved by other methods.

An upper bound for random probability measures in the case when S is Polish and $V = C_b(S)$ was proved in Deuschel and Stroock (1989, p. 189) by other methods. The assumption in that result implies assumption 2(i) of Theorem 3.1, as will be shown in Proposition 6.4; therefore the result follows from Theorem 3.1.

Example 3.4 is based on Exercise 6.2.24 in Dembo and Zeitouni (1998).

4

Identification and Reconciliation of Rate Functions

Our main objectives in this chapter are:

1. Identification (Section 4.1): to relate the rate functions $(\Lambda \mid V)^*$, Λ^* (Theorem 2.10) and $(\phi_M \mid V)^*$ (Theorem 3.2) to the analytically more explicit functions I and I_ψ, and in some cases to identify them completely in terms of I and I_ψ.
2. Reconciliation (Section 4.3): to find conditions under which $(\phi_M \mid V)^* = \Lambda^*$ and $I = \Lambda^*$.

We will also study the relationship between $(\Lambda \mid V)^*$ and Λ^* (Section 4.2) and conditions for the equality $\phi_M^* = I_\lambda$ in certain cases in which P may not be irreducible (Section 4.4).

In order to achieve these objectives we must strengthen the conditions V.1–V.3 imposed on a vector space $V \subset B(S)$ in Chapter 3: V will be assumed to satisfy

V.1′ V is a closed subalgebra of $B(S)$; that is, V is a vector subspace of $B(S)$, V is closed under multiplication, and V is closed for the uniform norm.

V.2 $1 \in V$.

V.3 The σ-algebra generated by V on S is $\mathcal{S}$.

V.4 V is P-stable; that is, if $g \in V$ then $Pg \in V$.

Of course, V.2 and V.3 are as in Chapter 3. If V satisfies V.1′, then it satisfies V.1; this is proved in Proposition J.3. Thus V.1′–V.4 strengthen V.1–V.3.

If S is Polish and $V = C_b(S)$, clearly V.1′–V.3 are satisfied, and V.4 is the standard Feller condition (Appendix B). For a general measurable space $(S, \mathcal{S})$, obviously $V = B(S)$ satisfies V.1′–V.4.

4.1 Identification of Rate Functions

In this section we will prove

1. $(\phi \mid V)^* = I$ and $(\phi_M \mid V)^* \geq I$; here ϕ is as in Chapter 1.
2. Assuming that P is irreducible, $\Lambda^* = I_\psi$ and $(\phi_M \mid V)^* \leq I_\psi$; the equality $\Lambda^* = I_\psi$ establishes the relationship between Λ^*, a rate function on $\mathcal{P}(S)$ generated by the convergence parameters of the irreducible transform kernels, and the Donsker–Varadhan rate function I.

We recall and extend some of the definitions presented in Chapter 1. For $x \in S$, we write as in Chapter 1,

$$\phi_x(g) = \overline{\lim_n}\, n^{-1} \log \mathbb{E}_x\left(\exp S_n(g)\right), \qquad g \in B(S),$$

and we define

$$\tilde{\phi}(g) = \sup_{x \in S} \phi_x(g),$$

and for $\mu \in \mathcal{P}(S)$,

$$\left(\tilde{\phi} \mid V\right)^*(\mu) = \sup\left\{\int g\, d\mu - \tilde{\phi}(g) : g \in V\right\}.$$

We recall that for $\mu \in \mathcal{P}(S)$,

$$I(\mu) = \sup\left\{\int \log\left(\frac{e^g}{Pe^g}\right) d\mu : g \in B(S)\right\},$$

and we introduce

$$I_V(\mu) = \sup\left\{\int \log\left(\frac{e^g}{Pe^g}\right) d\mu : g \in V\right\}.$$

Theorem 4.1

1. *Assume that V satisfies V.1–V.3. Then*

$$I_V = I. \tag{4.1}$$

2. *Assume that V satisfies V.1′–V.4. Then*

$$(\phi \mid V)^* = \left(\tilde{\phi} \mid V\right)^* = I. \tag{4.2}$$

For all $M \subset \mathcal{P}(S)$,

$$(\phi_M \mid V)^* \geq I. \tag{4.3}$$

Proof.

1. The inequality $I_V \le I$ is obvious. To prove the reverse inequality, let $\mu \in \mathcal{P}(S)$,
$$\mathcal{H}_\mu = \left\{g \in B(S) \colon \int g\, d\mu - \int \log(Pe^g)\, d\mu \le I_V(\mu)\right\}.$$
By monotone or dominated convergence, $\mathcal{H}_\mu$ is closed under monotone pointwise limits; also, $\mathcal{H}_\mu \supset V$. Therefore by Proposition I.1, $\mathcal{H}_\mu = B(S)$ and it follows that $I(\mu) \le I_V(\mu)$. This proves (4.1).
2. Since $\tilde{\phi} \le \phi$, we have
$$(\phi \mid V)^* \le \left(\tilde{\phi} \mid V\right)^* \le \tilde{\phi}^*.$$
We will prove
$$(\phi \mid V)^* \ge I \tag{4.4}$$
and
$$\tilde{\phi}^* \le I. \tag{4.5}$$
This will establish (4.2).

Let $\mathcal{U}(V) = \{e^f : f \in V\}$. For $u \in \mathcal{U}(V)$, let $z = Pu$, $g = \log(u/Pu)$. By Proposition J.3, $u \in V$, $z \in V$ and $g \in V$. We have
$$K_g z = P(e^g z) = P\left(u(Pu)^{-1}Pu\right) = Pu = z.$$
Let $c > 0$, $d > 0$ be such that $cz \le 1 \le dz$. Then for all $n \in \mathbb{N}$,
$$cz = K_g^n(cz) \le K_g^n 1 \le K_g^n(dz) = dz,$$
and it follows that $\phi(g) = 0$. Therefore for $\mu \in \mathcal{P}(S)$,
$$\int \log(u/Pu)\, d\mu = \int g\, d\mu - \phi(g) \le (\phi \mid V)^*(\mu),$$
and we have $I_V(\mu) \le (\phi \mid V)^*(\mu)$. Using (4.1), (4.4) is proved.

For $g \in B(S)$, $\rho > 0$, $x \in S$, $n \in \mathbb{N}$, let
$$u_n(\rho, x) = \sum_{k=0}^{n} \rho^k K_g^k 1(x), \qquad z_n(\rho, \cdot) = e^g u_n(\rho, \cdot).$$
Then
$$\frac{z_n(\rho, \cdot)}{Pz_n(\rho, \cdot)} = \frac{e^g u_n(\rho, \cdot)}{K_g u_n(\rho, \cdot)} \ge \rho e^g \left(\frac{u_n(\rho, \cdot)}{u_{n+1}(\rho, \cdot)}\right),$$
$$\log\left(\frac{z_n(\rho, \cdot)}{Pz_n(\rho, \cdot)}\right) \ge g + \log\rho + \log\left(\frac{u_n(\rho, \cdot)}{u_{n+1}(\rho, \cdot)}\right). \tag{4.6}$$

Let $w_n(\rho, \cdot) = \rho^n K_g^n 1$. We have

$$\frac{u_{n+1}(\rho,\cdot)}{u_n(\rho,\cdot)} = 1 + \frac{w_{n+1}(\rho,\cdot)}{u_n(\rho,\cdot)}$$
$$\leq 1 + \rho e^{\|g\|}\left(\frac{w_n(\rho,\cdot)}{u_n(\rho,\cdot)}\right) \leq 1 + \rho e^{\|g\|},$$

since $w_{n+1}(\rho, \cdot) = \rho(\rho^n K_g^n K_g 1) \leq \rho e^{\|g\|} w_n(\rho, \cdot)$, and it follows from (4.6) that

$$\log\left(\frac{z_n(\rho,\cdot)}{Pz_n(\rho,\cdot)}\right) \geq g + \log\rho + \log\left(1 + \rho e^{\|g\|}\right)^{-1}. \tag{4.7}$$

Now let $\rho < \exp(-\tilde{\phi}(g))$. Therefore for all $x \in S$,

$$\rho < R_x \triangleq \left[\overline{\lim_n}\, \left(K_g^n 1(x)\right)^{1/n}\right]^{-1},$$

and since R_x is the radius of convergence of the power series with coefficients $\{K_g^n 1(x)\}$, we have

$$\lim_n \left[\frac{u_n(\rho, x)}{u_{n+1}(\rho, x)}\right] = 1. \tag{4.8}$$

Since $z_n(\rho, \cdot) \in \mathcal{U}(B(S))$ by Proposition J.3, we have by (4.6)–(4.8) and Fatou's lemma,

$$\begin{aligned}
I(\mu) &= \sup\left\{\int \log\left(\frac{z}{Pz}\right) d\mu : z \in \mathcal{U}(B(S))\right\} \\
&\geq \underline{\lim_n} \int \log\left(\frac{z_n(\rho,\cdot)}{Pz_n(\rho,\cdot)}\right) d\mu \\
&\geq \int \underline{\lim_n} \log\left(\frac{z_n(\rho,\cdot)}{Pz_n(\rho,\cdot)}\right) d\mu \\
&\geq \int g\, d\mu + \log\rho.
\end{aligned}$$

It follows that for all $g \in B(S)$,

$$I(\mu) \geq \int g\, d\mu - \tilde{\phi}(g),$$

which implies

$$I(\mu) \geq \tilde{\phi}^*(\mu),$$

proving (4.5).

3. Since obviously $\phi_M \leq \phi$ we have $(\phi_M \mid V)^* \geq (\phi \mid V)^* = I$, which is (4.3). □

Theorem 4.2 *Let P be irreducible. Then*

$$\Lambda^* = I_\psi. \tag{4.9}$$

For all $M \subset \mathcal{P}(S)$, $\phi_M \geq \Lambda$, and therefore for all $V \subset B(S)$,

$$(\phi_M \mid V)^* \leq (\Lambda \mid V)^* \leq I_\psi. \tag{4.10}$$

If $\mu \ll \psi$, then for all $M \subset \mathcal{P}(S)$, V satisfying V.1′–V.4,

$$(\phi_M \mid V)^*(\mu) = I_\psi(\mu) = I(\mu). \tag{4.11}$$

We will need the following lemma. For $g \in F_b(S) \triangleq \{g \in F(S):\ g \text{ is bounded above}\}$, $0 \leq h \in B(S)$, let $R_{g,h}(x)$ be the radius of convergence of the power series with coefficients $\{K_g^n h(x)\}$:

$$R_{g,h}(x) = \left[\overline{\lim_n} \left(K_g^n h(x)\right)^{1/n}\right]^{-1}.$$

For $b > 0$, let $C_{g,h}(b) = \{x \in S : R_{g,h}(x) \geq b\}$. Also, for $\nu \in \mathcal{P}(S)$, let

$$\phi_{\nu,h}(g) = \overline{\lim_n}\, n^{-1} \log \nu K_g^n h.$$

Let us recall (Appendix B) that a nonempty set $F \in \mathcal{S}$ is P-closed if $P(x, F) = 1$ for all $x \in F$.

Lemma 4.3 *Let $g \in F_b(S)$, $0 \leq h \in B(S)$.*

1. *For $b > 0$, $C_{g,h}(b)$ is P-closed if it is nonempty.*
2. *For any $\nu \in \mathcal{P}(S)$,*

$$\nu\left(C_{g,h}\left(\exp\left[-\phi_{\nu,h}(g)\right]\right)\right) = 1. \tag{4.12}$$

Proof.

1. Let $D = (C_{g,h}(b))^c$, $D_k = \{x \in S : R_{g,h}(x) < b - 1/k\}$, $k \in \mathbb{N}$. Then

$$D = \bigcup_{k=1}^{\infty} D_k.$$

Next, let

$$D_{k,n} = \left\{x \in S :\ K_g^n h(x) > (b - 1/k)^{-n}\right\}.$$

Then

$$D_k \subset \overline{\lim_n}\, D_{k,n}.$$

For $x \in C_{g,h}(b)$,

$$K_g(x, D_{k,n}) \le (b - 1/k)^n \, K_g^{n+1} h(x),$$
$$\sum_n K_g(x, D_{k,n}) \le \sum_n (b - 1/k)^n \, K_g^{n+1} h(x) < \infty,$$

since $R_{g,h}(x) \ge b$. By the Borel–Cantelli lemma,

$$K_g\left(x, \overline{\lim_n} \, D_{k,n}\right) = 0$$

and therefore $K_g(x, D_k) = 0$, so finally $K_g(x, D) = 0$. But this implies $P(x, D) = 0$.

2. Let D, D_k, $D_{k,n}$ be as in step 1 with $b = \exp(-\phi_{\nu,h}(g))$. Then

$$\nu(D_{k,n}) \le (b - 1/k)^n \, \nu K_g^n h,$$
$$\sum_n \nu(D_{k,n}) \le \sum_n (b - 1/k)^n \, \nu K_g^n h < \infty,$$

since

$$\left[\overline{\lim_n} \, \left(\nu K_g^n h\right)^{1/n}\right]^{-1} = \exp\left(-\phi_{\nu,h}(g)\right) = b.$$

By the Borel–Cantelli lemma,

$$\nu\left(\overline{\lim_n} \, D_{k,n}\right) = 0$$

and therefore $\nu(D_k) = 0$, so finally $\nu(D) = 0$. □

We will also need the following elementary lemma.

Lemma 4.4 *Let* $h_n : \overline{\mathbb{R}^+} \to \mathbb{R}^+$ *be defined by*

$$h_n(t) = \frac{t}{1 + (t/n)} \quad \text{for } t \in \mathbb{R}^+, \quad h_n(\infty) = n, \quad n \in \mathbb{N}.$$

Then h_n *is increasing, concave,* $h_n \le n$, $h_n(t) \uparrow t$ *for each* $t \in \overline{\mathbb{R}^+}$, *and for each* $a, b, c \in \mathbb{R}^+$,

$$\frac{h_n(a) + c}{h_n(b) + c} \ge \min\{1, a/b\}. \tag{4.13}$$

Proof of Theorem 4.2.

1. To prove (4.9), we will show:

(i) If $\mu \not\ll \psi$, then

$$\Lambda^*(\mu) = \infty = I_\psi(\mu). \tag{4.14}$$

(ii) If $\mu \ll \psi$, then

$$\Lambda^*(\mu) \le I(\mu). \tag{4.15}$$

Since obviously $\phi \geq \Lambda$, which implies $\phi^* \leq \Lambda^*$, and $\phi^* = I$ by Theorem 4.1, it will follow that

$$\Lambda^*(\mu) = I(\mu). \tag{4.16}$$

(i) Let ν be a small measure. Then $\nu \ll \psi$ (Section B.2). Assume that $\psi(A) = 0$. Using the fact that $\psi P \ll \psi$ (Proposition B.2(1)), we have: for all $n \in \mathbb{N}$, $\nu P^n(A) = 0$. This implies: for all $t > 0$, $\Lambda(t\mathbf{1}_A) \leq 0$. For, let (s, ν) be a P-small pair with $s \leq 1$. Then if $g = t\mathbf{1}_A$,

$$\begin{aligned} \nu K_g^n s &= \mathbb{E}_\nu \left\{ \left(\exp \sum_{j=1}^n t\mathbf{1}_A(X_j) \right) s(X_n) \right\} \\ &\leq \mathbb{E}_\nu \left\{ \left(\exp \sum_{j=1}^n t\mathbf{1}_A(X_j) \right) \prod_{i=1}^n \mathbf{1}_{A^c}(X_i) \right\} \\ &\quad + \sum_{i=1}^n \mathbb{E}_\nu \left\{ \left(\exp \sum_{j=1}^n t\mathbf{1}_A(X_j) \right) \mathbf{1}_A(X_i) \right\} \\ &\leq 1 + e^{nt} \sum_{i=1}^n \nu P^i(A) = 1, \end{aligned}$$

and therefore

$$\Lambda(t\mathbf{1}_A) = \overline{\lim_n}\, n^{-1} \log \nu K_g^n s \leq 0.$$

Next, for all $t > 0$,

$$t\mu(A) = \int (t\mathbf{1}_A)\, d\mu \leq \Lambda(t\mathbf{1}_A) + \Lambda^*(\mu) \leq \Lambda^*(\mu),$$

and therefore $\Lambda^*(\mu) < \infty$ implies $\mu(A) = 0$. We have shown: $\Lambda^*(\mu) < \infty$ implies $\mu \ll \psi$, proving (4.14).

(ii) If $g \in F(S)$, $\mu \in \mathcal{P}(S, \psi)$, and $\int |g|\, d\mu < \infty$, then

$$I(\mu) \geq \int g\, d\mu - \Lambda(g). \tag{4.17}$$

This statement is somewhat more general than what is needed to prove (4.15); it will be useful in Section 11.2. The proof is similar to that of $\tilde{\phi}^* \leq I$ in Theorem 4.1 but more subtle.

Assume that $g \in F_b(S)$. Let (s, ν) be a K_g-small pair. By Lemma 4.3, if

$$b = \exp(-\Lambda(g)) = \exp(-\phi_{\nu,s}(g)) \quad \text{and} \quad F = C_{g,s}(b),$$

then $\nu(F) = 1$ (so $F \neq \emptyset$) and F is P-closed. For $\rho > 0$, $x \in S$, let

$$u(\rho, x) = \sum_{k=0}^{\infty} \rho^k K_g^k s(x), \quad w(\rho, x) = e^{g(x)} u(\rho, x).$$

By the irreducibility of K_g, we have $w(\rho, x) > 0$ for all $\rho > 0$, $x \in S$. Fix $\rho < b$. Then for all $x \in F$, $w(\rho, x) < \infty$,

$$Pw(\rho, \cdot)(x) = K_g u(\rho, \cdot)(x) \leq \rho^{-1} u(\rho, x) < \infty \tag{4.18}$$

and

$$\log\left(\frac{w(\rho, x)}{Pw(\rho, \cdot)(x)}\right) \geq \log\left(\frac{e^{g(x)} u(\rho, x)}{\rho^{-1} u(\rho, x)}\right) = g(x) + \log \rho. \tag{4.19}$$

We will now define a sequence $\{w_n\} \subset \mathcal{U}(B(S))$ (defined in Theorem 4.1) such that $\{w_n / Pw_n\}$ approximates $w(\rho, \cdot)/Pw(\rho, \cdot)$ in a suitable way. Let h_n be as in Lemma 4.4 and define on S

$$w_n = h_n \circ w(\rho, \cdot) + 1/n.$$

Then:

(a) $w_n \in \mathcal{U}(B(S))$.

(b) For each $x \in F$,

$$\lim_n \left(\frac{w_n(x)}{Pw_n(x)}\right) = \frac{w(\rho, x)}{Pw(\rho, \cdot)(x)}. \tag{4.20}$$

For, obviously $w_n(x) \to w(\rho, x)$ and

$$\begin{aligned} Pw_n(x) &= \int P(x, dy)\, [h_n\, (w(\rho, y)) + 1/n] \\ &\longrightarrow \int P(x, dy) w(\rho, y) = Pw(\rho, \cdot)(x) \end{aligned}$$

by monotone convergence.

(c) For each $x \in F$, $n \in \mathbb{N}$,

$$\log\left(\frac{w_n(x)}{Pw_n(x)}\right) \geq \min\{0, g(x) + \log \rho\}. \tag{4.21}$$

To prove this, let $\lambda = P(x, \cdot)$, $f = w(\rho, \cdot)$. Then

$$\lambda(\{y\colon f(y) < \infty\}) \geq \lambda(F) = 1,$$

and by (4.18),

$$\int f\, d\lambda = Pw(\rho, \cdot)(x) < \infty.$$

By Jensen's inequality, recalling that h_n is concave,

$$\begin{aligned} Pw_n(x) &= \int P(x, dy)\,[h_n\,(w(\rho, y)) + 1/n] \\ &= \int \lambda(dy) h_n\,(f(y)) + 1/n \\ &\le h_n\left(\int f\,d\lambda\right) + 1/n = h_n\,(Pw(\rho, \cdot)(x)) + 1/n, \end{aligned}$$

and therefore by (4.13) and (4.19),

$$\begin{aligned} \log\left(\frac{w_n(x)}{Pw_n(x)}\right) &\ge \log\left(\frac{h_n\,(w(\rho, x)) + 1/n}{h_n\,(Pw(\rho, \cdot)(x)) + 1/n}\right) \\ &\ge \log\left(\min\left\{1, \frac{w(\rho, x)}{Pw(\rho, \cdot)(x)}\right\}\right) \ge \min\,\{0, g(x) + \log\rho\}, \end{aligned}$$

showing that (4.21) holds for $g \in F_b(S)$. Next, since F is P-closed, we have $\psi(F^c) = 0$ (Proposition B.2(2)). Since $\mu \ll \psi$, $\mu(F^c) = 0$. If $\int |g|\,d\mu < \infty$, then by (4.21) Fatou's lemma applies, and we have by (4.20),

$$\begin{aligned} I(\mu) &= \sup\left\{\int \log\left(\frac{z}{Pz}\right)\,d\mu \colon z \in \mathcal{U}\,(B(S))\right\} \\ &\ge \underline{\lim}_n \int_F \log\left(\frac{w_n}{Pw_n}\right)\,d\mu \\ &\ge \int_F \underline{\lim}_n \log\left(\frac{w_n}{Pw_n}\right)\,d\mu = \int_F \log\left(\frac{w(\rho, \cdot)}{Pw(\rho, \cdot)}\right)\,d\mu, \end{aligned}$$

and by (4.19)

$$> \int g\,d\mu + \log\rho,$$

for all $\rho < \exp(-\Lambda(g))$. Therefore $I(\mu) \ge \int g\,d\mu - \Lambda(g)$ for $g \in F_b(S)$ such that $\int |g|\,d\mu < \infty$.

Now let $g \in F(S)$, $\int |g|\,d\mu < \infty$, $g_a \triangleq g \wedge a$ for $a > 0$. Then $\int |g_a|\,d\mu < \infty$, $g_a \in F_b(S)$, and therefore

$$I(\mu) \ge \int g_a\,d\mu - \Lambda(g_a) \ge \int g_a\,d\mu - \Lambda(g).$$

Letting $a \to \infty$, (4.17) follows.

From (4.17) we obtain

$$I(\mu) \ge \sup\left\{\int g\,d\mu - \Lambda(g) \colon g \in B(S)\right\} = \Lambda^*(\mu),$$

proving (4.15), hence (4.16).

2. To prove (4.10), we first note that for any $\lambda \in \mathcal{P}(S)$, $q \in \mathbb{N}$, we have $\phi_{\lambda P^q} \leq \phi_\lambda$. For: if $g \in B(S)$, since $e^{-\|g\|}P \leq K_g$,

$$\begin{aligned}\phi_{\lambda P^q}(g) &= \overline{\lim_n}\, n^{-1} \log(\lambda P^q K_g^n 1)\\ &\leq \overline{\lim_n}\, n^{-1} \log(e^{q\|g\|} \lambda K_g^{n+q} 1)\\ &= \phi_\lambda(g).\end{aligned}$$

Next, let (s, ν) be a small pair, say $P^m \geq s \otimes \nu$ for some $m \in \mathbb{N}$. Let $\mu \in \mathcal{P}(S)$. By irreducibility, there exists $k \in \mathbb{N}$ such that $c = \mu P^k s > 0$. Since $\mu P^q \geq c\nu$, where $q = k + m$, it follows that $\Lambda \leq \phi_\nu \leq \phi_{\mu P^q} \leq \phi_\mu$ and therefore $\phi_M \geq \Lambda$ for any $M \subset \mathcal{P}(S)$. Therefore

$$(\phi_M \mid V)^* \leq (\Lambda \mid V)^* \leq \Lambda^* = I_\psi.$$

3. If $\mu \ll \psi$, then $I_\psi(\mu) = I(\mu)$. Also, if V satisfies V.1′–V.4, then by Theorem 4.1, $(\phi \mid V)^* = I$. It follows that for any $M \subset \mathcal{P}(S)$,

$$I(\mu) = (\phi \mid V)^*(\mu) \leq (\phi_M \mid V)^*(\mu) \leq I_\psi(\mu) = I(\mu).$$

which is (4.11). □

4.2 On the Relationship between $(\Lambda|V)^*$ and I_ψ

Proposition 4.5 *Assume that V satisfies V.1′–V.4 and P is irreducible. Then the following conditions are equivalent:*

$$\operatorname{dom}(\Lambda \mid V)^* \subset \mathcal{P}(S, \psi); \tag{4.22}$$

$$(\Lambda \mid V)^* = I_\psi. \tag{4.23}$$

Proof. From the definition of I_ψ, (4.23) $\Longrightarrow$ (4.22) is obvious. To establish (4.22) $\Longrightarrow$ (4.23), first suppose $\mu \not\ll \psi$. Then

$$(\Lambda \mid V)^*(\mu) = \infty = I_\psi(\mu).$$

On the other hand, if $\mu \ll \psi$, by Theorems 4.1 and 4.2,

$$\begin{aligned}I(\mu) = (\phi \mid V)^*(\mu) &\leq (\Lambda \mid V)^*(\mu)\\ &\leq \Lambda^*(\mu) = I_\psi(\mu) = I(\mu),\end{aligned}$$

and therefore $(\Lambda \mid V)^*(\mu) = I_\psi(\mu)$. □

The first statement of Proposition 4.6 extends and strengthens 2(ii) $\Longrightarrow$ 2(i) in Corollary 2.19; note that here we exploit the equality (4.9).

Proposition 4.6 *Assume that V satisfies V.1′–V.4 and P is irreducible, and that I_ψ is V-tight. Then:*

$$\text{If } 0 \leq g_k \in V,\ \{g_k\} \text{ is decreasing, and } \int g_k \, d\psi \to 0,$$
$$\text{then } \lim_k \Lambda(g_k) \leq 0. \tag{4.24}$$
$$(\Lambda \mid V)^* = I_\psi. \tag{4.25}$$

Proof.

1. Since $\Lambda\colon B(S) \to \mathbb{R}$ satisfies conditions 1–3 of Proposition F.3 and $\Lambda^* = I_\psi$ by Theorem 4.2, we have by Theorem 2.13 and (F.4) of Proposition F.3,
$$\begin{aligned}\Lambda(g_k) &= \sup\left\{\int g_k \, d\mu - I_\psi(\mu)\colon \mu \in \mathcal{P}(S)\right\}\\ &\leq \max\left\{\sup_{\mu\in L_b} \int g_k \, d\mu, c - b\right\},\end{aligned}$$
where
$$L_b = \{\mu \in \mathcal{P}(S)\colon I_\psi(\mu) \leq b\} \quad \text{and} \quad c = \sup g_1.$$
L_b is V-compact and $L_b \subset \mathcal{P}(S, \psi)$. Therefore by Proposition H.3,
$$\sup_{\mu\in L_b} \int g_k \, d\mu \longrightarrow 0.$$
It follows that $\lim_k \Lambda(g_k) \leq \max\{0, c - b\}$. But b is arbitrary.
2. We will show that (4.24) $\implies$ (4.25). We first prove: (4.24) implies $\operatorname{dom}(\Lambda \mid V)^* \subset \mathcal{P}(S, \psi)$. Let $\mu \in \mathcal{P}(S)$, and assume that $a = (\Lambda \mid V)^*(\mu) < \infty$. Let $f \geq 0$, $f \in C(V)$, where $C(V)$ is as in Appendix J, and assume that $\int f \, d\psi = 0$. Then there exists a decreasing sequence $\{g_k\}$ such that $g_k \geq 0$, $g_k \in V$, $f = \lim_k g_k$, and therefore by dominated convergence $\int g_k \, d\psi \to 0$. For all $t > 0$, we have by step 1,
$$\begin{aligned}t\int f \, d\mu &= \lim_k \int t g_k \, d\mu\\ &= \lim_k \left\{\left[\int t g_k \, d\mu - \Lambda(t g_k)\right] + \Lambda(t g_k)\right\} \leq a.\end{aligned}$$
Since t is arbitrary, it follows that $\int f \, d\mu = 0$. Applying Proposition J.1(2), we have $\mu \ll \psi$. By Proposition 4.5, (4.25) follows. □

4.3 Reconciliation of Rate Functions

Assume that P is irreducible. In this section we will establish

I. Conditions under which $(\phi_M \mid V)^* = I_\psi$ (by Theorem 4.2, we know now that $\Lambda^* = I_\psi$).
II. Conditions under which $I = I_\psi$.

In Proposition 4.9 we present several equivalent conditions which will play an important role in establishing I in the case $V = B(S)$. In preparation for it, we prove two lemmas.

Lemma 4.7 *Let $M \subset \mathcal{P}(S)$, and let F be a P-closed set. Then the following conditions are equivalent:*

$$\text{For all } b > 0, \quad \lim_n e^{bn} \sup_{\nu \in M} \nu P^n(F^c) = 0. \tag{4.26}$$

$$\text{For all } b > 0, \quad \phi_M(b\mathbf{1}_{F^c}) = 0. \tag{4.27}$$

Proof. (4.26) $\Longrightarrow$ (4.27): For $t > 0$, let $g = t\mathbf{1}_{F^c}$. We claim that for all $n \in \mathbb{N}$,

$$K_g^n 1 \leq 1 + \sum_{j=1}^{n} e^{jt} P^j \mathbf{1}_{F^c}. \tag{4.28}$$

We proceed by induction. For $n = 1$,

$$K_g 1 = P(e^g) = P(\mathbf{1}_F e^g) + P(\mathbf{1}_{F^c} e^g) \leq 1 + e^t P\mathbf{1}_{F^c}.$$

Induction step:

$$\begin{aligned} K_g^{n+1} 1 = K_g(K_g^n 1) &\leq K_g\left(1 + \sum_{j=1}^{n} e^{jt} P^j \mathbf{1}_{F^c}\right) \\ &\leq 1 + e^t P\mathbf{1}_{F^c} + \sum_{j=1}^{n} e^{jt} K_g(P^j \mathbf{1}_{F^c}). \end{aligned}$$

Now

$$\begin{aligned} K_g(P^j \mathbf{1}_{F^c}) &= P(\mathbf{1}_F e^g P^j \mathbf{1}_{F^c}) + P(\mathbf{1}_{F^c} e^g P^j \mathbf{1}_{F^c}) \\ &\leq e^t P^{j+1} \mathbf{1}_{F^c}, \end{aligned}$$

since for $x \in F$, $P^j(x, F^c) = 0$. Therefore

$$K_g^{n+1} 1 \leq 1 + \sum_{j=1}^{n+1} e^{jt} P^j \mathbf{1}_{F^c},$$

proving (4.28). Next, we have: for all $t > 0$,

$$\sum_{j=1}^{\infty} e^{jt} \sup_{\nu \in M} \nu P^j(F^c) < \infty.$$

This follows from (4.26) by taking $b = t + 1$. Therefore

$$\begin{aligned}
\phi_M(g) &= \overline{\lim_n}\, n^{-1} \log \sup_{\nu \in M} \nu K_g^n 1 \\
&\le \overline{\lim_n}\, n^{-1} \log \left(1 + \sum_{j=1}^{n} e^{jt} \sup_{\nu \in M} \nu P^j(F^c) \right) = 0.
\end{aligned}$$

Since $g \ge 0$, we have $\phi_M(g) \ge 0$, hence $\phi_M(g) = 0$.

(4.27) $\Longrightarrow$ (4.26): For $x \in S$, $n \in \mathbb{N}$, $t > 0$,

$$\begin{aligned}
K_g^n 1(x) &= \mathbb{E}_x \left(\exp \sum_{j=1}^{n} t \mathbf{1}_{F^c}(X_j) \right) \\
&\ge e^{nt} \mathbb{P}_x[X_1 \in F^c, \ldots, X_n \in F^c] = e^{nt} \mathbb{P}_x[X_n \in F^c],
\end{aligned}$$

and therefore

$$\begin{aligned}
\sup_{\nu \in M} \nu K_g^n 1 &\ge e^{nt} \sup_{\nu \in M} \nu P^n(F^c), \\
0 = \phi_M(g) &= \overline{\lim_n}\, n^{-1} \log \sup_{\nu \in M} \nu K_g^n 1 \\
&\ge t + \overline{\lim_n}\, n^{-1} \log \sup_{\nu \in M} \nu P^n(F^c).
\end{aligned}$$

Since $t > 0$ is arbitrary, this implies (4.26). □

Lemma 4.8 *Let $M \subset \mathcal{P}(S)$, $\lambda \in \mathcal{P}(S)$. Then the following conditions are equivalent:*

$$\text{For every } t > 0, \text{ every } A \in \mathcal{S} \text{ with } \lambda(A) = 0,\ \phi_M(t\mathbf{1}_A) = 0. \tag{4.29}$$

$$\text{For every } 0 \le g \in B(S) \text{ such that } \textstyle\int g\, d\lambda = 0,\ \phi_M(g) = 0. \tag{4.30}$$

Proof. (4.29) $\Longrightarrow$ (4.30): We have, for $\delta \in (0, 1)$,

$$g \le \delta + c\mathbf{1}_{[g \ge \delta]},$$

where $c = \sup g$,

$$= \delta + (1 - \delta) \left[(1 - \delta)^{-1} c \mathbf{1}_{[g \ge \delta]} \right].$$

By the convexity of ϕ_M,

$$\phi_M(g) \le \delta \phi_M(1) + (1 - \delta) \phi_M \left[(1 - \delta)^{-1} c \mathbf{1}_{[g \ge \delta]} \right].$$

But $\phi_M(1) = 1$ and the second term vanishes, since

$$\lambda\left([g \geq \delta]\right) \leq \delta^{-1} \int g \, d\lambda = 0.$$

Therefore $\phi_M(g) \leq \delta$. But δ is arbitrary in $(0, 1)$.

(4.30) $\Longrightarrow$ (4.29): Take $g = t\mathbf{1}_A$. □

Proposition 4.9 *Let P be irreducible, $M \subset \mathcal{P}(S)$. Then the following conditions are equivalent:*

For every P-closed set F, every $t > 0$,

$$\lim_n e^{tn} \sup_{\nu \in M} \nu P^n(F^c) = 0. \tag{4.31}$$

For every $A \in \mathcal{S}$ such that $\psi(A) = 0$, every $t > 0$,

$$\lim_n e^{tn} \sup_{\nu \in M} \nu P^n(A) = 0. \tag{4.32}$$

For every P-closed set F, every $t > 0$,

$$\phi_M(t\mathbf{1}_{F^c}) = 0. \tag{4.33}$$

For every $A \in \mathcal{S}$ such that $\psi(A) = 0$, every $t > 0$,

$$\phi_M(t\mathbf{1}_A) = 0. \tag{4.34}$$

For all $g \in B(S)$ such that $g \geq 0$ and $\int g \, d\psi = 0$,

$$\phi_M(g) = 0. \tag{4.35}$$

Proof. We will need the following facts, proved in Proposition B.2:

1. $\psi P \ll \psi$.
2. $\psi(F^c) = 0$ for every P-closed set F.
3. If $A \in \mathcal{S}$ and $\psi(A) = 0$, then there exists a P-closed set F such that $A \subset F^c$.

To prove (4.32) $\Longrightarrow$ (4.31) and (4.34) $\Longrightarrow$ (4.33): use fact 2. (4.31) $\Longrightarrow$ (4.32) and (4.33) $\Longrightarrow$ (4.34) follow from fact 3. The equivalence of (4.31) and (4.33) (and, respectively, of (4.34) and (4.35)) follows from Lemma 4.7 (respectively, Lemma 4.8). □

In the next result we give a necessary and sufficient condition for the equality $(\phi_M \mid V)^* = I_\psi$.

Proposition 4.10 *Assume that V satisfies V.1′–V.4 and P is irreducible. Let $M \subset \mathcal{P}(S)$. Then the following conditions are equivalent:*

$$\mathrm{dom}(\phi_M \mid V)^* \subset \mathcal{P}(S, \psi); \tag{4.36}$$

$$(\phi_M \mid V)^* = I_\psi. \tag{4.37}$$

Proof. (4.37) $\Longrightarrow$ (4.36): Obvious from the definition of I_ψ.

(4.36) $\Longrightarrow$ (4.37): Same argument as in Proposition 4.5, i.e., if $\mu \notin \mathcal{P}(S, \psi)$, then

$$(\phi_M \mid V)^*(\mu) = \infty = I_\psi(\mu),$$

and if $\mu \in \mathcal{P}(S, \psi)$, then by (4.11) of Theorem 4.2,

$$(\phi_M \mid V)^*(\mu) = I_\psi(\mu).$$

□

Next, we present sufficient conditions for (4.36) in the case $V = B(S)$. Under an additional assumption, the sufficient conditions are necessary.

Proposition 4.11 *Let P be irreducible, $M \subset \mathcal{P}(S)$.*

1. *If any of the equivalent conditions* (4.31)–(4.35) *of Proposition 4.9 holds, then*

$$\mathrm{dom}\, \phi_M^* \subset \mathcal{P}(S, \psi), \tag{4.38}$$

   and therefore $\phi_M^* = I_\psi$.
2. *If* (4.38) *holds and ϕ_M is $\sigma(B(S), \mathcal{P}(S))$-lower semicontinuous, then all the equivalent conditions of Proposition 4.9 hold.*

Proof.

1. Suppose $\mu \in \mathcal{P}(S)$, $\phi_M^*(\mu) < \infty$ and let $A \in \mathcal{S}$ satisfy $\psi(A) = 0$. Then for all $t > 0$,

$$t\mu(A) = \int (t\mathbf{1}_A)\, d\mu \le \phi_M(t\mathbf{1}_A) + \phi_M^*(\mu).$$

   Since $\phi_M(t\mathbf{1}_A) = 0$, it follows that $\mu(A) = 0$.
2. Since $\phi_M : B(S) \to \mathbb{R}$ satisfies conditions 1–3 of Proposition F.3, by (F.4) of Proposition F.3 we have

$$\phi_M(g) = \sup\left\{\int g\, d\mu - \phi_M^*(\mu) : \mu \in \mathcal{P}(S)\right\}.$$

Then by (4.38),

$$\phi_M(g) = \sup\left\{\int g\, d\mu - \phi_M^*(\mu) \colon \mu \ll \psi\right\}.$$

If $g \geq 0$, $\int g\, d\psi = 0$, then $\int g\, d\mu = 0$ for all $\mu \ll \psi$, and therefore $\phi_M(g) = 0$.

□

We will now study under what conditions $I_\psi = I$. By Theorems 4.1 and 4.2, this will imply: for all $M \subset \mathcal{P}(S)$, all V satisfying V.1′–V.4,

$$(\phi_M \mid V)^* = I_\psi = I.$$

Theorem 4.12 *Let P be irreducible. Assume: for every $x \in S$, every P-closed set F, every $t > 0$,*

$$\lim_n e^{tn} P^n(x, F^c) = 0. \tag{4.39}$$

Then

$$I_\psi = I. \tag{4.40}$$

For every $M \subset \mathcal{P}(S)$, every V satisfying V.1′–V.4,

$$(\phi_M \mid V)^* = I_\psi = I. \tag{4.41}$$

Conversely, if (4.40) *holds and $\tilde{\phi}$ is $\sigma(B(S), \mathcal{P}(S))$-lower semicontinuous, then* (4.39) *holds.*

Remark 4.13 Assumption (4.39) amounts to stating that condition (4.31) of Proposition 4.9 holds for $M = \{\delta_x\}$ for each $x \in S$. Therefore (4.39) may be replaced by any of conditions (4.32)–(4.35) with the same specification.

Proof of Theorem 4.12.

1. From the definition of I_ψ, $I_\psi = I$ if and only if $\operatorname{dom} I \subset \mathcal{P}(S, \psi)$. As noted in Remark 4.13, by (4.39) we have: for every $A \in \mathcal{S}$ with $\psi(A) = 0$, every $t > 0$,

$$\tilde{\phi}(t\mathbf{1}_A) = \sup_{x \in S} \phi_x(t\mathbf{1}_A) = 0. \tag{4.42}$$

The next steps are similar to the proof of Proposition 4.11. Assume that $I(\mu) < \infty$ and $\psi(A) = 0$. Then for all $t > 0$,

$$t\mu(A) = \int (t\mathbf{1}_A)\, d\mu \leq \tilde{\phi}(t\mathbf{1}_A) + \tilde{\phi}^*(\mu) = I(\mu),$$

by (4.42) and (4.2) of Theorem 4.1. Therefore $\mu(A) = 0$. Hence $\mu \in \operatorname{dom} I$ implies $\mu \ll \psi$.

2. Next, (4.41) follows from Theorems 4.1 and 4.2.
3. Last, since $\tilde{\phi}\colon B(S) \to \mathbb{R}$ satisfies conditions 1–3 of Proposition F.3, by (F.4) of Proposition F.3 we have

$$\begin{aligned}\tilde{\phi}(g) &= \sup\left\{\int g\, d\mu - \tilde{\phi}^*(\mu) : \mu \in \mathcal{P}(S)\right\}\\ &= \sup\left\{\int g\, d\mu - I(\mu) : \mu \ll \psi\right\},\end{aligned}$$

by (4.2) and (4.40). Therefore $0 \le g$, $\int g\, d\psi = 0$ imply $\tilde{\phi}(g) = 0$. By Proposition 4.9, (4.39) holds. □

Remark 4.14

1. Suppose that P is irreducible and for some $m \in \mathbb{N}$, $P^m(x, \cdot) \ll \psi$ for all $x \in S$. Then $I_\psi = I$. For, let F be a P-closed set. Then $\psi(F^c) = 0$ (Proposition B.2(2)) and for $k \ge 1$, $x \in S$,

$$P^{m+k}(x, F^c) = \int P^k(x, dy) P^m(y, F^c) = 0.$$

Apply now Theorem 4.12.

2. Suppose that there exist $m, n \in \mathbb{N}$, $c > 0$, such that for all $x, y \in S$,

$$P^m(x, \cdot) \le c \sum_{j=1}^{n} P^j(y, \cdot). \tag{4.43}$$

Then:

(i) P is irreducible.
(ii) $I_\psi = I$.
(iii) S is petite.

For, 2(i) is clear since for any $x \in S$, $P^m(x, \cdot)$ is an irreducibility measure by (4.43). Also, since $P^m(x, \cdot) \ll \psi$, 2(ii) follows from (1). Claim 2(iii) is obvious.

3. Suppose:

(a) S is Polish, P is irreducible, and $\psi(G) > 0$ for every nonempty open set $G \subset S$.
(b) For some $m \in \mathbb{N}$, $P^m \mathbf{1}_K$ is lower semicontinuous for every compact set $K \subset S$.

Then $I_\psi = I$. For, by item 1 and the regularity of probability measures on S, it suffices to prove: if K is compact and $\psi(K) = 0$, then $P^m(x, K) = 0$ for all $x \in S$. Since $\psi P^m \ll \psi$, which easily follows from the fact that $\psi P \ll \psi$ (Proposition B.2(1)), if $\psi(K) = 0$ then

$$0 = \psi P^m(K) = \int (P^m 1_K)\, d\psi.$$

Let $G = [P^m 1_K > 0]$. If $G \neq \emptyset$, then by 3(a) and 3(b) $\psi(G) > 0$, which is impossible. Therefore $G = \emptyset$, proving the claim.

4.4 Additional Results on the Equality of Rate Functions

The aim of Propositions 4.15 and 4.17 is to establish conditions under which $\phi_M^* = I_\lambda$ when P is not necessarily irreducible and $\lambda \in \mathcal{P}(S)$ satisfies the condition $\lambda P \ll \lambda$. This will be applied in Theorem 7.17.

Actually, the stated property of λ is not needed in the next result, which is similar to Proposition 4.10.

Proposition 4.15 *Let $M \subset \mathcal{P}(S)$, $\lambda \in M$. Then the following conditions are equivalent:*

$$\operatorname{dom} \phi_M^* \subset \mathcal{P}(S, \lambda); \tag{4.44}$$

$$\phi_M^* = I_\lambda. \tag{4.45}$$

Proof. (4.45) $\Longrightarrow$ (4.44): Obvious from the definition of I_λ.

(4.44) $\Longrightarrow$ (4.45): If $\mu \notin \mathcal{P}(S, \lambda)$, then

$$\phi_M^*(\mu) = \infty = I_\lambda(\mu).$$

It remains to prove that if $\mu \ll \lambda$, then

$$\phi_M^*(\mu) = I_\lambda(\mu) = I(\mu).$$

Since $\phi_M \leq \phi$, we have $\phi_M^* \geq \phi^* = I$ by Theorem 4.1. Therefore we only have to prove: $\mu \ll \lambda$ implies $I(\mu) \geq \phi_M^*$. The proof is similar to that of $I \geq \tilde{\phi}^*$ in Theorem 4.1, but slightly more involved. Let $g \in B(S)$ and let u_n, z_n be as in the proof of Theorem 4.1; then inequalities (4.6) and (4.7) hold. Let

$$F = C_g \left[\exp\left(-\phi_\lambda(g)\right)\right],$$

where $C_g(b) = C_{g,1}(b)$ for $b \geq 0$ in the notation of Lemma 4.3. Then by that lemma, F is P-closed and $\lambda(F) = 1$, which implies $\mu(F) = 1$. Now let

$\rho < \exp(-\phi_\lambda(g))$. Then for $x \in F$, $\{u_n(\rho, x)\}$ converges and

$$\lim_n \left[u_n(\rho, x)/u_{n+1}(\rho, x)\right] = 1. \tag{4.46}$$

By (4.6), (4.7), (4.46), and Fatou's lemma we have

$$\begin{aligned} I(\mu) &\geq \underline{\lim}_n \int_F \log\left(\frac{z_n}{Pz_n}\right) d\mu \\ &\geq \int_F \underline{\lim}_n \log\left(\frac{z_n}{Pz_n}\right) d\mu \geq \int g\, d\mu + \log \rho. \end{aligned}$$

As in the proof of (4.2), since $\lambda \in M$, it follows that

$$I(\mu) \geq \phi_\lambda^*(\mu) \geq \phi_M^*(\mu).$$

□

In Proposition 4.17 we will present sufficient conditions for (4.44). Under an additional assumption, the conditions are also necessary.

We state first the following extension of Proposition 4.9.

Proposition 4.16 *Assume that $\lambda \in \mathcal{P}(S)$ satisfies $\lambda P \ll \lambda$. Let $M \subset \mathcal{P}(S)$. Then the following conditions are equivalent:*

For every P-closed set F with $\lambda(F) = 1$, every $t > 0$,

$$\lim_n e^{tn} \sup_{\nu \in M} \nu P^n(F^c) = 0. \tag{4.47}$$

For every $A \in \mathcal{S}$ such that $\lambda(A) = 0$, every $t > 0$,

$$\lim_n e^{tn} \sup_{\nu \in M} \nu P^n(A) = 0. \tag{4.48}$$

For every P-closed set F with $\lambda(F) = 1$, every $t > 0$,

$$\phi_M(t\mathbf{1}_{F^c}) = 0. \tag{4.49}$$

For every $A \in \mathcal{S}$ such that $\lambda(A) = 0$, every $t > 0$,

$$\phi_M(t\mathbf{1}_A) = 0. \tag{4.50}$$

For all $g \in B(S)$ such that $g \geq 0$ and $\int g\, d\lambda = 0$,

$$\phi_M(g) = 0. \tag{4.51}$$

Proof. This is just a rerun of the proof of Proposition 4.9. Fact 3 among the background facts for that proof is now rephrased as follows: if $A \in \mathcal{S}$ and $\lambda(A) = 0$, then there exists a P-closed set F such that $A \subset F^c$ and $\lambda(F^c) = 0$. This is proved as in Proposition B.2(3), using the assumption $\lambda P \ll \lambda$. □

Proposition 4.17 *Assume that $\lambda \in \mathcal{P}(S)$ satisfies $\lambda P \ll \lambda$. Let $M \subset \mathcal{P}(S)$, $\lambda \in M$.*

1. *If any of the equivalent conditions* (4.47)–(4.51) *of Proposition 4.16 holds, then*

$$\operatorname{dom} \phi_M^* \subset \mathcal{P}(S, \lambda), \tag{4.52}$$

and therefore $\phi_M^* = I_\lambda$.

2. *If* (4.52) *holds and ϕ_M is $\sigma(B(S), \mathcal{P}(S))$-lower semicontinuous, then all the equivalent conditions of Proposition 4.16 hold.*

Proof. Like the proof of Proposition 4.11, with ψ replaced by λ. Take into account Proposition 4.15. □

Finally, we extend (4.40) of Theorem 4.12. We omit the proof of the next result, which is entirely similar to that of (4.40).

Proposition 4.18 *Suppose that $\lambda \in \mathcal{P}(S)$ satisfies $\lambda P \ll \lambda$. Assume for every $x \in S$, every P-closed set F such that $\lambda(F) = 1$, every $t > 0$,*

$$\lim_n e^{tn} P^n(x, F^c) = 0. \tag{4.53}$$

Then $I_\lambda = I$. Conversely, if $I_\lambda = I$ and $\tilde{\phi}$ is $\sigma(B(S), \mathcal{P}(S))$-lower semicontinuous, then (4.53) *holds.*

4.5 Notes

Theorem 4.1 extends a result in Deuschel and Stroock (1989). The equality $(\phi \mid C_b(S))^* = I$ when S is Polish and P is Feller (Section B.3) was previously proved by Stroock (1984) and de Acosta (1985).

Theorem 4.2 (4.9) states in a concise formulation some results in de Acosta (1988, 1990). The proof given here is somewhat different. The convenient notation I_λ appears in Wu (2000a).

Sufficient conditions for the equality $\Lambda^* = I$ more restrictive than (4.39) were given in de Acosta (1988) and, in a more relaxed form, in de Acosta (1990).

Theorem 6.5.2 in Dembo and Zeitouni (1998) follows from Proposition 4.17 with λ an invariant probability measure, $M = \{\lambda\}$.

In Wu (2000a) a result equivalent to Proposition 4.18 is stated under the additional assumption of λ-essential irreducibility.

5

Necessary Conditions: Bounds on the Rate Function, Invariant Measures, Irreducibility, and Recurrence

Before embarking on a fuller discussion of upper bounds (Chapters 6 and 7) and the large deviation principle (Chapter 8), in this chapter we change the point of view and seek necessary conditions: we assume that $\{\mathbb{P}_\nu[L_n \in \cdot]\}$ satisfies the large deviation principle in the V topology with a (V-lower semicontinuous) rate function J which is not known a priori, and derive several consequences.

In Section 5.1 we show that J is always bounded by the rate functions $(\phi_\nu \mid V)^*$ and, if P is irreducible, $(\Lambda \mid V)^*$. Also, if J is convex, then $J = (\phi_\nu \mid V)^*$. If $V = B(S)$, we can give additional information about the identification of J.

In Section 5.2, it is shown that if J is V-tight, then P has an invariant probability measure π; the key fact is Lemma 5.5. If in addition P is irreducible, then π is unique and it is a P-maximal irreducibility probability measure.

In Section 5.3 we study the situation when the large deviation principle in the τ topology is assumed to hold for $\{\mathbb{P}_x[L_n \in \cdot]\}$ for all $x \in S$ with a common a priori unknown τ-tight rate function J. Irreducibility is not assumed in Theorem 5.8; rather, the irreducibility and a recurrence property of P are part of the conclusion. Also, J is completely identified.

In more detail, we prove:

1. P is irreducible and the unique P-invariant probability measure π is a P-maximal irreducibility probability measure.
2. $J = I_\pi$ and $I_\pi = I$.
3. P is positive Harris recurrent.

This result highlights the central role of irreducibility.

5.1 Bounds on the Rate Function J

In order to provide the proper context for the bounds proved below, we recall the inequalities $I \le (\phi_V \mid V)^* \le (\Lambda \mid V)^* \le I_\psi$ when P is irreducible and V satisfies V.1′–V.4, proved in Theorems 4.1 and 4.2.

Theorem 5.1 *Assume that V satisfies V.1′–V.4.*

1. *Assume that $\{\mathbb{P}_v[L_n \in \cdot]\}$ satisfies the lower bound in the V topology with rate function J. Then*

$$(\phi_V \mid V)^* \le J. \tag{5.1}$$

2. (i) *Assume that $\{\mathbb{P}_v[L_n \in \cdot]\}$ satisfies the upper bound in the V topology with rate function J and J is convex. Then*

$$J \le (\phi_V \mid V)^*. \tag{5.2}$$

   (ii) *Assume that P is irreducible and $\{\mathbb{P}_v[L_n \in \cdot]\}$ satisfies the upper bound in the V topology with rate function J. Then*

$$J \le (\Lambda \mid V)^* \qquad (\le \Lambda^* = I_\psi). \tag{5.3}$$

3. *Assume that $\{\mathbb{P}_v[L_n \in \cdot]\}$ satisfies the large deviation principle in the V topology with rate function J and J is convex. Then*

$$J = (\phi_V \mid V)^*. \tag{5.4}$$

Proof. Let $\mu \in \mathcal{P}(S)$, and assume $(\phi_V \mid V)^*(\mu) < \infty$. Given $\epsilon > 0$, let $f \in V$ be such that

$$\int f\, d\mu - \phi_V(f) > (\phi_V \mid V)^*(\mu) - \epsilon,$$

and let

$$U = \left\{\rho \in \mathcal{P}(S) \colon \int f\, d\rho - \phi_V(f) > (\phi_V \mid V)^*(\mu) - \epsilon\right\}.$$

Then

$$\begin{aligned}\mathbb{P}_v[L_n \in U] &\le \exp\left(-n\left((\phi_V \mid V)^*(\mu) - \epsilon\right)\right)\\ &\quad\times \mathbb{E}_v \exp\left(n \int f\, dL_n - n\phi_V(f)\right)\\ &= \exp\left(-n\left((\phi_V \mid V)^*(\mu) - \epsilon\right)\right)\\ &\quad\times \exp\left(-n\phi_V(f)\right) \mathbb{E}_v\left(\exp \sum_{j=0}^{n-1} f(X_j)\right),\end{aligned}$$

$$\overline{\lim_n}\, n^{-1} \log \mathbb{P}_v[L_n \in U] \le -(\phi_V \mid V)^*(\mu) + \epsilon.$$

On the other hand, since U is V-open and $\mu \in U$,

$$\underline{\lim}_n n^{-1} \log \mathbb{P}_\nu[L_n \in U] \geq -J(\mu).$$

Therefore

$$-J(\mu) \leq -(\phi_\nu \mid V)^*(\mu) + \epsilon.$$

Since ϵ is arbitrary, we have $(\phi_\nu \mid V)^*(\mu) \leq J(\mu)$, establishing (5.1) if $(\phi_\nu \mid V)^*(\mu) < \infty$. If $(\phi_\nu \mid V)^*(\mu) = \infty$, a similar argument yields $J(\mu) = \infty$.

The assumptions in part 2 of Theorem 5.1 are not comparable. By Theorem 4.2, the conclusion in (5.2) is stronger.

To prove (5.2): by Corollary E.4, we have for $g \in V$,

$$\phi_\nu(g) \leq \sup\left\{\int g \, d\mu - J(\mu) : \mu \in \mathcal{P}(S)\right\},$$

and therefore

$$(\phi_\nu \mid V)^*(\mu) \geq J(\mu),$$

for all $\mu \in \mathcal{P}(S)$ by Corollary F.2.

To obtain (5.3), let $\mu \in \mathcal{P}(S)$, $\epsilon > 0$. Assume that $J(\mu) < \infty$. Since

$$U = \{\rho \in \mathcal{P}(S) : J(\rho) > J(\mu) - \epsilon\}$$

is V-open, there exist $\delta > 0$, H finite, $H \subset V$ such that

$$F \triangleq \left\{\rho \in \mathcal{P}(S) : \left|\int f \, d\rho - \int f \, d\mu\right| \leq \delta \text{ for all } f \in H\right\} \subset U.$$

Since F is V-closed,

$$\overline{\lim}_n n^{-1} \log \mathbb{P}_\nu[L_n \in F] \leq -\inf_{\rho \in F} J(\rho) \leq -J(\mu) + \epsilon.$$

On the other hand, by Theorem 2.10,

$$\underline{\lim}_n n^{-1} \log \mathbb{P}_\nu[L_n \in G] \geq -(\Lambda \mid V)^*(\mu),$$

where

$$G = \left\{\rho \in \mathcal{P}(S) : \left|\int f \, d\rho - \int f \, d\mu\right| < \delta \text{ for all } f \in H\right\},$$

a V-open set. Therefore

$$-(\Lambda \mid V)^*(\mu) \leq -J(\mu) + \epsilon.$$

Since ϵ is arbitrary, we have $J(\mu) \le (\Lambda \mid V)^*(\mu)$, proving (5.3) if $J(\mu) < \infty$. If $J(\mu) = \infty$, a similar argument yields the inequality.

Part 3 of Theorem 5.1 follows at once from (5.1) and (5.2). □

Remark 5.2 In Theorem 2.12 for $M = \{\nu\}$, it was assumed that P is irreducible and $\{\mathbb{P}_\nu[L_n \in \cdot]\}$ satisfies the upper bound in the V topology with rate function $J = (\Lambda \mid V)^*$. Under this assumption, which is stronger than the assumptions in part 2 of Theorem 5.1, it was proved that $(\phi_\nu \mid V) = (\Lambda \mid V)$. This conclusion is stronger than the conclusions (5.2)–(5.4).

Proposition 5.3 *Let P be irreducible, $\nu \in \mathcal{P}(S)$. Assume that $\{\mathbb{P}_\nu[L_n \in \cdot]\}$ satisfies the large deviation principle in the τ topology with rate function J. Then the following conditions are equivalent:*

$$J = I_\psi. \tag{5.5}$$

$$\operatorname{dom} J \subset \mathcal{P}(S, \psi). \tag{5.6}$$

$$\textit{If } 0 \le g \in B(S) \textit{ and } \textstyle\int g\, d\psi = 0, \textit{ then } \phi_\nu(g) = 0. \tag{5.7}$$

$$\phi_\nu^* = I_\psi. \tag{5.8}$$

Proof. Condition (5.5) ⟹ (5.6): Obvious.

(5.6) ⟹ (5.7): By Corollary E.4 with $V = B(S)$ and (5.6), for all $g \in B(S)$,

$$\begin{aligned}\phi_\nu(g) &\le \sup\left\{\int g\, d\mu - J(\mu) : \mu \in \mathcal{P}(S)\right\} \\ &= \sup\left\{\int g\, d\mu - J(\mu) : \mu \in \mathcal{P}(S, \psi)\right\}.\end{aligned}$$

Therefore if $g \ge 0$ and $\int g\, d\psi = 0$, we have $\int g\, d\mu = 0$ for $\mu \ll \psi$, and consequently $\phi_\nu(g) = 0$.

(5.7) ⟹ (5.8): This follows from Proposition 4.11(1).

(5.8) ⟹ (5.5): By Theorem 5.1, $\phi_\nu^* \le J \le I_\psi$, so (5.8) implies $J = I_\psi$.

The implication $J = I_\psi \implies \phi_\nu^* = I_\psi$ also follows from (5.4), since I_ψ is convex. □

5.2 The V-Tightness of J and Invariant Measures

We recall that a measure μ is *P-invariant*, or simply *invariant*, if $\mu P = P$. The state space S is *P-indecomposable* if there do not exist two disjoint P-closed sets. If P is irreducible, then S is P-indecomposable: for, as shown in Proposition B.2, $\psi(F) = 1$ for every P-closed set F.

Theorem 5.4 *Assume that V satisfies V.1′–V.4 and that $\{\mathbb{P}_\nu[L_n \in \cdot]\}$ satisfies the large deviation principle in the V topology with V-tight rate function J. Then:*

1. *There exists an invariant probability measure.*
2. *If S is P-indecomposable, then P has a unique invariant probability measure π, $\pi(F) = 1$ for every P-closed set F, and $J(\mu) = 0$ if and only if $\mu = \pi$.*
3. *If P is irreducible then, moreover, $\pi \equiv \psi$.*

The following lemma states an important property of the rate function I.

Lemma 5.5 *Let $\mu \in \mathcal{P}(S)$. Then $I(\mu) = 0$ if and only if μ is invariant.*

Proof. Suppose $I(\mu) = 0$. Then by the definition of I for all $f \in B(S)$, $t \in \mathbb{R}$,

$$\int tf \, d\mu \le \int \log(Pe^{tf}) \, d\mu.$$

For $t > 0$,

$$\int f \, d\mu \le t^{-1} \int \log(Pe^{tf}) \, d\mu,$$

so

$$\begin{aligned}\int f \, d\mu &\le \lim_{t \to 0^+} t^{-1} \int \log(Pe^{tf}) \, d\mu \\ &= \int (Pf) \, d\mu = \int f \, d(\mu P).\end{aligned}$$

Similarly,

$$\int f \, d\mu \ge \lim_{t \to 0^-} t^{-1} \int \log(Pe^{tf}) \, d\mu = \int f \, d(\mu P).$$

Therefore $\int f \, d\mu = \int f \, d(\mu P)$ for all $f \in B(S)$, hence $\mu = \mu P$. Conversely, by Jensen's inequality, $Pe^f \ge e^{Pf}$ for $f \in B(S)$. Therefore

$$\int \log(Pe^f) \, d\mu \ge \int (Pf) \, d\mu = \int f \, d\mu,$$

hence

$$I(\mu) = \sup_{f \in B(S)} \int \log\left(\frac{e^f}{Pe^f}\right) d\mu \le 0.$$

Taking $f \equiv 0$, we have $I(\mu) = 0$. □

Lemma 5.6 *Suppose that S is P-indecomposable and π is an invariant probability measure. Then π is unique and $\pi(F) = 1$ for every P-closed set F.*

Proof. Suppose that π and π' are invariant probability measures and $\pi \neq \pi'$. Let $\rho = (\pi - \pi')^+$, $\sigma = (\pi - \pi')^-$. Then $\rho \neq 0$, $\sigma \neq 0$, they are mutually singular, and by Hernández-Lerma and Lasserre (2003, p. 26) both are invariant and there exist disjoint P-closed sets B_ρ and B_σ such that

$$(\rho(S))^{-1}\rho(B_\rho) = 1, \qquad (\sigma(S))^{-1}\sigma(B_\sigma) = 1.$$

But this is impossible. Therefore $\pi = \pi'$.

Next, by Hernández-Lerma and Lasserre (2003, p. 35), π is ergodic; that is, $\pi(F) = 0$ or 1 for every P-closed set F. Let F be a P-closed set and let

$$H = \{x \in S\colon G1_F(x) = 0\},$$

where $G = \sum_{n=0}^{\infty} P^n$, the potential kernel of P. Then $F \cap H = \emptyset$. If $H \neq \emptyset$, then, as in the proof of Proposition B.2, H is P-closed. But this is impossible. Therefore $H^c = S$ and

$$0 < \int (G1_F)\, d\pi = \pi G(F),$$

which implies $\pi(F) > 0$. Since π is ergodic, it follows that $\pi(F) = 1$. □

The following statement is a general fact about Markov kernels. Only part 1 of Lemma 5.7 is needed for Theorem 5.4; part 2 will be used in Theorem 8.20.

Lemma 5.7 *Assume that P has an invariant probability measure π. Then*

1. *If P is irreducible, then π is unique and $\pi \equiv \psi$.*
2. *Suppose that π is ergodic. Then the following conditions are equivalent:*
   (i) *P is irreducible.*
   (ii) *If $B \in \mathcal{S}$ and $\pi(B) = 1$, then for every $x \in S$ there exists $m \in \mathbb{N}$ such that $P^m(x, B) > 0$.*

   If either condition above holds, then π is a P-maximal irreducibility probability measure.

Proof. The uniqueness of π follows from Lemma 5.6. For the second assertion, we first prove that $\psi \ll \pi$. Suppose that $A \in \mathcal{S}$, $\psi(A) > 0$. Then for all $x \in S$,

$$\sum_{n=1}^{\infty} 2^{-n} P^n(x, A) > 0,$$

and therefore

$$\pi(A) = \sum_{n=1}^{\infty} 2^{-n}\pi P^n(A) = \int \left[\sum_{n=1}^{\infty} 2^{-n} P^n(\cdot, A)\right] d\pi > 0,$$

so $\psi \ll \pi$.

We prove next that $\pi \ll \psi$. We will use the following fact: if $\mu \in \mathcal{P}(S)$ and $\mu(F^c) = 0$ for every P-closed set F, then $\mu \ll \psi$. To prove this, suppose $\psi(A) = 0$. Then by Proposition B.2 there exists a P-closed set F such that $A \subset F^c$. But then $\mu(A) \leq \mu(F^c) = 0$. By Lemma 5.6, $\pi(F^c) = 0$ for every P-closed set F. By the previous argument, it follows that $\pi \ll \psi$.

2(i) $\Longrightarrow$ 2(ii): By part 1, if $\pi(B) = 1$ then $\psi(B) = 1$, and 2(ii) follows.

2(ii) $\Longrightarrow$ 2(i): We will prove that π is an irreducibility measure. Let $A \in \mathcal{S}$, $\pi(A) > 0$, and let

$$B = \{x \in S \colon PG\mathbf{1}_A(x) > 0\},$$

where G is the potential kernel of P. We must prove that $B = S$. We will assume that $B^c \neq \emptyset$ and reach a contradiction. We have

1. B^c is P-closed. For, if $x \in B^c$, then
$$\begin{aligned}\int P(x, dy) PG\mathbf{1}_A(y) &= P^2 G\mathbf{1}_A(x) \\ &\leq PG\mathbf{1}_A(x) = 0,\end{aligned}$$
and therefore $P(x, B) = 0$.
2. $\pi(B^c) = 0$. For, since
$$\int (PG\mathbf{1}_A)\, d\pi = \sum_{n=1}^{\infty} \pi P^n(A) = \infty,$$
we have $\pi(B) > 0$, and since π is ergodic it follows that $\pi(B) = 1$.

Let $B_k = \{x \in S \colon PG\mathbf{1}_A(x) > k^{-1}\}$. By condition 2(ii), given $x \in S$ there exist $m \in \mathbb{N}$, $k \in \mathbb{N}$ such that

$$P^m(x, B_k) > 0.$$

Therefore

$$\begin{aligned}0 < P^m \mathbf{1}_{B_k}(x) &\leq k P^m PG\mathbf{1}_A(x) \\ &\leq kPG\mathbf{1}_A(x),\end{aligned}$$

which is impossible if $x \in B^c$. Therefore $B^c = \emptyset$.

The final assertion of Lemma 5.7 follows from part 1. □

Proof of Theorem 5.4. To prove part 1, we have

$$\begin{aligned}0 &= \overline{\lim_n}\, n^{-1} \log \mathbb{P}_\nu\,[L_n \in \mathcal{P}(S)] \\ &\leq -\inf\{J(\mu) \colon \mu \in \mathcal{P}(S)\},\end{aligned}$$

so $\inf\{J(\mu) : \mu \in \mathcal{P}(S)\} = 0$. Since J is V-tight,

$$Z \triangleq \{\mu \in \mathcal{P}(S) : J(\mu) = 0\} \neq \emptyset.$$

By (4.3) and (5.1), we have $J \geq (\phi_\nu \mid V)^* \geq I$. Therefore by Lemma 5.5 if $\mu \in Z$ then μ is invariant.

Part 2 follows from Lemma 5.6, except for the last statement. But since $Z \neq \emptyset$ and π is unique, we must have $Z = \{\pi\}$.

Part 3 follows from Lemma 5.7. □

5.3 Irreducibility and Recurrence

Theorem 5.8 *Assume that there exists a τ-tight function $J : \mathcal{P}(S) \to \overline{\mathbb{R}^+}$ such that for every $x \in S$, $\{\mathbb{P}_x[L_n \in \cdot]\}$ satisfies the large deviation principle in the τ topology with rate function J. Then:*

1. *There exists a unique invariant probability measure π and $\pi(F) = 1$ for every P-closed set F.*
2. *P is irreducible and π is a P-maximal irreducibility probability measure.*
3. *$J = I_\pi$ and $I_\pi = I$.*
4. *P is positive Harris recurrent.*

Lemma 5.9 *Under the assumptions of Theorem 5.8, for every $\nu \in \mathcal{P}(S)$, $\{\mathbb{P}_\nu[L_n \in \cdot]\}$ satisfies the large deviation lower bound with rate function J.*

Proof. By Fatou's lemma and Hölder's inequality, for any measurable set $B \subset \mathcal{P}(S)$,

$$\begin{aligned}
\underline{\lim}_n \{\mathbb{P}_\nu[L_n \in B]\}^{1/n} &= \underline{\lim}_n \left\{\int \nu(dx)\mathbb{P}_x[L_n \in B]\right\}^{1/n} \\
&\geq \underline{\lim}_n \int \nu(dx) \{\mathbb{P}_x[L_n \in B]\}^{1/n} \\
&\geq \int \underline{\lim}_n \{\mathbb{P}_x[L_n \in B]\}^{1/n} \nu(dx) \\
&\geq \int \exp\left(-\inf_{\mu \in G} J(\mu)\right) \nu(dx) = \exp\left(-\inf_{\mu \in G} J(\mu)\right),
\end{aligned}$$

where $G = \mathrm{int}_\tau(B)$. □

Proof of Theorem 5.8.

1. By (5.1), $J \geq \phi_x^*$ for all $x \in S$. Arguing as in the proof of Theorem 5.4, we have: the equation $J(\mu) = 0$, $\mu \in \mathcal{P}(S)$, has solutions and any solution is invariant.
2. We claim that the equation $J(\mu) = 0$ has a unique solution. Let π_1, π_2 be two solutions; by step 1, π_1 and π_2 are invariant.

   Suppose $\pi_1 \neq \pi_2$, and let $\rho = (\pi_1 - \pi_2)^+$, $\sigma = (\pi_1 - \pi_2)^-$. Then $\rho \neq 0$, $\sigma \neq 0$, they are mutually singular, and by Hernández-Lerma and Lasserre (2003, p. 26), both ρ and σ are invariant measures.

   Let $\tilde{\rho} = (\rho(S))^{-1}\rho$, $\tilde{\sigma} = (\sigma(S))^{-1}\sigma$. By Lemma 5.9, $\{\mathbb{P}_{\tilde{\rho}}[L_n \in \cdot]\}$ satisfies the large deviation lower bound with rate function J, so by (5.1),

$$\phi_{\tilde{\rho}}^*(\pi_2) \leq J(\pi_2) = 0.$$

   Then $\pi_2 \ll \tilde{\rho}$. For, (4.47) is satisfied for $\lambda = \tilde{\rho}$ and $M = \{\tilde{\rho}\}$. Since $\pi_2 \in \operatorname{dom} \phi_{\tilde{\rho}}^*$, by (4.52) it follows that $\pi_2 \in \mathcal{P}(S, \tilde{\rho})$. Next, since $\sigma \leq \pi_2$, we have $\tilde{\sigma} \ll \tilde{\rho}$. Inverting the roles of π_1 and π_2, we obtain $\tilde{\sigma} \equiv \tilde{\rho}$, which is impossible. Therefore $\pi_1 = \pi_2$.
3. Let π be the unique probability measure such that $J(\pi) = 0$. We claim that π is a P-irreducibility measure. For, let $f \in B(S)$, $f \geq 0$, and assume $\pi(f) = \int f \, d\pi > 0$. For $\delta > 0$, let

$$U_\delta = \{\rho \in \mathcal{P}(S) \colon |\rho(f) - \pi(f)| < \delta\}.$$

   Then for all $x \in S$,

$$\begin{aligned} \left| n^{-1} \sum_{j=0}^{n-1} P^j f(x) - \pi(f) \right| &= |\mathbb{E}_x L_n(f) - \pi(f)| \\ &\leq \mathbb{E}_x \, |L_n(f) - \pi(f)| \\ &= \mathbb{E}_x \, |L_n(f) - \pi(f)| \, \mathbf{1}[L_n \in U_\delta] \\ &\quad + \mathbb{E}_x \, |[L_n(f) - \pi(f)| \, \mathbf{1}[L_n \in U_\delta^c] \\ &\leq \delta + c\mathbb{P}_x[L_n \in U_\delta^c], \end{aligned} \tag{5.9}$$

   where $c = 2\|f\|$. Next, we have

$$\lim_n \mathbb{P}_x[L_n \in U_\delta^c] = 0. \tag{5.10}$$

   For, by the large deviation upper bound,

$$\overline{\lim_n} \, n^{-1} \log \mathbb{P}_x[L_n \in U_\delta^c] \leq -\inf\left\{J(\mu) \colon \mu \in U_\delta^c\right\}. \tag{5.11}$$

But $\inf\{J(\mu)\colon \mu \in U_\delta^c\} > 0$. In fact, since U_δ^c is τ-closed and J is τ-tight, the infimum is attained, say at $\rho \in U_\delta^c$, and by step 2, $J(\rho) > 0$ because $\rho \neq \pi$. Now (5.10) follows from (5.11).

From (5.9) and (5.10) we have: for all $\delta > 0$, $x \in S$,

$$\overline{\lim_n} \left| n^{-1} \sum_{j=0}^{n-1} P^j f(x) - \pi(f) \right| \leq \delta,$$

so

$$\lim_n n^{-1} \sum_{j=0}^{n-1} P^j f(x) = \pi(f).$$

Therefore, for all $x \in S$, $P^j f(x) > 0$ for some $j \in \mathbb{N}$, proving the claim.

4. π is a P-maximal irreducibility probability measure. For, suppose μ is a P-irreducibility probability measure and $A \in \mathcal{S}$, $\mu(A) > 0$. Arguing as in the first part of the proof of Lemma 5.7, we have $\pi(A) > 0$. Therefore $\mu \ll \pi$.

   Since P is irreducible and π is invariant, by Lemma 5.7 π is the unique invariant probability measure. Also, $\pi(F) = 1$ for every P-closed set F by Lemma 5.6 or Proposition B.2.

5. We show next that $J = I_\pi$. First, since P is irreducible, by (5.3) we have $J \leq I_\pi$. Next, from Lemma 5.9 and (5.1), it follows that $\phi_\pi^* \leq J$. Finally, $\phi_\pi^* = I_\pi$. For, (4.31) is satisfied for $M = \{\pi\}$. Therefore, by Proposition 4.11, we have $\operatorname{dom} \phi_\pi^* \subset \mathcal{P}(S, \pi)$, which implies $\phi_\pi^* = I_\pi$. We have proved that $J = I_\pi$.

6. We prove that $I_\pi = I$. By Corollary E.4, for each $x \in S$, $g \in B(S)$,

$$\begin{aligned}\phi_x(g) &\leq \sup\left\{\int g\, d\mu - I_\pi(\mu)\colon \mu \in \mathcal{P}(S)\right\}\\ &= \sup\left\{\int g\, d\mu - I(\mu)\colon \mu \ll \pi\right\}.\end{aligned}$$

   If $g \geq 0$ and $\int g\, d\pi = 0$, then $\int g\, d\mu = 0$ for all $\mu \ll \pi$. Therefore $\phi_x(g) = 0$. By Theorem 4.12 and Remark 4.13, it follows that $I_\pi = I$.

7. We show last that P is positive Harris recurrent (Section B.3). Since P is irreducible and π is invariant and a P-maximal irreducibility probability measure, it only remains to show that for all $x \in S$, $A \in \mathcal{S}$ with $\pi(A) > 0$,

$$\mathbb{P}_x[V_A = \infty] = 1, \tag{5.12}$$

where $V_A = \sum_{j=1}^{\infty} \mathbf{1}_A(X_j)$. In turn, to prove (5.12) it suffices, by a well-known argument using the strong Markov property (see, e.g., de Acosta and Ney, 2014, p. 87) to show: for all $x \in S$,

$$\mathbb{P}_x[\tau_A < \infty] = 1, \tag{5.13}$$

where $\tau_A = \inf\{n \geq 1 : X_n \in A\}$. Let $0 < \epsilon < \pi(A)$, $B = \{\mu \in \mathcal{P}(S) : \mu(A) \leq \epsilon\}$. Since $\pi \notin B$, B is τ-closed, and I is τ-tight,

$$a = \inf\{I(\mu) : \mu \in B\} > 0.$$

Let $n_0 \in \mathbb{N}$, $n_0 > \epsilon^{-1}$. Then for $n \geq n_0$,

$$\begin{aligned}
\mathbb{P}_x[\tau_A > n] &\leq \mathbb{P}_x[L_n(A) < \epsilon] \\
&\leq \mathbb{P}_x[L_n \in B], \\
\overline{\lim_n}\, n^{-1} \log \mathbb{P}_x[\tau_A > n] &\leq \overline{\lim_n}\, n^{-1} \log \mathbb{P}_x[L_n \in B] \\
&\leq -a,
\end{aligned}$$

and therefore $\lim_n \mathbb{P}_x[\tau_A > n] = 0$, which implies (5.13). □

5.4 Notes

Theorem 5.1.3 is proved in Dinwoodie (1993) for S Polish, P Feller (see Appendix B), and $V = C_b(S)$. Lemma 5.5 is due to Donsker and Varadhan (1975, 1976); see also Lemma 4.1.45, Deuschel and Stroock (1989).

In the context of Theorem 5.8, under the assumption that S is Polish and $J = I$—in contrast to Theorem 5.8, in which J is not known a priori—the irreducibility of P, the existence and uniqueness of π, and $I = I_\pi$ are proved in Wu (2000a), Proposition 2.5.

6

Upper Bounds II: Equivalent Analytic Conditions

In Section 6.1 we study in a more abstract setting the condition 2(i), which played a key role in Theorems 3.1 and 3.2. We present several analytic conditions which are equivalent to a generalization of 2(i). The results apply to the functions $\Gamma_A \mid V$, $\phi_M \mid V$, $\tilde{\phi} \mid V$, $\Lambda \mid V$, where V is a vector subspace of $B(S)$ satisfying V.1–V.3.

Condition (6.3) of Theorem 6.1 with $\Gamma = \phi_M \mid V$ will play an important role in Theorems 8.1 and 8.3. Condition (6.4) of Theorem 6.1 with $\Gamma = \Gamma_A$ provides the basis for a different proof of the upper bound in Theorem 3.1 (hence in Theorem 3.2).

In Section 6.2 a more comprehensive form of Theorem 3.1.2 is presented, including the alternative proof of the upper bound. In Section 6.3 a more comprehensive form of Theorem 3.2.2 is presented. In order to relate the rate function $(\phi_M \mid V)^*$ to I, we will assume there that V satisfies V.1′–V.4, as was done in Chapter 4.

Under the assumption that P is irreducible, in Section 6.4 we derive necessary conditions for a set $M \subset \mathcal{P}(S)$ to be a uniformity set for the upper bound in the V topology. A stronger result is proved in the case $V = B(S)$.

6.1 Analytic Properties of Certain Real-Valued Functions on V

The functions Γ_A, ϕ_M, $\tilde{\phi}$, Λ are real-valued functions of a special form defined on $B(S)$. We will consider a more general function $\Gamma\colon V \to \mathbb{R}$, where V satisfies V.1–V.3, which captures that form.

We define

$$\Gamma^*(\ell) = \sup\{\ell(g) - \Gamma(g)\colon g \in V\}, \qquad \ell \in V^*;$$

in particular, for $\mu \in \mathcal{P}(S)$,

$$\Gamma^*(\mu) = \sup\left\{\ell_\mu(g) - \Gamma(g) \colon g \in V\right\}$$
$$= \sup\left\{\int g\, d\mu - \Gamma(g) \colon g \in V\right\},$$

where ℓ_μ is as in Chapter 3. For $a \geq 0$, we define

$$L_a = \{\mu \in \mathcal{P}(S) \colon \Gamma^*(\mu) \leq a\}.$$

For $d \in \mathbb{N}$, V^d is the Cartesian product of d copies of V. For $f = (f_1, \dots, f_d) \in V^d$ and $\xi = (\xi_1, \dots, \xi_d) \in \mathbb{R}^d$, we write

$$\langle f, \xi\rangle = \sum_{i=1}^{d} \xi_i f_i \quad \text{and} \quad \Gamma_f(\xi) = \Gamma(\langle f, \xi\rangle).$$

For $u \in \mathbb{R}^d$,

$$\Gamma_f^*(u) = \sup\left\{\langle u, \xi\rangle - \Gamma_f(\xi) \colon \xi \in \mathbb{R}^d\right\}.$$

For $f \in V^d$, as in Theorem 2.10, $\Phi_f \colon \mathcal{P}(S) \to \mathbb{R}^d$ is defined by $\Phi_f(\mu) = \int f\, d\mu$.

The topology $\sigma(V, \mathcal{P}(S))$ is the smallest topology on V such that each map $g \mapsto \int g\, d\mu$, $\mu \in \mathcal{P}(S)$, is continuous: a net $\{g_\delta\}_{\delta \in D} \subset V$ converges in this topology to $g \in V$ if and only if $\lim_\delta \int g_\delta\, d\mu = \int g\, d\mu$ for all $\mu \in \mathcal{P}(S)$.

Theorem 6.1 *Assume that V satisfies V.1–V.3 and $\Gamma \colon V \to \mathbb{R}$ satisfies:*

G.1 Γ is convex.

G.2 Γ is $\|\cdot\|$-continuous, where $\|\cdot\|$ is the supremum norm.

G.3 $\Gamma(c) = \Gamma(0) + c$ for all $c \in \mathbb{R}$.

G.4 If $f, g \in V$, $f \leq g$, then $\Gamma(f) \leq \Gamma(g)$.

Then the conditions (6.1)–(6.4) *are equivalent:*

$$\textit{If } 0 \leq g_k \in V \textit{ and } g_k \downarrow 0 \textit{ pointwise, then } \Gamma(g_k) \to 0. \tag{6.1}$$

If $\ell \in V^*$ and $\Gamma^*(\ell) < \infty$, then $\ell \in \mathcal{P}(S)$:

$$\textit{there exists a unique } \mu \in \mathcal{P}(S) \textit{ such that } \ell = \ell_\mu. \tag{6.2}$$

The following two conditions hold:

$$\Gamma \textit{ is } \sigma(V, \mathcal{P}(S))\textit{-lower semicontinuous.} \tag{6.3a}$$

$$\textit{For all } a \geq 0,\ L_a \textit{ is } V\textit{-compact.} \tag{6.3b}$$

The following two conditions hold:

For every $d \in \mathbb{N}$, $f \in V^d$, $u \in \mathbb{R}^d$,

$$\Gamma_f^*(u) = \begin{cases} \inf\left\{\Gamma^*(\mu) : \mu \in \Phi_f^{-1}(u)\right\} & \text{if } u \in \Phi_f(\mathcal{P}(S)), \\ \infty & \text{otherwise.} \end{cases} \tag{6.4a}$$

For all $a \geq 0$, L_a *is* V*-compact.* (6.4b)

Moreover, if (6.4) *is satisfied, then for every* V*-closed set* $C \subset \mathcal{P}(S)$,

$$\inf_{\mu \in C} \Gamma^*(\mu) = \sup_{f \in \mathcal{V}} \inf\left\{\Gamma_f^*(u) : u \in \overline{\Phi_f(C)}\right\}, \tag{6.5}$$

where

$$\mathcal{V} = \bigcup\{V^d : d \in \mathbb{N}\}.$$

In particular, for $\mu \in \mathcal{P}(S)$,

$$\Gamma^*(\mu) = \sup\left\{\Gamma_f^*\left(\Phi_f(\mu)\right) : f \in \mathcal{V}\right\}.$$

Proof. We prove (6.1) $\Longrightarrow$ (6.2) $\Longrightarrow$ (6.3) $\Longrightarrow$ (6.1), (6.3) $\Longleftrightarrow$ (6.4).

(6.1) $\Longrightarrow$ (6.2): This is proved like (3.5) in the proof of Theorem 3.1. Note that in Theorem 3.1, Γ_A satisfies $\Gamma_A(0) = 0$, but the proof works under assumption G.3 of Theorem 6.1.

(6.2) $\Longrightarrow$ (6.3a): Since Γ is convex and $\|\cdot\|$-continuous, for all $a \in \mathbb{R}$, $\{g \in V : \Gamma(g) \leq a\}$ is convex and $\|\cdot\|$-closed, hence $\sigma(V, V^*)$-closed. Therefore Γ is $\sigma(V, V^*)$-lower semicontinuous. By Proposition F.1 and (6.2),

$$\begin{aligned} \Gamma(g) &= \sup\{\ell(g) - \Gamma^*(\ell) : \ell \in V^*\} \\ &= \sup\left\{\int g \, d\mu - \Gamma^*(\mu) : \mu \in \mathcal{P}(S)\right\}, \end{aligned}$$

showing that Γ is $\sigma(V, \mathcal{P}(S))$-lower semicontinuous. Note that if $V = B(S)$, P is irreducible, and $\Gamma = \Lambda$, (6.3a) holds automatically by Theorem 2.13.

(6.2) $\Longrightarrow$ (6.3b): As in the compactness proof (a) in Theorem 3.1.

(6.3) $\Longrightarrow$ (6.1): By (F.4) of Proposition F.3, condition (6.3a) implies, for all $g \in V$,

$$\Gamma(g) = \sup\left\{\int g \, d\mu - \Gamma^*(\mu) : \mu \in \mathcal{P}(S)\right\}. \tag{6.6}$$

We can now argue as in the proof of 2(ii) $\Longrightarrow$ 2(i), Corollary 2.19. Assume that $0 \le g_k \in V$, $g_k \downarrow 0$ pointwise. By (6.6), for all $a \ge 0$,

$$\Gamma(g_k) \le \max\left\{\sup_{\mu\in L_a} \int g_k \, d\mu, c - a\right\},$$

where $c = \sup g_1$. By (6.3b) and Proposition H.1,

$$\sup_{\mu\in L_a} \int g_k \, d\mu \longrightarrow 0.$$

Therefore $\lim_k \Gamma(g_k) \le \max\{0, c - a\}$. But a is arbitrary. This proves (6.1).

(6.3) $\Longrightarrow$ (6.4a): For $u \in \mathbb{R}^d$, let

$$\widetilde{\Gamma_f}(u) = \begin{cases} \inf\left\{\Gamma^*(\mu) : \mu \in \Phi_f^{-1}(u)\right\} & \text{if } u \in \Phi_f(\mathcal{P}(S)), \\ \infty & \text{otherwise.} \end{cases}$$

The proof that $\widetilde{\Gamma_f} = \Gamma_f^*$ is as in Theorem 2.17 with $E = \mathbb{R}^d$ and Γ instead of Λ; the $\sigma(V, \mathcal{P}(S))$-lower semicontinuity of Γ is an assumption here, while in Theorem 2.17 the lower semicontinuity of Λ follows from Theorem 2.13. Note that for $f \in V^d$ and $\xi \in \mathbb{R}^d$, we have

$$\langle f, \xi\rangle = \sum_{i=1}^{d} \xi_i f_i \in V.$$

(6.4) $\Longrightarrow$ (6.3a): Let $g \in V$. By (6.4a), since Γ_g is convex and finite on $\mathbb{R}$, hence continuous,

$$\begin{aligned} \Gamma(g) = \Gamma_g(1) &= \Gamma_g^{**}(1) \\ &= \sup\left[u \cdot 1 - \Gamma_g^*(u) : u \in \mathbb{R}\right] \\ &= \sup\left[u \cdot 1 - \widetilde{\Gamma}_g(u) : u \in \mathbb{R}\right]; \end{aligned}$$

arguing as in the proof of Theorem 2.17 with $f = g$, $\xi = 1$,

$$= \sup\left[\int g \, d\mu - \Gamma^*(\mu) : \mu \in \mathcal{P}(S)\right].$$

This proves (6.3a).

We prove now (6.5). Let $f \in V^d$. If $u \in \mathbb{R}^d$, $\mu \in \mathcal{P}(S)$, and $u = \Phi_f(\mu)$, then

$$\begin{aligned} \Gamma_f^*(u) &= \sup\left\{\left\langle \Phi_f(\mu), \xi\right\rangle - \Gamma_f(\xi) : \xi \in \mathbb{R}^d\right\} \\ &= \sup\left\{\int \langle f, \xi\rangle \, d\mu - \Gamma(\langle f, \xi\rangle) : \xi \in \mathbb{R}^d\right\} \\ &\le \sup\left\{\int g \, d\mu - \Gamma(g) : g \in V\right\} = \Gamma^*(\mu). \end{aligned}$$

Therefore

$$\inf\left\{\Gamma_f^*(u) : u \in \overline{\Phi_f(C)}\right\} \le \inf\left\{\Gamma_f^*(u) : u \in \Phi_f(C)\right\}$$
$$\le \inf\left\{\Gamma^*(\mu) : \mu \in C\right\}$$

and

$$\sup_{f \in \mathcal{V}} \inf\left\{\Gamma_f^*(u) : u \in \overline{\Phi_f(C)}\right\} \le \inf\left\{\Gamma^*(\mu) : \mu \in C\right\}.$$

To prove the opposite inequality, let $c = \inf\{\Gamma^*(\mu) : \mu \in C\}$. Since the right-hand side of (6.5) is nonnegative, if $c = 0$ there is nothing to prove. For $c > 0$, let $0 < b < c$. We have $C \cap L_b = \emptyset$ and

$$C = \bigcap \left\{\Phi_f^{-1}\left(\overline{\Phi_f(C)}\right) : f \in \mathcal{V}\right\}.$$

To prove this equality, we first note that obviously

$$C \subset \Phi_f^{-1}\left(\Phi_f(C)\right) \subset \Phi_f^{-1}\left(\overline{\Phi_f(C)}\right)$$

for all $f \in \mathcal{V}$. On the other hand, if $\mu \notin C$ then, since C is V-closed, there exist $d \in \mathbb{N}$, $f \in V^d$, U open in $\mathbb{R}^d$, such that $\mu \in \Phi_f^{-1}(U)$ and $C \subset \Phi_f^{-1}(U^c)$. Therefore

$$\Phi_f(C) \subset \Phi_f\left(\Phi_f^{-1}(U^c)\right) \subset U^c,$$
$$\overline{\Phi_f(C)} \subset U^c,$$
$$\Phi_f^{-1}\left(\overline{\Phi_f(C)}\right) \subset \Phi_f^{-1}(U^c),$$

and it follows that

$$\mu \notin \Phi_f^{-1}\left(\overline{\Phi_f(C)}\right).$$

By the V-compactness of L_b, there exist $k \in \mathbb{N}$ and $f_i \in V^{d_i}$, $i = 1, \ldots, k$, such that

$$\bigcap_{i=1}^{k} \Phi_{f_i}^{-1}\left(\overline{\Phi_{f_i}(C)}\right) \cap L_b = \emptyset.$$

Let $h = (f_1, \ldots, f_k) \in V^d$, where $d = \sum_{i=1}^{k} d_i$. Then

$$\Phi_h^{-1}\left(\overline{\Phi_h(C)}\right) \cap L_b = \emptyset. \tag{6.7}$$

Therefore by (6.4a) and (6.7),

$$\inf\left\{\Gamma_h^*(u) : u \in \overline{\Phi_h(C)}\right\} = \inf\left\{\inf\left\{\Gamma^*(\mu) : \mu \in \Phi_h^{-1}(u)\right\} : u \in \overline{\Phi_h(C)}\right\}$$
$$= \inf\left\{\Gamma^*(\mu) : \mu \in \Phi_h^{-1}\left(\overline{\Phi_h(C)}\right)\right\} \ge b.$$

Therefore

$$\sup_{f \in \mathcal{V}} \inf \left\{ \Gamma_f^*(u) : u \in \overline{\Phi_f(C)} \right\} \geq b.$$

Letting $b \to c$, (6.5) is proved. □

Definition 6.2 Let V satisfy V.1–V.3. A function $\Gamma\colon V \to \mathbb{R}$ is *V-regular* if it satisfies assumptions G.1–G.4 of Theorem 6.1 and any, hence all, of the equivalent conditions (6.1)–(6.4).

Corollary 6.3 *Let V satisfy V.1–V.3 and let $\Gamma\colon V \to \mathbb{R}$ be V-regular. Then*

1. *If $g_n \in V$, $g_n \uparrow g \in V$ pointwise, then $\Gamma(g_n) \uparrow \Gamma(g)$.*
2. *If $g_n \in V$, $g_n \downarrow g \in V$ pointwise, then $\Gamma(g_n) \downarrow \Gamma(g)$.*

Proof. 1 (respectively, 2) is proved like statement 1 (respectively, statement 3) of Corollary 2.19. □

The following is a sufficient condition for (6.1) in the case when S is Polish and $V = C_b(S)$.

Proposition 6.4 *Assume that S is Polish and $\Gamma\colon C_b(S) \to \mathbb{R}$ satisfies assumptions G.1–G.4 of Theorem 6.1, $\Gamma(0) = 0$, and:*

$$\begin{gathered}\text{For every } a > 0, \text{ there exists a compact set } K \subset S \text{ such that}\\ \text{if } f \in C_b(S),\ f \mid K = 0, \text{ and } \|f\| \leq a, \text{ then } \Gamma(f) \leq 1.\end{gathered} \tag{6.8}$$

Then Γ is $C_b(S)$-regular.

Proof. Let $0 \leq f_k \in C_b(S)$, $f_k \downarrow 0$ pointwise. Let $\lambda > 1$, $a = 2\lambda\|f_1\|$. Let K be a compact set as in assumption (6.8). By Dini's theorem, $f_k \downarrow 0$ uniformly on K. Let $k_0 \in \mathbb{N}$ be such that for $k \geq k_0$,

$$\sup_{x \in K} f_k(x) \leq \frac{1}{4\lambda}.$$

Let

$$U = \left\{ x \in S\colon f_{k_0}(x) < \frac{1}{2\lambda} \right\};$$

then $K \subset U$. Let $g \in C_b(S)$ be such that $\mathbf{1}_K \leq g \leq \mathbf{1}_U$. For $k \geq k_0$, we have

$$2\lambda f_k g \leq 2\lambda f_{k_0} \mathbf{1}_U < 1,$$

and therefore

$$\Gamma(2\lambda f_k g) \leq \Gamma(1) = 1. \tag{6.9}$$

Also

$$\|2\lambda f_k(1-g)\| \le 2\lambda\|f_1\| = a \quad \text{and} \quad 2\lambda f_k(1-g) \mid K = 0,$$

which implies

$$\Gamma(2\lambda f_k(1-g)) \le 1. \tag{6.10}$$

Therefore, for $k \ge k_0$,

$$\begin{aligned}\Gamma(f_k) = \Gamma\left(\frac{1}{\lambda}(\lambda f_k) + \left(1 - \frac{1}{\lambda}\right)0\right) &\le \frac{1}{\lambda}\Gamma(\lambda f_k) \\ &= \frac{1}{\lambda}\Gamma\left(\lambda f_k g + \lambda f_k(1-g)\right) \\ &\le \frac{1}{2\lambda}\Gamma(2\lambda f_k g) + \frac{1}{2\lambda}\Gamma\left(2\lambda f_k(1-g)\right) \\ &\le \frac{1}{\lambda}\end{aligned}$$

by (6.9)–(6.10). Therefore $\lim_k \Gamma(f_k) \le 1/\lambda$. Let $\lambda \to \infty$. □

6.2 Upper Bounds for Random Probability Measures II

Let $(\Omega', \mathcal{A})$, $\{\mathbb{P}_\alpha\}_{\alpha\in A}$, $\{M_n\}$, Γ_A, $(\Gamma_A \mid V)^*$ be as in Chapter 3. We note that $\Gamma_A\colon V \to \mathbb{R}$ satisfies assumptions G.1–G.4. G.1 follows easily from Hölder's inequality; G.3 and G.4 are immediate. We prove assumption G.2: for $f, g \in V$,

$$\begin{aligned}e^{-n\|f-g\|}\mathbb{E}_\alpha\left\{\exp\left(n\int f\, dM_n\right)\right\} &\le \mathbb{E}_\alpha\left\{\exp\left(n\int g\, dM_n\right)\right\} \\ &\le e^{n\|f-g\|}\mathbb{E}_\alpha\left\{\exp\left(n\int f\, dM_n\right)\right\},\end{aligned}$$

and therefore $|\Gamma_A(f) - \Gamma_A(g)| \le \|f-g\|$. The following result is a more inclusive statement of Theorem 3.1.2.

Theorem 6.5 *Assume that V satisfies V.1–V.3. Then the conditions* (6.11)–(6.13) *are equivalent:*

$$\Gamma_A \mid V \textit{ is } V\textit{-regular.} \tag{6.11}$$

The following two conditions hold:

For every measurable set $B \subset \mathcal{P}(S)$,

$$\overline{\lim_n}\, n^{-1} \log \sup_{\alpha\in A} \mathbb{P}_\alpha[M_n \in B] \le -\inf\{(\Gamma_A \mid V)^*(\mu) \colon \mu \in \mathrm{cl}_V(B)\}. \tag{6.12a}$$

$$(\Gamma_A \mid V)^* \textit{ is } V\textit{-tight.} \tag{6.12b}$$

There exists a V-tight function $J: \mathcal{P}(S) \to \overline{\mathbb{R}^+}$ such that for every measurable set $B \subset \mathcal{P}(S)$,

$$\overline{\lim_n}\, n^{-1} \log \sup_{\alpha\in A} \mathbb{P}_\alpha[M_n \in B] \le -\inf\{J(\mu): \mu \in \mathrm{cl}_V(B)\}. \tag{6.13}$$

Lemma 6.6 *Let $Y_n: \Omega' \to \mathbb{R}^d$, $n \in \mathbb{N}$, be random vectors such that $\rho = \sup_n \|Y_n\| < \infty$. Let*

$$\theta(\xi) = \overline{\lim_n}\, n^{-1} \log \sup_{\alpha\in A} \mathbb{E}_\alpha\left(\exp\langle nY_n, \xi\rangle\right), \qquad \xi \in \mathbb{R}^d,$$

$$\theta^*(u) = \sup\left\{\langle u, \xi\rangle - \theta(\xi): \xi \in \mathbb{R}^d\right\}, \qquad u \in \mathbb{R}^d.$$

Then for every closed set $F \subset \mathbb{R}^d$,

$$\overline{\lim_n}\, n^{-1} \log \sup_{\alpha\in A} \mathbb{P}_\alpha[Y_n \in F] \le -\inf_{u\in F} \theta^*(u). \tag{6.14}$$

Proof. This is a rerun of the proof of Theorem 3.1.1. We will indicate the first steps. Let

$$B_\rho = \left\{u \in \mathbb{R}^d: \|u\| \le \rho\right\}, \qquad a = \inf\left\{\theta^*(u): u \in F \cap B_\rho\right\},$$

and for $\xi \in \mathbb{R}^d$, $\epsilon > 0$,

$$H(\xi) = \left\{u \in \mathbb{R}^d: \langle u, \xi\rangle - \theta(\xi) > a - \epsilon\right\}.$$

Then

$$F \cap B_\rho \subset \left\{u \in \mathbb{R}^d: \theta^*(u) > a - \epsilon\right\} = \bigcup\left\{H(\xi): \xi \in \mathbb{R}^d\right\}.$$

Since $F \cap B_\rho$ is compact and $H(\xi)$ is open, there exist $\xi_1, \ldots, \xi_k \in \mathbb{R}^d$ such that

$$F \cap B_\rho \subset \bigcup_{i=1}^k H(\xi_i).$$

Since $\mathbb{P}_\alpha[Y_n \in F] = \mathbb{P}_\alpha[Y_n \in F \cap B_\rho]$, proceeding as in the proof of Theorem 3.1.1 we arrive at (6.14). □

Proof of Theorem 6.5. (6.11) $\Longrightarrow$ (6.12a): Let $\mathcal{V}$, Φ_f ($f \in \mathcal{V}$) be as in the proof of Theorem 6.1. Let B be a measurable set $\subset \mathcal{P}(S)$, $C = \mathrm{cl}_V(B)$. For all $f \in \mathcal{V}$, $f: S \to \mathbb{R}^d$ for some $d \in \mathbb{N}$,

$$\overline{\lim_n}\, n^{-1} \log \sup_{\alpha\in A} \mathbb{P}_\alpha[M_n \in B] \le \overline{\lim_n}\, n^{-1} \log \sup_{\alpha\in A} \mathbb{P}_\alpha\left[\int f\, dM_n \in \overline{\Phi_f(C)}\right];$$

by Lemma 6.6, letting $Y_n: \Omega' \to \mathbb{R}^d$ be defined by $Y_n = \int f\, dM_n$,

$$\le -\inf\left\{\theta^*(u): u \in \overline{\Phi_f(C)}\right\},$$

where, for $\xi \in \mathbb{R}^d$, $u \in \mathbb{R}^d$,

$$\theta(\xi) = \overline{\lim_n} \, n^{-1} \log \sup_{\alpha \in A} \mathbb{E}_\alpha \exp\langle nY_n, \xi\rangle,$$
$$\theta^*(u) = \sup_{\xi \in \mathbb{R}^d} [\langle u, \xi\rangle - \theta(\xi)] .$$

But, in the notation of Theorem 6.1,

$$\theta(\xi) = \Gamma_A(\langle f, \xi\rangle) = \Gamma_{A,f}(\xi) \quad \text{and} \quad \theta^*(u) = \Gamma^*_{A,f}(u).$$

Therefore

$$\begin{aligned}\overline{\lim_n} \, n^{-1} \log \sup_{\alpha \in A} \mathbb{P}_\alpha[M_n \in B] &\leq \inf_{f \in \mathcal{V}} \left\{- \inf \left[\Gamma^*_{A,f}(u) \colon u \in \overline{\Phi_f(C)}\right]\right\} \\ &= - \sup_{f \in \mathcal{V}} \inf \left\{\Gamma^*_{A,f}(u) \colon u \in \overline{\Phi_f(C)}\right\} \\ &= - \inf \{(\Gamma_A \mid V)^*(\mu) \colon \mu \in C\}\end{aligned}$$

by Theorem 6.1. Completing the step, (6.11) $\Longrightarrow$ (6.12b) is obvious. (6.12) $\Longrightarrow$ (6.13) is trivial.

(6.13) $\Longrightarrow$ (6.11): We will show that Γ_A satisfies (6.1). By Corollary E.4, applied to $\gamma_{\alpha,n} = \mathbb{P}_\alpha[M_n \in \cdot]$, we have

$$\Gamma_A(g) \leq \sup_{\mu \in \mathcal{P}(S)} \left[\int g \, d\mu - J(\mu)\right], \qquad g \in V. \tag{6.15}$$

The proof is completed as in (6.3) $\Longrightarrow$ (6.1) of Theorem 6.1. □

Remark 6.7

1. We will compare the proofs of Theorems 2.10 and 6.5 ((6.11) $\Longrightarrow$ (6.12a)). The following facts are used in Theorem 2.10:
   (a) Any V-open set $G \subset \mathcal{P}(S)$ is the union of sets of the form $\Phi_f^{-1}(U)$, $f \in V^d$, U open in $\mathbb{R}^d$, $d \in \mathbb{N}$.
   (b) By Theorem 2.1, Λ_f^* is a rate function for the lower bound for $\{\inf_{\nu \in M} \mathbb{P}_\nu [\Phi_f(L_n) \in \cdot]\}$, where $f \colon S \to \mathbb{R}^d$ is a bounded measurable function; note that $\Phi_f(L_n) = n^{-1} S_n(f)$.

   In proving Theorem 6.5, (6.11) $\Longrightarrow$ (6.12a), we use the following facts, applying Theorem 6.1:

   (a)′ Any V-closed set C can be expressed as an intersection of sets of the form $\Phi_f^{-1}(K)$, $f \in V^d$, K closed in $\mathbb{R}^d$, $d \in \mathbb{N}$:

$$C = \bigcap \left\{\Phi_f^{-1}\left(\overline{\Phi_f(C)}\right) \colon f \in \mathcal{V}\right\}.$$

(b)′ By Lemma 6.6, applied to $\Phi_f(M_n) = \int f\,dM_n$, $\Gamma^*_{A,f}$ is a rate function for the upper bound for $\{\sup_{\alpha\in A}\mathbb{P}_\alpha[\Phi_f(M_n)\in\cdot]\}$, where $f\colon S\to\mathbb{R}^d$ is a bounded measurable function.

2. By Theorem 6.5, if there exists a V-tight rate function J for the upper bound for $\{\sup_{\alpha\in A}\mathbb{P}_\alpha[M_n\in\cdot]\}$ in the V topology, then $(\Gamma_A \mid V)^*$ is also a V-tight rate function for the upper bound for $\{\sup_{\alpha\in A}\mathbb{P}_\alpha[M_n\in\cdot]\}$ in the V topology. If, moreover, J is convex then, by (6.15) and Corollary F.2, $J\le(\Gamma_A \mid V)^*$. This shows that $(\Gamma_A \mid V)^*$ is optimal among V-tight convex rate functions for the upper bound for $\{\sup_{\alpha\in A}\mathbb{P}_\alpha[M_n\in\cdot]\}$ in the V topology.

6.3 Upper Bounds for Empirical Measures II

As a consequence of Theorem 6.5, we have the following more complete statement of Theorem 3.2.2.

Theorem 6.8 *Assume that V satisfies V.1′–V.4. Let $M\subset\mathcal{P}(S)$.*

1. *The conditions* (6.16)–(6.18) *are equivalent:*

$$\phi_M \mid V \text{ is } V\text{-regular.} \tag{6.16}$$

The following two conditions hold:

For every measurable set $B\subset\mathcal{P}(S)$,

$$\begin{aligned}\overline{\lim_n}\, n^{-1}\log\sup_{\nu\in M}\mathbb{P}_\nu[L_n\in B] &\le -\inf\{(\phi_M \mid V)^*(\mu)\colon \mu\in \mathrm{cl}_V(B)\}.\\ &\le -\inf\{I(\mu)\colon \mu\in\mathrm{cl}_V(B)\}.\end{aligned}\tag{6.17a}$$

$$(\phi_M \mid V)^* \text{ is } V\text{-tight.} \tag{6.17b}$$

There exists a V-tight function $J\colon\mathcal{P}(S)\to\overline{\mathbb{R}^+}$ such that for every measurable set $B\subset\mathcal{P}(S)$,

$$\overline{\lim_n}\, n^{-1}\log\sup_{\nu\in M}\mathbb{P}_\nu[L_n\in B]\le -\inf\{J(\mu)\colon\mu\in\mathrm{cl}_V(B)\}. \tag{6.18}$$

2. *If any (hence all) of conditions* (6.16)–(6.18) *is satisfied, then P has an invariant probability measure π. If, moreover, S is P-indecomposable, then π is unique, $\pi(F)=1$ for every P-closed set F, and $(\phi_M \mid V)^*(\mu)=0$ if and only if $\mu=\pi$.*

Proof. Statement 1 follows from Theorem 6.5, except for the second inequality in (6.17a). This inequality follows from the fact that, by Theorem 4.1,

$$(\phi_M \mid V)^*\ge(\phi \mid V)^*=I. \tag{6.19}$$

Statement 2 follows from (6.19) and the proof of Theorem 5.4. As in that result, (6.17) implies that

$$Z \triangleq \{\mu \in \mathcal{P}(S) \colon (\phi_M \mid V)^*(\mu) = 0\} \neq \emptyset.$$

By (6.19) and Lemma 5.5, if $\mu \in Z$ then μ is invariant. If, moreover, S is P-indecomposable, then by Lemma 5.6 there is a unique invariant probability measure π. It follows that $Z = \{\pi\}$. □

Remark 6.9

1. Just as in Remark 6.7(2), $(\phi_M \mid V)^*$ is optimal among V-tight convex rate functions for the upper bound for $\{\sup_{\nu \in M} \mathbb{P}_\nu[L_n \in \cdot]\}$ in the V topology.
2. A sufficient condition for the V-tightness of I is: $0 \le f_k \in V$, $f_k \downarrow 0$ pointwise imply $\tilde{\phi}(f_k) \to 0$. For, $I = (\tilde{\phi} \mid V)^*$ by Theorem 4.1 and by Theorem 6.1 $(\tilde{\phi} \mid V)^*$ is V-tight. In the case $M = S$, I is V-tight since $(\phi \mid V)^*$ is V-tight and $(\phi \mid V)^* = I$ by Theorem 4.1.

By (6.17a), applied to $M = \{\delta_x\}$, for every $x \in S$ I is a rate function for the upper bound for $\{\mathbb{P}_x[L_n \in \cdot]\}$ in the V topology. However, in general I is not necessarily V-tight. It turns out that the sufficient condition for the V-tightness of I given in Remark 6.9(2) is actually necessary for the existence of a V-tight rate function for the upper bound for $\{\mathbb{P}_x[L_n \in \cdot]\}$ in the V topology, independent of $x \in S$.

Theorem 6.10 *Assume that V satisfies V.1′–V.4. Then the following conditions are equivalent:*

1. *$\tilde{\phi} \mid V$ is V-regular.*
2. *For every $x \in S$, $\{\mathbb{P}_x[L_n \in \cdot]\}$ satisfies the upper bound in the V topology with V-tight rate function I.*
3. *There exists a V-tight function $J \colon \mathcal{P}(S) \to \overline{\mathbb{R}^+}$ such that for every $x \in S$, $\{\mathbb{P}_x[L_n \in \cdot]\}$ satisfies the upper bound in the V topology with rate function J.*

Proof. Condition 1 $\Longrightarrow$ 2: This follows from Theorem 6.8 with $M = \{\delta_x\}$, $x \in S$, and Remark 6.9(2). Condition 2 $\Longrightarrow$ 3 is trivial.

Condition 3 $\Longrightarrow$ 1: The proof is like that of (6.13) $\Longrightarrow$ (6.11) in the proof of Theorem 6.5, with one more step. By Corollary E.4 with $A = \{\delta_x\}$ ($x \in S$, fixed), $\gamma_{\alpha,n} = \mathbb{P}_x[L_n \in \cdot]$, and $\Gamma_A = \phi_x$, we have for $g \in V$,

$$\phi_x(g) \le \sup\left\{\int g \, d\mu - J(\mu) \colon \mu \in \mathcal{P}(S)\right\}.$$

Therefore

$$\tilde{\phi}(g) \leq \sup\left\{\int g\,d\mu - J(\mu) \colon \mu \in \mathcal{P}(S)\right\}.$$

Proceeding now as in the proof of Theorem 6.5, we obtain $\lim_k \tilde{\phi}(g_k) = 0$ whenever $0 \leq g_k \in V$, $g_k \downarrow 0$ pointwise. □

It turns out that if condition 2 of Theorem 6.10 is upgraded to asserting the simultaneous validity of the upper bound for $\{\mathbb{P}_\nu[L_n \in \cdot]\}$ for all $\nu \in \mathcal{P}(S)$, then this stronger condition is equivalent to the uniformity of S for the upper bound.

Theorem 6.11

$$\sup\{\phi_\nu \colon \nu \in \mathcal{P}(S)\} = \phi \text{ on } B(S). \tag{6.20}$$

Assume that V satisfies V.1′–V.4. Then the following conditions are equivalent:

1. $\sup\{\phi_\nu \colon \nu \in \mathcal{P}(S)\} \mid V$ *is V-regular.*
2. *For every $\nu \in \mathcal{P}(S)$, $\{\mathbb{P}_\nu[L_n \in \cdot]\}$ satisfies the upper bound in the V topology with V-tight rate function I.*
3. *There exists a V-tight function $J \colon \mathcal{P}(S) \to \overline{\mathbb{R}^+}$ such that for every $\nu \in \mathcal{P}(S)$, $\{\mathbb{P}_\nu[L_n \in \cdot]\}$ satisfies the upper bound in the V topology with rate function J.*
4. *S is a uniformity set for the upper bound in the V topology with V-tight rate function I.*

Proof. Let $\hat{\phi} = \sup\{\phi_\nu \colon \nu \in \mathcal{P}(S)\}$. Clearly $\hat{\phi} \leq \phi$. Suppose that for some $g \in B(S)$, $\hat{\phi}(g) < \phi(g)$, and let $\hat{\phi}(g) < \lambda < \phi(g)$, $\rho = e^{-\lambda}$. Similarly to the proof of Theorem 4.1, since $\exp(-\phi_\nu(g))$ is the radius of convergence of the power series with coefficients $\{\nu K_g^n 1\}$, we have: for every $\nu \in \mathcal{P}(S)$,

$$\sum_{n\geq 1} \rho^n \left(\nu K_g^n 1\right) < \infty,$$

or $\int h\,d\nu < \infty$, where

$$h(x) = \sum_{n\geq 1} \rho^n K_g^n 1(x), \qquad x \in S.$$

But this easily implies that $h \in B(S)$.

Next, since $\rho > \exp(-\phi(g))$, we have

$$\overline{\lim_n}\, \rho^n \sup_{x\in S} K_g^n 1(x) = \infty.$$

Therefore there exists $\{x_k\} \subset S$ such that $\rho^k K_g^k 1(x_k) \to \infty$. Since $h(x_k) \geq \rho^k K_g^k 1(x_k)$, this is impossible. Therefore $\hat{\phi} = \phi$.

The proofs of condition 1 $\Longrightarrow$ 2 $\Longrightarrow$ 3 $\Longrightarrow$ 1 are the same as in Theorem 6.10. Next, by (6.20), $\phi \mid V$ is V-regular if condition 1 holds. Then by Theorem 6.8, S is a uniformity set for the upper bound in the V topology with V-tight rate function $(\phi \mid V)^* = I$ (Theorem 4.1).

The fact that condition 4 implies condition 2 is obvious. □

6.4 Necessary Conditions for the Uniformity of a Set of Initial Distributions for the Upper Bound

First we will prove a general result for V satisfying V.1′–V.4, and then a stronger result in the case $V = B(S)$.

We will need a special assumption on V and P, as follows. Let P be irreducible and

$$\mathcal{G}(V) = \{G \in \mathcal{S}^+ : 1_G = \lim f_k, \text{ where } \{f_k\} \subset V \text{ is an increasing sequence}\}.$$

The assumption is:

$$\text{There exists a petite set } G \in \mathcal{G}(V). \tag{6.21}$$

Assumption (6.21) is satisfied when

1. $V = B(S)$.
2. S is a Polish space, $V = C_b(S)$, and there exists an open petite set in $\mathcal{S}^+$; note that every open set belongs to $\mathcal{G}(C_b(S))$. If P is strong Feller (Appendix B) then this latter existence condition is satisfied (Proposition B.11).

Theorem 6.12 *Let P be irreducible. Assume that V satisfies V.1′–V.4 and that* (6.21) *holds. Let $M \subset \mathcal{P}(S)$ be a uniformity set for the upper bound in the V topology with V-tight rate function J. Then:*

1. *There exists a petite set $C \in \mathcal{S}^+$ such that*

$$\inf\{\nu(C) : \nu \in M\} > 0. \tag{6.22}$$

2. *In particular, if $M \subset S$ then M is petite.*

Proof. By Theorem 6.8, the assumption implies that M is a uniformity set for the upper bound in the V topology with V-tight rate function $(\phi_M \mid V)^*$ and that P has an invariant probability measure π. Moreover, by Lemma 5.7, since P is irreducible, π is unique and $\pi \equiv \psi$. Also, by Theorem 6.8,

$$(\phi_M \mid V)^*(\mu) = 0 \text{ if and only if } \mu = \pi. \tag{6.23}$$

We will show:

$$\lim_n \sup_{\nu \in M} \mathbb{P}_\nu[\tau_G > n] = 0, \tag{6.24}$$

where G is as in (6.21) and $\tau_G = \inf\{n \geq 1 : X_n \in G\}$. Since $\psi(G) > 0$ and $\pi \equiv \psi$, we have $\pi(G) > 0$. Let $0 < \epsilon < \pi(G)$,

$$B = \{\mu \in \mathcal{P}(S) : \mu(G) \leq \epsilon\}.$$

Then

$$B = \bigcap_k \left\{\mu \in \mathcal{P}(S) : \int f_k \, d\mu \leq \epsilon\right\},$$

where $\{f_k\} \subset V$ is increasing and $\mathbf{1}_G = \lim_k f_k$, and it follows that B is a V-closed subset of $\mathcal{P}(S)$. Since $\pi \notin B$, by the V-tightness of $(\phi_M \mid V)^*$ and (6.23),

$$\beta = \inf\{(\phi_M \mid V)^*(\mu) : \mu \in B\} > 0.$$

Let $n_0 \in \mathbb{N}$, $n_0 > \epsilon^{-1}$. Then for $n \geq n_0$,

$$\mathbb{P}_\nu[\tau_G > n] \leq \mathbb{P}_\nu[L_n(G) < \epsilon] \leq \mathbb{P}_\nu[L_n \in B],$$

and therefore

$$\begin{aligned}\overline{\lim_n}\, n^{-1} \log \sup_{\nu \in M} \mathbb{P}_\nu[\tau_G > n] &\leq \overline{\lim_n}\, n^{-1} \log \sup_{\nu \in M} \mathbb{P}_\nu[L_n \in B] \\ &\leq -\beta,\end{aligned}$$

which implies (6.24). The conclusion follows now from Proposition B.6. □

In the next result we show that if $V = B(S)$, then the conclusion in Theorem 6.12 can be sharpened. For the definition of uniform set, see Appendix B. This notion is more restrictive than (6.22); see Corollary B.8. In particular, if $M \subset S$ and M is uniform, then M is petite.

Theorem 6.13 *Let P be irreducible. Let $M \subset \mathcal{P}(S)$ be a uniformity set for the upper bound in the τ topology with τ-tight rate function J. Then M is uniform.*

Proof. Since in the present case for any $A \in \mathcal{S}^+$ the set $B = \{\mu \in \mathcal{P}(S) : \mu(A) \leq \epsilon\}$ is τ-closed, (6.24) with A instead of G is proved by the argument in Theorem 6.12. □

Remark 6.14 By Theorem 6.13, if P is irreducible and S is a uniformity set for the upper bound in the τ topology with τ-tight rate function J, then P is uniformly recurrent. For, S is petite; see Proposition B.9.

6.5 Notes

Theorems 6.1, 6.5 and 6.8 are partly based on de Acosta (1990), but the material there has been reorganized, expanded, and reformulated here.

A result similar to Theorem 6.1 and part of (6.16)–(6.18) in Theorem 6.8 for S Polish, $V = C_b(S)$ or $B(S)$, and certain specifications of Γ is stated in Wu (2000a). The list of conditions in that work does not include (6.3) or (6.4) of Theorem 6.1 but includes a different condition.

The assumption in Proposition 6.4 appears in Deuschel and Stroock (1989); see the notes in Section 3.3.

Equality (6.20) for S Polish is proved in Wu (2000a). The proof given here is based on that paper.

7

Upper Bounds III: Sufficient Conditions

In this chapter we will present different sufficient conditions for the upper bound for $\{\sup_{\nu \in M} \mathbb{P}_\nu[L_n \in \cdot]\}$ in the V topology; V will be assumed to satisfy V.1′–V.4. In view of Theorem 6.8, in each case the task will be to show that the stated condition implies: $\phi_M \mid V$ (or $\tilde{\phi} \mid V$) is V-regular.

7.1 Sufficient Conditions for the Upper Bound in the V Topology

Theorem 7.1 *Assume that V satisfies V.1′–V.4 and the following condition holds: there exists $m \in \mathbb{N}$ such that $\{P^m(x, \cdot): x \in S\}$ is V-relatively compact. Then S is a uniformity set for the upper bound in the V topology with V-tight rate function $(\phi \mid V)^* = I$.*

Proof. Let $g \in B(S)$, $g \geq 0$. We claim that

$$\phi(g) \leq m^{-1} \log \sup_{x \in S} \int P^m(x, dy) \exp(mg(y)) . \tag{7.1}$$

For, by the convexity of the exponential function, for $p \in \mathbb{N}$,

$$\begin{aligned} \exp\left(\sum_{j=1}^{pm} g(X_j)\right) &= \exp\left(\sum_{i=1}^{m}\sum_{j=0}^{p-1} g(X_{i+mj})\right) \\ &\leq m^{-1} \sum_{i=1}^{m} \exp\left(\sum_{j=0}^{p-1} mg(X_{i+mj})\right) . \end{aligned} \tag{7.2}$$

Next, for $x \in S$, $i = 1, \ldots, m$, we have by the Markov property,

$$\mathbb{E}_x\left[\exp\left(\sum_{j=0}^{p-1} mg(X_{i+mj})\right)\right] \leq \mathbb{E}_x\left[\exp\left(\sum_{j=0}^{p-2} mg(X_{i+mj})\right)\right]$$
$$\times \sup_{y\in S} \mathbb{E}_y\left[\exp\left(mg(X_m)\right)\right] \tag{7.3}$$
$$\leq e^{m\|g\|}\left\{\sup_{y\in S} \mathbb{E}_y\left[\exp\left(mg(X_m)\right)\right]\right\}^{p-1}.$$

For $n \in \mathbb{N}$, $p = [n/m] + 1$, by (7.2) and (7.3),

$$\mathbb{E}_x\left[\exp\left(\sum_{j=1}^{n} g(X_j)\right)\right] \leq \mathbb{E}_x\left[\exp\left(\sum_{j=1}^{pm} g(X_j)\right)\right]$$
$$\leq e^{m\|g\|}\left\{\sup_{y\in S} \mathbb{E}_y\left[\exp\left(mg(X_m)\right)\right]\right\}^{p-1},$$

and therefore

$$\phi(g) = \overline{\lim_n}\, n^{-1} \log \sup_{x\in S} \mathbb{E}_x\left[\exp\left(\sum_{j=1}^{n} g(X_j)\right)\right]$$
$$\leq m^{-1} \log \sup_{y\in S} \mathbb{E}_y\left[\exp\left(mg(X_m)\right)\right],$$

proving (7.1). If $0 \leq g_k \in V$, $g_k \downarrow 0$ pointwise, then by Proposition H.1 the right-hand side of (7.1) with $g = g_k$ converges to 0, hence $\phi(g_k) \to 0$.

The upper bound for $\{\sup_{x\in S} \mathbb{P}_x[L_n \in \cdot]\}$ in the V topology now follows from Theorem 6.8, and the equality $(\phi \mid V)^* = I$ follows from Theorem 4.1. □

Corollary 7.2 *Assume that the following condition holds: there exists $m \in \mathbb{N}$ and a finite measure λ such that $\sup_{x\in S} P^m(x,\cdot) \leq \lambda$. Then S is a uniformity set for the upper bound in the τ topology with τ-tight rate function I.*

Proof. Take $V = B(S)$ in Theorem 7.1; the compactness condition follows from Proposition H.1. Or, directly, assume that $0 \leq g_k \in B(S)$, $g_k \downarrow 0$ pointwise. By (7.1),

$$\phi(g_k) \leq m^{-1} \log \int \exp(mg_k)\, d\lambda \longrightarrow 0$$

by dominated convergence. □

Theorem 7.8 (Corollary 7.9) is a substantial improvement of Theorem 7.1. In order to state it, we need the following definition.

Definition 7.3 A set $C \subset S$ is *(P-V)-tight* if there exists $m \in \mathbb{N}$ such that $\{P^m(x,\cdot)\colon x \in C\}$ is V-relatively compact.

Remark 7.4 If P is irreducible and C is (P-τ)-tight, then C is petite. For, by Proposition B.5, there exists an increasing sequence of petite sets $\{D_j\}_{j\in\mathbb{N}}$ such that $\bigcup_{j=1}^{\infty} D_j = S$. By Proposition H.1,

$$\inf_{x\in C} P^m(x, D_j) = 1 - \sup_{x\in C} P^m(x, D_j^c) \uparrow 1.$$

Therefore there exists $j_0 \in \mathbb{N}$ such that

$$P^m \mathbf{1}_{D_{j_0}} \geq \frac{1}{2}\mathbf{1}_C. \tag{7.4}$$

Let $q \in \mathbb{N}$, $\nu \in \mathcal{P}(S)$, $\alpha > 0$ be such that

$$\sum_{k=1}^{q} P^k \geq \alpha \mathbf{1}_{D_{j_0}} \otimes \nu.$$

Then

$$P^m \sum_{j-1}^{q} P^j \geq \alpha \left(P^m \mathbf{1}_{D_{j_0}}\right) \otimes \nu \geq \left(\frac{\alpha}{2}\right) \mathbf{1}_C \otimes \nu, \tag{7.5}$$

showing that C is petite.

Remark 7.5 If P is Harris recurrent (Section B.3) and C is (P-τ)-tight, then C is uniform (Definition B.7). Let m be as in the definition of a (P-τ)-tight set, $k > m$, and let $A \in \mathcal{S}^+$, $\tau_A = \inf\{n \geq 1 : X_n \in A\}$. Since on $[\tau_A > m]$ we have

$$\tau_A = m + \tau_A \circ \Theta^m,$$

where Θ is the canonical shift, by the Markov property it follows that

$$\begin{aligned} \mathbb{P}_x[\tau_A \geq k] = \mathbb{P}_x[m + \tau_A \circ \Theta^m \geq k] &= \mathbb{E}_x \mathbf{1}_{[k,\infty)}(m + \tau_A \circ \Theta^m) \\ &= \mathbb{E}_x \mathbb{E}_{X_m} \mathbf{1}_{[k,\infty)}(m + \tau_A) \\ &= \int P^m(x, dy) \mathbb{P}_y[\tau_A \geq k - m]. \end{aligned} \tag{7.6}$$

By Harris recurrence,

$$g_k(y) = \mathbb{P}_y[\tau_A \geq k - m] \downarrow 0$$

for all $y \in S$. Therefore by Proposition H.1,

$$\sup_{x\in C} \mathbb{P}_x[\tau_A \geq k] = \sup_{x\in C} P^m g_k(x) \downarrow 0. \tag{7.7}$$

Note that by Corollary B.8, the conclusion of Remark 7.5 is stronger than that of Remark 7.4.

Remark 7.6 A sufficient condition for a set $C \in \mathcal{S}$ to be (P-τ)-tight is as follows. Let $\varphi\colon \mathbb{R}^+ \to \mathbb{R}^+$ be a nondecreasing convex function such that

$$\varphi(0) = 0 \quad \text{and} \quad \lim_{t\to\infty} t^{-1}\varphi(t) = \infty. \tag{7.8}$$

Suppose there exist $k \in \mathbb{N}$, $\lambda \in \mathcal{P}(S)$ such that for all $x \in C$,

$$P^k(x,\cdot) \ll \lambda \quad \text{and} \quad a = \sup_{x\in C} \int \varphi\left(\frac{dP^k(x,\cdot)}{d\lambda}\right) d\lambda < \infty. \tag{7.9}$$

Then $\{P^k(x,\cdot)\colon x \in C\}$ is τ-relatively compact. For, define $\varphi^*\colon \mathbb{R}^+ \to \mathbb{R}^+$ by

$$\varphi^*(t) = \sup\{st - \varphi(s)\colon s \geq 0\}.$$

Then it is easily proved that φ^* satisfies:

1. $\varphi^*(0) = 0$.
2. For all $t \geq 0$, $0 \leq \varphi^*(t) < \infty$.
3. φ^* is increasing, convex, and continuous.
4. For all $s \geq 0$, $t \geq 0$, $st \leq \varphi(s) + \varphi^*(t)$.

Suppose $0 \leq g_j \in B(S)$, $g_j \downarrow 0$ pointwise. Then for all $p > 0$,

$$\begin{aligned} \int g_j\, dP^k(x,\cdot) &= \int (pg_j)\left(p^{-1}\frac{dP^k(x,\cdot)}{d\lambda}\right) d\lambda \\ &\leq \int \varphi^*(pg_j)\, d\lambda + \int \varphi\left(p^{-1}\frac{dP^k(x,\cdot)}{d\lambda}\right) d\lambda. \end{aligned} \tag{7.10}$$

By dominated convergence,

$$\lim_j \int \varphi^*(pg_j)\, d\lambda = 0.$$

Therefore, since $\varphi(p^{-1}u) \leq p^{-1}\varphi(u)$ for $u \geq 0$, $p > 1$,

$$\limsup_j \sup_{x\in C} \int g_j\, dP^k(x,\cdot) \leq p^{-1}a \longrightarrow 0 \tag{7.11}$$

as $p \to \infty$. By Proposition H.1, the assertion follows.

Remark 7.7 Suppose that S is a Polish space and for some $m \in \mathbb{N}$ the map $x \mapsto P^m(x,\cdot)$ is V-continuous. Then every compact set $K \subset S$ is (P-V)-tight.

For a set $C \in \mathcal{S}$, let $\tau_C = \inf\{n \geq 1\colon X_n \in C\}$.

Theorem 7.8 *Assume that V satisfies V.1′–V.4 and let $M \subset \mathcal{P}(S)$. Suppose that for every $b > 0$ there exists a (P-V)-tight set $C \in \mathcal{S}$ such that*

$$\sup_{x\in C} \mathbb{E}_x e^{b\tau} < \infty, \quad \textit{where } \tau = \tau_C; \tag{7.12}$$

$$\sup_{\nu\in M} \mathbb{E}_\nu e^{b\tau} < \infty. \tag{7.13}$$

Then $\phi_M \mid V$ is V-regular, and therefore (6.17) *of Theorem 6.8 holds.*

Proof. Let $0 \le g_k \in V$, and assume that $g_k \downarrow 0$ pointwise. We must prove that $\phi_M(g_k) \to 0$. What is needed is an inequality resembling (7.1). That will be (7.27).

Let $b > 2\|g_1\| + 1$, let $C \in \mathcal{S}$ be a $(P\text{-}V)$-tight set such that (7.12)–(7.13) hold, and let $m \in \mathbb{N}$ be as in Definition 7.3. Let

$$\tau(1) = \tau = \inf\{n \ge 1 : X_n \in C\},$$

and for $j \ge 1$, inductively,

$$\tau(j+1) = \inf\{n \ge 1 + \tau(j) : X_n \in C\}.$$

Then we have: on $[\tau(j) < \infty]$,

$$\tau(j+1) = \tau(j) + \tau \circ \Theta^{\tau(j)}, \tag{7.14}$$

where $\Theta : \Omega \to \Omega$ is the canonical shift, so

$$\Theta^{\tau(j)}(y) = \left(y_{k+\tau(j)(y)}\right)_{k\ge 0}, \qquad \text{for } y = (y_k)_{k\ge 0} \in \Omega,$$

and by (7.12), (7.14), and the strong Markov property,

$$\mathbb{P}_x\left[\tau(j) < \infty \text{ for all } j \ge 1\right] = 1, \qquad \text{for all } x \in C, \tag{7.15}$$

and

$$\sup_{x\in C} \mathbb{E}_x e^{b\tau(j)} \le \left(\sup_{x\in C} \mathbb{E}_x e^{b\tau}\right)^j < \infty, \qquad j \in \mathbb{N}. \tag{7.16}$$

Let $\sigma = \inf\{n \ge m : X_n \in C\}$, $\sigma(0) = 0$ (for convenience), $\sigma(1) = \sigma$, and for $j \ge 1$, inductively,

$$\sigma(j+1) = \inf\{n \ge m + \sigma(j) : X_n \in C\}.$$

Since $\tau(m) \ge m$, we have $\sigma \le \tau(m)$ and

$$\sup_{x\in C} \mathbb{E}_x e^{b\sigma} \le \sup_{x\in C} \mathbb{E}_x e^{b\tau(m)} < \infty.$$

Similarly to (7.14)–(7.16), we have:

$$\sigma(j+1) = \sigma(j) + \sigma \circ \Theta^{\sigma(j)}, \qquad \text{on } [\sigma(j) < \infty], \tag{7.17}$$

$$\mathbb{P}_x\left[\sigma(j) < \infty \text{ for all } j \geq 1\right] = 1, \qquad \text{for all } x \in C,$$

$$\sup_{x\in C} \mathbb{E}_x e^{b\sigma(j)} \leq \big(\sup_{x\in C} \mathbb{E}_x e^{b\sigma}\big)^j < \infty, \qquad j \in \mathbb{N}. \tag{7.18}$$

Moreover, by (7.13) and (7.17), again by the strong Markov property,

$$\begin{aligned} \mathbb{P}_\nu\left[\sigma(j) < \infty \text{ for all } j \geq 1\right] = 1, &\qquad \text{for all } \nu \in M, \\ \sup_{\nu\in M} \mathbb{E}_\nu e^{b\sigma(j)} < \infty, &\qquad \text{for all } j \in \mathbb{N}. \end{aligned} \tag{7.19}$$

For $n \in \mathbb{N}$, let $p(n) = [n/m] + 1$ or $[n/m] + 2$, whichever one is even. Then

$$n \leq p(n)m \leq \sigma\left(p(n)\right).$$

Let $f \in V$, $0 \leq f \leq \|g_1\|$, and let $Y_j = f(X_j)$. We have

$$\begin{aligned} \mathbb{E}_\nu\left(\exp \sum_{j=1}^{n} Y_j\right) &\leq \mathbb{E}_\nu\left(\exp \sum_{j=1}^{\sigma(p(n))} Y_j\right) \\ &= \mathbb{E}_\nu\left(\exp \sum_{i=0}^{p(n)-1} \sum_{j=\sigma(i)+1}^{\sigma(i+1)} Y_j\right). \end{aligned} \tag{7.20}$$

By a rearrangement of the sums in the exponent, to be exploited below, and the convexity of the exponential function, we have: setting $q(n) = (1/2)p(n) - 1$,

$$\begin{aligned} &\mathbb{E}_\nu\left(\exp \sum_{i=0}^{p(n)-1} \sum_{j=\sigma(i)+1}^{\sigma(i+1)} Y_j\right) \\ &= \mathbb{E}_\nu\left[\exp\left(\sum_{i=0}^{q(n)} \sum_{\substack{j=\\ \sigma(2i)+1}}^{\sigma(2i+1)} Y_j + \sum_{i=0}^{q(n)} \sum_{\substack{j=\\ \sigma(2i+1)+1}}^{\sigma(2i+2)} Y_j\right)\right] \\ &\leq \frac{1}{2}\mathbb{E}_\nu\left(\exp \sum_{i=0}^{q(n)} \sum_{\substack{j=\\ \sigma(2i)+1}}^{\sigma(2i+1)} 2Y_j\right) + \frac{1}{2}\mathbb{E}_\nu\left(\exp \sum_{i=0}^{q(n)} \sum_{\substack{j=\\ \sigma(2i+1)+1}}^{\sigma(2i+2)} 2Y_j\right). \end{aligned} \tag{7.21}$$

We will use below the following straightforward generalization of (7.17): for $p, q \in \mathbb{N}$, $p > q$,

$$\sigma(p) = \sigma(q) + \sigma(p-q) \circ \Theta^{\sigma(q)} \qquad \text{on } [\sigma(q) < \infty]. \tag{7.22}$$

By (7.22) we have, for $k > 1$,

$$\sum_{j=\sigma(k)+1}^{\sigma(k+1)} Y_j = \sum_{\substack{j=\sigma(k-1)\\+a(k)+1}}^{\substack{\sigma(k-1)\\+b(k)}} Y_j = \sum_{j=a(k)+1}^{b(k)} Y_{j+\sigma(k-1)}$$

$$= \sum_{j=a(k)+1}^{b(k)} Y_j \circ \Theta^{\sigma(k-1)},$$

where $a(k) = \sigma \circ \Theta^{\sigma(k-1)}$, $b(k) = \sigma(2) \circ \Theta^{\sigma(k-1)}$,

$$= \Phi \circ \Theta^{\sigma(k-1)},$$

where

$$\Phi(y) = \sum_{j=\sigma(1)(y)+1}^{\sigma(2)(y)} Y_j(y) \qquad \text{for } y = (y_j)_{j\geq 0} \in \Omega.$$

Therefore by the strong Markov property,

$$\begin{aligned} \mathbb{E}_\nu\left[\left(\exp \sum_{j=\sigma(k)+1}^{\sigma(k+1)} 2Y_j\right) \middle| \mathcal{F}_{\sigma(k-1)}\right] &= \mathbb{E}_{X_{\sigma(k-1)}}\left(\exp \sum_{j=\sigma(1)+1}^{\sigma(2)} 2Y_j\right) \quad \text{a.s. } [\mathbb{P}_\nu] \\ &\leq \sup_{x\in C} \mathbb{E}_x\left(\exp \sum_{j=\sigma(1)+1}^{\sigma(2)} 2Y_j\right) \quad \text{a.s. } [\mathbb{P}_\nu]. \end{aligned} \tag{7.23}$$

Applying (7.23) to the first term in (7.21) with $k = 2q(n)$, we have

$$\begin{aligned} &\mathbb{E}_\nu\left[\left(\exp \sum_{i=0}^{q(n)} \sum_{\substack{j=\\ \sigma(2i)+1}}^{\sigma(2i+1)} 2Y_j\right) \middle| \mathcal{F}_{\sigma(2q(n)-1)}\right] \\ &= \left(\exp \sum_{i=0}^{q(n)-1} \sum_{\substack{j=\\ \sigma(2i)+1}}^{\sigma(2i+1)} 2Y_j\right) \mathbb{E}_\nu\left[\exp \sum_{\substack{j=\\ \sigma(2q(n))+1}}^{\sigma(2q(n)+1)} 2Y_j \middle| \mathcal{F}_{\sigma(2q(n)-1)}\right] \\ &\leq \left(\exp \sum_{i=0}^{q(n)-1} \sum_{\substack{j=\\ \sigma(2i)+1}}^{\sigma(2i+1)} 2Y_j\right) \times \sup_{x\in C} \mathbb{E}_x\left(\exp \sum_{j=\sigma(1)+1}^{\sigma(2)} 2Y_j\right) \quad \text{a.s. } [\mathbb{P}_\nu]. \end{aligned}$$

Iterating by repeated applications of (7.23), we obtain

$$
\begin{aligned}
\mathbb{E}_\nu\left(\exp\sum_{i=0}^{q(n)}\sum_{\substack{j=\\ \sigma(2i)+1}}^{\sigma(2i+1)} 2Y_j\right) &\le \mathbb{E}_\nu\left(\exp\sum_{j=1}^{\sigma(1)} 2Y_j\right) \\
&\quad\times\left[\sup_{x\in C}\mathbb{E}_x\left(\exp\sum_{j=\sigma(1)+1}^{\sigma(2)} 2Y_j\right)\right]^{q(n)}.
\end{aligned}
\tag{7.24}
$$

Proceeding similarly with the second term in (7.21), we have

$$
\begin{aligned}
\mathbb{E}_\nu\left(\exp\sum_{i=0}^{q(n)}\sum_{\substack{j=\\ \sigma(2i+1)+1}}^{\sigma(2i+2)} 2Y_j\right) &\le \mathbb{E}_\nu\left(\exp\sum_{j=\sigma(1)+1}^{\sigma(2)} 2Y_j\right) \\
&\quad\times\left[\sup_{x\in C}\mathbb{E}_x\left(\exp\sum_{j=\sigma(1)+1}^{\sigma(2)} 2Y_j\right)\right]^{q(n)}.
\end{aligned}
\tag{7.25}
$$

From (7.20), (7.21), (7.24), and (7.25), we have

$$
\begin{aligned}
\mathbb{E}_\nu\left(\exp\sum_{j=1}^{n} f(X_j)\right) &\le \mathbb{E}_\nu\left(\exp\sum_{j=1}^{\sigma(2)} 2f(X_j)\right) \\
&\quad\times\left[\sup_{x\in C}\mathbb{E}_x\left(\exp\sum_{j=\sigma+1}^{\sigma(2)} 2f(X_j)\right)\right]^{q(n)}.
\end{aligned}
\tag{7.26}
$$

Now

$$
\sup_{\nu\in M}\mathbb{E}_\nu\left(\exp\sum_{j=1}^{\sigma(2)} 2f(X_j)\right) \le \sup_{\nu\in M}\mathbb{E}_\nu e^{b\sigma(2)} < \infty
$$

by (7.19). Therefore from (7.26),

$$
\begin{aligned}
\phi_M(f) &= \overline{\lim_n}\, n^{-1}\log\sup_{\nu\in M}\mathbb{E}_\nu\left(\exp\sum_{j=1}^{n} f(X_j)\right) \\
&\le \frac{1}{2m}\log\sup_{x\in C}\mathbb{E}_x\left(\exp\sum_{j=\sigma+1}^{\sigma(2)} 2f(X_j)\right).
\end{aligned}
\tag{7.27}
$$

Next, for $l \in \mathbb{N}$, $l \ge m+1$, setting $f = g_k$,

$$
\begin{aligned}
\mathbb{E}_x\left(\exp\sum_{j=\sigma+1}^{\sigma(2)} 2g_k(X_j)\right) &\le \mathbb{E}_x\left(\exp\sum_{j=m+1}^{l} 2g_k(X_j)\right) \\
&\quad+\mathbb{E}_x\left\{\left[\exp\left(2\|g_1\|\sigma(2)\right)\right]\mathbf{1}\left[\sigma(2)>l\right]\right\}.
\end{aligned}
\tag{7.28}
$$

Let

$$f_k(y) = \mathbb{E}_y\left(\exp\sum_{j=1}^{l-m} 2g_k(X_j)\right).$$

By the Markov property,

$$\mathbb{E}_x\left(\exp\sum_{j=m+1}^{l} 2g_k(X_j)\right) = \int P^m(x,dy)f_k(y). \tag{7.29}$$

Since $f_k \downarrow 1$ pointwise by dominated convergence and $f_k \in V$ by Proposition J.3, by the definition of a $(P\text{-}V)$-tight set and Proposition H.1,

$$\sup_{x\in C}\int P^m(x,dy)f_k(y) = 1 + \sup_{x\in C}\int P^m(x,dy)\left(f_k(y)-1\right) \longrightarrow 1. \tag{7.30}$$

On the other hand, since $b > 2\|g_1\| + 1$,

$$\mathbb{E}_x\left\{\left[\exp\left(2\|g_1\|\sigma(2)\right)\right]\mathbf{1}\left[\sigma(2) > l\right]\right\} \le e^{-l}\mathbb{E}_x\left[\exp\left(b\sigma(2)\right)\right]. \tag{7.31}$$

By (7.27)–(7.31),

$$\lim_k \phi_M(g_k) \le \frac{1}{2m}\log\left\{1 + e^{-l}\sup_{x\in C}\mathbb{E}_x\left[\exp\left(b\sigma(2)\right)\right]\right\}.$$

Letting $l \to \infty$, by (7.18) we finally have $\lim_k \phi_M(g_k) = 0$. □

Corollary 7.9 *Assume that V satisfies V.1′–V.4. Suppose that for every $b > 0$ there exists a $(P\text{-}V)$-tight set $C \in \mathcal{S}$ such that*

$$\sup_{x\in S}\mathbb{E}_x e^{b\tau} < \infty, \quad \textit{where } \tau = \tau_C.$$

Then S is a uniformity set for the upper bound in the V topology with V-tight rate function $(\phi \mid V)^* = I$.

Proof. Apply Theorem 7.8 with $M = S$. The equality $(\phi \mid V)^* = I$ follows from Theorem 4.1. □

Corollary 7.10 *Assume that V satisfies V.1′–V.4. Suppose that for every $b > 0$ there exists a $(P\text{-}V)$-tight set $C \in \mathcal{S}$ such that*

$$\sup_{x\in C}\mathbb{E}_x e^{b\tau} < \infty, \quad \textit{where } \tau = \tau_C;$$

$$\mathbb{E}_x e^{b\tau} < \infty, \quad \textit{for every } x \in S.$$

Then for every $x \in S$, $\{\mathbb{P}_x[L_n \in \cdot]\}$ satisfies the upper bound in the V topology with V-tight rate function I.

Proof. By Theorem 6.10, it suffices to prove that $\tilde{\phi} \mid V$ is V-regular. By the present assumptions and the proof of Theorem 7.8, (7.27) holds with $M = \{\delta_x\}$ for every $x \in S$. Therefore the left-hand side of (7.27) can be replaced by $\tilde{\phi}(f)$. By the argument following (7.27), $\tilde{\phi} \mid V$ is V-regular. □

7.2 Some Results When S Is a Polish Space

The following result is a sufficient condition for the upper bound for random probability measures when S is Polish and $V = C_b(S)$.

Theorem 7.11 *Assume that $(\Omega', \mathcal{A})$, $\{\mathbb{P}_\alpha\}_{\alpha \in A}$ and $\{M_n\}$, Γ_A, $(\Gamma_A \mid V)^*$ are as in Chapter 3, and that S is Polish and $V = C_b(S)$.*

Suppose that there exists a measurable function $h\colon S \to \mathbb{R}^+$ such that

$$\text{For all } a \geq 0,\ K_a \triangleq \{x \in S\colon h(x) \leq a\} \text{ is relatively compact.} \tag{7.32a}$$

$$\Gamma_A(h) \triangleq \overline{\lim_n}\, n^{-1} \log \sup_{\alpha \in A} \mathbb{E}_\alpha \left[\exp\left(n \int h \, dM_n\right)\right] < \infty. \tag{7.32b}$$

Then $\Gamma_A \mid C_b(S)$ is $C_b(S)$-regular and therefore (6.12) *of Theorem 6.5 holds with $V = C_b(S)$.*

Proof. Let $0 \leq f_k \in C_b(S)$, $f_k \downarrow 0$ pointwise. Let $c = \sup f_1$. Given $a > 0$, $\epsilon > 0$, by (7.32a) and Dini's theorem there exists $k_0 \in \mathbb{N}$ such that, for $k \geq k_0$,

$$\begin{aligned} f_k &= f_k \mathbf{1}_{K_a} + f_k \mathbf{1}_{K_a^c} \\ &\leq \epsilon + c\mathbf{1}_{K_a^c} \leq \epsilon + \frac{c}{a} h. \end{aligned}$$

Therefore for $k \geq k_0$, $a > c$,

$$\mathbb{E}_\alpha \left[\exp\left(n \int f_k \, dM_n\right)\right] \leq e^{n\epsilon} \left\{\mathbb{E}_\alpha \left[\exp\left(n \int h \, dM_n\right)\right]\right\}^{c/a},$$

$$\Gamma_A(f_k) \leq \epsilon + \frac{c}{a} \Gamma_A(h),$$

and therefore

$$\lim_k \Gamma_A(f_k) \leq \epsilon + \frac{c}{a} \Gamma_A(h).$$

Since $\epsilon > 0$ and $a > c$ are arbitrary, it follows that $\lim_k \Gamma_A(f_k) = 0$. □

Theorem 7.11 implies the following consequence for $\{L_n\}$.

Theorem 7.12 *Assume that S is Polish and P is Feller (Section B.4). Let $M \subset \mathcal{P}(S)$.*

1. *Suppose that there exists a measurable function* $h\colon S \to \mathbb{R}^+$ *such that:*

   For all $a \geq 0$, $K_a \triangleq \{x \in S\colon h(x) \leq a\}$ *is relatively compact.* (7.33a)

   There exists $m \in \mathbb{N}_0$ *such that*

$$\phi_M^{(m)}(h) \triangleq \overline{\lim_n}\, n^{-1} \log \sup_{\nu \in M} \mathbb{E}_\nu \left(\exp \sum_{j=m}^{n} h(X_j) \right) < \infty. \tag{7.33b}$$

   Then (6.17) *of Theorem 6.8 holds with* $V = C_b(S)$.

2. *Suppose also that*

$$\tilde{\phi}^{(m)}(h) \triangleq \sup_{x \in S} \phi_x^{(m)}(h) < \infty.$$

   Then I *is* $C_b(S)$*-tight.*

Proof.

1. Since P is Feller, $V = C_b(S)$ satisfies V.1′–V.4. Therefore by Theorem 6.8 it suffices to show that $\phi_M \mid C_b(S)$ is $C_b(S)$-regular.

   Let $M_n = n^{-1} \sum_{j=m}^{n-1} \delta_{X_j}$. Then (7.33b) reads: $\Gamma_M(h) < \infty$. By Theorem 7.11, $\Gamma_M \mid C_b(S)$ is $C_b(S)$-regular. Since for $0 \leq f \in C_b(S)$, $c = \sup f$, $n > m$,

$$\mathbb{E}_\nu \left(\exp \sum_{j=0}^{n-1} f(X_j) \right) \leq e^{mc} \mathbb{E}_\nu \left(\exp \sum_{j=m}^{n-1} f(X_j) \right),$$

   we have

$$\phi_M(f) \leq \Gamma_M(f).$$

   It follows that $\phi_M \mid C_b(S)$ is also $C_b(S)$-regular.

2. To prove that I is $C_b(S)$-tight, by Remark 6.9(2) it suffices to show that $\tilde{\phi} \mid C_b(S)$ is $C_b(S)$-regular. But as in part 1 of the proof, this follows from the assumption $\tilde{\phi}^{(m)}(h) < \infty$. □

Remark 7.13 In general, the assumptions in Theorems 7.8 and 7.12 are not comparable. In one particular case, however, a comparison is possible.

Assume that S is Polish and (7.33) of Theorem 7.12 holds with $M = S$. Then for all $b > 0$ there exists a compact set K such that

$$\sup_{x \in S} \mathbb{E}_x e^{b\tau} < \infty, \quad \text{where } \tau = \tau_K.$$

For, let

$$c = \sup_n n^{-1} \log \sup_{x\in S} \mathbb{E}_x \left(\exp \sum_{j=m}^{n} h(X_j) \right).$$

We have, for $\tau = \tau_{K_a}$,

$$\begin{aligned}
\sum_{n=m+1}^{\infty} e^{bn}\mathbb{P}_x[\tau = n] &\le \sum_{n=m+1}^{\infty} e^{bn}\mathbb{P}_x\left[h(X_i) > a,\ i = 1,\dots,n-1\right] \\
&\le \sum_{n=m+1}^{\infty} e^{bn}\mathbb{P}_x\left[\sum_{j=m}^{n-1} h(X_j) > (n-m)a\right] \\
&\le \sum_{n=m+1}^{\infty} e^{bn}e^{-(n-m)a}\mathbb{E}_x\left(\exp\sum_{j=m}^{n} h(X_j)\right) \\
&\le e^{ma}\sum_{n=m+1}^{\infty} e^{-(a-b-c)n}.
\end{aligned}$$

Choosing $a > b + c$, $K = K_a$, the assertion is proved.

Corollary 7.14 *Assume that S is Polish, P is Feller, and there exists a measurable function $u\colon S \to \mathbb{R}^+$ such that*

1. $\inf u > 0$.
2. *For some $m \in \mathbb{N}_0$, $P^m u$ is bounded on compact sets.*
3. *For all $b \ge 0$,*

$$\left\{x\colon \frac{u(x)}{Pu(x)} \le b\right\}$$

is relatively compact.

Let $M \subset \mathcal{P}(S)$ be such that there exists a compact set K satisfying $\nu(K) = 1$ for all $\nu \in M$. Then (6.17) of Theorem 6.8 holds with $V = C_b(S)$ and I is $C_b(S)$-tight.

Proof. We have, by the Markov property, for $n > m$,

$$\begin{aligned}
\mathbb{E}_\nu\left[\prod_{j=m}^{n-1}\frac{u(X_j)}{Pu(X_j)}\right]u(X_n) &= \mathbb{E}_\nu\left[\prod_{j=m}^{n-1}\frac{u(X_j)}{Pu(X_j)}\right]\mathbb{E}_{X_{n-1}}u(X_1) \\
&= \mathbb{E}_\nu\left[\prod_{j=m}^{n-2}\frac{u(X_j)}{Pu(X_j)}\right]u(X_{n-1}) = \dots \\
&= \mathbb{E}_\nu u(X_m) = \int (P^m u)\, d\nu.
\end{aligned}$$

Let $h = \log\{u/Pu\}$. Then h satisfies (7.33a) and

$$\begin{aligned}\mathbb{E}_\nu\left(\exp\sum_{j=m}^{n} h(X_j)\right) &= \mathbb{E}_\nu\left[\prod_{j=m}^{n}\frac{u(X_j)}{Pu(X_j)}\right]\\ &\le c\mathbb{E}_\nu\left[\prod_{j=m}^{n-1}\frac{u(X_j)}{Pu(X_j)}\right]u(X_n)\\ &= c\int (P^m u)\,d\nu,\end{aligned}$$

where $c = (\inf u)^{-1}$. Let $a = \sup_{x\in K} P^m u(x)$. Then

$$\sup_{\nu\in M}\mathbb{E}_\nu\left(\exp\sum_{j=m}^{n} h(X_j)\right) \le ca,$$

and $\phi_M^{(m)}(h) = 0$, where $\phi_M^{(m)}$ is as in Theorem 7.12. Therefore the first assertion follows from Theorem 7.12.1. Also, taking $\nu = \delta_x$, $x \in S$, we have $\phi_x^{(m)} = 0$, and the second assertion follows from Theorem 7.12.2. □

Under suitable additional assumptions, it is possible to prove converses to Theorem 7.8 and Corollary 7.9.

Theorem 7.15 *Assume that S is Polish. Suppose that $M \subset \mathcal{P}(S)$ and there exists a V-tight function $J\colon \mathcal{P}(S) \to \overline{\mathbb{R}^+}$ such that M is a uniformity set for the upper bound in the V topology with rate function J. If either*

1. *S is locally compact and $V = C_b(S)$, or*
2. *$V = B(S)$,*

then for every $b > 0$ there exists a compact set $K \subset S$ such that

$$\sup_{\nu\in M}\mathbb{E}_\nu e^{b\tau} < \infty, \quad \textit{where } \tau = \tau_K.$$

Proof.

1. By Garling (2018, Thm. 3.4.1), there exists an increasing sequence $\{K_j\}$ of compact subsets of S such that $K_j \uparrow S$ and $K_j \subset K_{j+1}^\circ$ for all j.

   Let $\epsilon \in (0,1)$, and for $j \in \mathbb{N}$ let

   $$M_j = \left\{\mu \in \mathcal{P}(S)\colon \mu\left((K_j^\circ)^c\right) \ge 1-\epsilon\right\}.$$

   Since $((K_j^\circ)^c) \subset K_{j-1}^c$, we have

   $$\bigcap_j M_j = \emptyset.$$

Let $b > 0$ and $L = \{\mu \in \mathcal{P}(S) \colon J(\mu) \leq 2b\}$. Since L is $C_b(S)$-compact, M_j is $C_b(S)$-closed, $\{M_j\}$ is decreasing, and

$$\bigcap_j (M_j \cap L) = \emptyset,$$

there exists $j_0 \in \mathbb{N}$ such that $M_{j_0} \cap L = \emptyset$, and therefore

$$\inf\left\{J(\mu) \colon \mu \in M_{j_0}\right\} \geq 2b.$$

Let $n_0 \in \mathbb{N}$, $n_0 > \epsilon^{-1}$, $K = K_{j_0}$, $\tau = \tau_K$. Then for $n \geq n_0$,

$$\begin{aligned}[\tau > n] &\subset \left[X_i \in (K^\circ)^c,\ i = 1, \ldots, n\right] \\ &\subset \left[L_n\left((K^\circ)^c\right) \geq 1 - \epsilon\right] = [L_n \in M_{j_0}],\end{aligned}$$

and therefore

$$\begin{aligned}\overline{\lim_n}\, n^{-1} \log \sup_{\nu \in M} \mathbb{P}_\nu[\tau > n] &\leq \overline{\lim_n}\, n^{-1} \log \sup_{\nu \in M} \mathbb{P}_\nu[L_n \in M_{j_0}] \\ &\leq -\inf\left\{J(\mu) \colon \mu \in M_{j_0}\right\} \leq -2b,\end{aligned}$$

and it follows that $\sup_{\nu \in M} \mathbb{E}_\nu e^{b\tau} < \infty$.

2. The proof is similar to the previous one. Let b, L be as above. Since L is $B(S)$-compact, it is $C_b(S)$-compact; by Prohorov's theorem (Parthasarathy, 1967, Ch. 2), there exists an increasing sequence of compact sets $\{K_j\}$ such that for all $\mu \in L$, $\mu(K_j) \uparrow 1$. Let $M_j = \{\mu \in \mathcal{P}(S) \colon \mu(K_j^c) \geq 1 - \epsilon\}$, where $\epsilon \in (0, 1)$. Clearly

$$\bigcap_j (M_j \cap L) = \emptyset,$$

and since L is $B(S)$ compact, M_j is $B(S)$-closed, and $\{M_j\}$ is decreasing, there exists $j_0 \in \mathbb{N}$ such that $M_{j_0} \cap L = \emptyset$. The proof is completed as before. □

Remark 7.16 If, in Theorem 7.15, $M = S$ or it is assumed that every compact set is a uniformity set, then the proof shows that $\sup_{x \in K} \mathbb{E}_x e^{b\tau} < \infty$.

7.3 Another Sufficient Condition for the Upper Bound in the τ Topology

The next result provides a different type of sufficient condition for the upper bound in the τ topology. For a fixed $\lambda \in \mathcal{P}(S)$, $r > 1$, $a > 1$, we introduce

$$M(r, a) = \left\{\nu \in \mathcal{P}(S) \colon \nu \ll \lambda \text{ and } \left\|\frac{d\nu}{d\lambda}\right\|_r \leq a\right\}.$$

Note that only $a > 1$ is meaningful here since, if $h = d\nu/d\lambda$, then

$$1 = \int h \, d\lambda \le \left(\int h^r \, d\lambda \right)^{1/r}.$$

In Theorem 7.17, the sets $M(r, a)$ are uniformity sets for the upper bound in the τ topology.

We recall (Chapter 1) that for $\lambda \in \mathcal{P}(S)$ and $\mu \in \mathcal{P}(S)$,

$$I_\lambda(\mu) = \begin{cases} I(\mu) & \mu \ll \lambda, \\ \infty & \text{otherwise.} \end{cases}$$

We also recall (Chapter 5, Section 5.2) that $\lambda \in \mathcal{P}(S)$ is ergodic if $\lambda(F) = 0$ or 1 for every P-closed set F.

Theorem 7.17 *Let $\lambda \in \mathcal{P}(S)$, and let $\varphi\colon \mathbb{R}^+ \to \mathbb{R}^+$ be a nondecreasing function such that*

$$\varphi(0) = 0 \quad \text{and} \quad \lim_{t \to \infty} t^{-1} \varphi(t) = \infty.$$

Assume: there exist $m \in \mathbb{N}$, $p > 1$ such that

$$P \text{ is a bounded linear operator on } L^p(\lambda). \tag{7.34}$$

$$C = \sup \left\{ \int \varphi\left((P^m f)^p\right) \, d\lambda \colon f \in B_p^+ \right\} < \infty, \tag{7.35}$$

where

$$B_p^+ = \left\{ f \in L^p(\lambda) \colon f \ge 0, \|f\|_p \le 1 \right\}.$$

Then:

1. *For all $r > 1$, $a > 1$, setting $M = M(r, a)$, and for every measurable set $B \subset \mathcal{P}(S)$,*

$$\overline{\lim_n}\, n^{-1} \log \sup_{\nu \in M} \mathbb{P}_\nu[L_n \in B] \le -\inf \{\phi_M^*(\mu) \colon \mu \in \text{cl}_\tau(B)\}, \tag{7.36}$$

   and ϕ_M^* is τ-tight.
2. *If $\lambda P \ll \lambda$, then $\phi_M^* = I_\lambda$ and P has an invariant probability measure π such that $\pi \ll \lambda$.*
3. *If $\lambda P \ll \lambda$ and λ is ergodic, then there is exactly one invariant probability measure π such that $\pi \ll \lambda$. Moreover, $\pi \equiv \lambda$, so π is ergodic and $\phi_M^* = I_\pi$.*

Remark 7.18

1. Assumption (7.35) is equivalent to the condition

$$\left\{(P^m f)^p : f \in B_p^+\right\}$$

being uniformly integrable, as is well known (see, e.g., Meyer, 1966).
2. $M(r, a)$ is τ-compact. For:

(i) $M(r, a)$ is τ-closed. In fact, let $\{\nu_\alpha\}$ be a net in $M(r, a)$, $\nu_\alpha \underset{\tau}{\to} \nu$. Then clearly $\nu \ll \lambda$. Let s be the conjugate exponent of r. Then for all $b > 0$, $f \in L^s(\lambda)$, if $f_b = f\mathbf{1}[|f| \le b]$, we have

$$\begin{aligned}\int |f_b| \left(\frac{d\nu}{d\lambda}\right) d\lambda = \int |f_b|\, d\nu &= \lim_\alpha \int |f_b|\, d\nu_\alpha \\ &= \lim_\alpha \int |f_b| \left(\frac{d\nu_\alpha}{d\lambda}\right) d\lambda \\ &\le \lim_\alpha \left(\int |f_b|^s\, d\lambda\right)^{1/s} \left(\int \left|\frac{d\nu_\alpha}{d\lambda}\right|^r d\lambda\right)^{1/r} \\ &\le a\|f_b\|_s.\end{aligned}$$

Let $b \to \infty$. Then for all $f \in L^s(\lambda)$,

$$\int |f| \left(\frac{d\nu}{d\lambda}\right) d\lambda \le a\|f\|_s,$$

and it follows that

$$\left\|\frac{d\nu}{d\lambda}\right\|_r = \sup\left\{\int f \left(\frac{d\nu}{d\lambda}\right) d\lambda : \|f\|_s \le 1\right\} \le a,$$

showing $\nu \in M(r, a)$.

(ii) $M(r, a)$ is τ-relatively compact. To show this, let $0 \le g_k \in B(S)$, $g_k \downarrow 0$ pointwise; then for $\nu \in M(r, a)$,

$$\begin{aligned}\int g_k\, d\nu &= \int g_k \left(\frac{d\nu}{d\lambda}\right) d\lambda \\ &\le \left(\int g_k^s\, d\lambda\right)^{1/s} \left(\int \left|\frac{d\nu}{d\lambda}\right|^r d\lambda\right)^{1/r} \le \|g_k\|_s a.\end{aligned}$$

But $\|g_k\|_s \downarrow 0$ by dominated convergence, and therefore

$$\sup_{\nu \in M(r,a)} \int g_k\, d\nu \longrightarrow 0.$$

The claim follows now from Proposition H.1.

Lemma 7.19 *Let $q \geq p$. Then, under assumptions* (7.34) *and* (7.35),

$$P \text{ is a bounded linear operator on } L^q(\lambda). \tag{7.37}$$

$$\sup\left\{\int \varphi((P^m g)^q)\, d\lambda \colon g \in B_q^+\right\} \leq C. \tag{7.38}$$

Proof. To prove (7.37), let $g \in L^q(\lambda)$, $f = |g|^{q/p}$. Then $f \in L^p(\lambda)$:

$$\|f\|_p^p = \int \left(|g|^{q/p}\right)^p\, d\lambda = \int |g|^q\, d\lambda = \|g\|_q^q. \tag{7.39}$$

By Hölder's inequality,

$$Pf^{p/q} = \int P(\cdot, dy) f^{p/q}(y) \leq \left(\int P(\cdot, dy) f(y)\right)^{p/q} = (Pf)^{p/q},$$

so

$$\int |P|g||^q\, d\lambda = \int \left|Pf^{p/q}\right|^q\, d\lambda \leq \int |Pf|^p\, d\lambda < \infty, \tag{7.40}$$

since $f \in L^p(\lambda)$ and assumption (7.34) of Theorem 7.17 holds. Therefore $Pg \in L^q(\lambda)$. Moreover, by (7.34) and (7.39)–(7.40),

$$\sup\left\{\int |Pg|^q\, d\lambda \colon \|g\|_q \leq 1\right\} \leq \sup\left\{\int |Ph|^p\, d\lambda \colon \|h\|_p \leq 1\right\} < \infty,$$

showing that P is a bounded linear operator on $L^q(\lambda)$.

For the proof of (7.38), let $g \geq 0$, $f = g^{q/p}$. Since $(P^m g)^q \leq (P^m f)^p$ by the argument in the proof of (7.37), and φ is nondecreasing, we have

$$\int \varphi((P^m g)^q)\, d\lambda \leq \int \varphi((P^m f)^p)\, d\lambda.$$

Therefore by (7.39),

$$\sup\left\{\int \varphi((P^m g)^q)\, d\lambda \colon g \in B_q^+\right\} \leq \sup\left\{\int \varphi((P^m h)^p)\, d\lambda \colon h \in B_p^+\right\} = C. \quad \square$$

For the next lemma, we recall that if T is a bounded linear operator on a Banach space E, then the limit

$$\lim_n \|T^n\|^{1/n} = \ell \tag{7.41}$$

exists,

$$\ell = \inf_n \|T^n\|^{1/n},$$

and $\ell = r_\sigma(T)$, the spectral radius of T (see, e.g., Dunford and Schwartz, 1958). For a bounded linear operator T on $L^q(\lambda)$, its norm will be denoted

$$||T||_{q\to q}.$$

Let $g \in B(S)$. Then K_g is a bounded linear operator on $L^q(\lambda)$ for $q \geq p$ since, by (7.37), if $f \in L^q(\lambda)$,

$$\begin{aligned}\int |K_g f|^q \, d\lambda &= \int \left| \int P(x, dy) e^{g(y)} f(y) \right|^q \, d\lambda(x) \\ &\leq e^{q||g||} \int |Pf|^q \, d\lambda \\ &\leq e^{q||g||} ||P||^q_{q\to q} ||f||^q_q.\end{aligned}$$

For $g \in B(S)$, we define

$$\phi^{(q)}(g) = \lim_n n^{-1} \log ||K^n_g||_{q\to q},$$

which exists by (7.41).

Lemma 7.20 *Let $q \geq p$. Then under assumption* (7.34) *of Theorem 7.17, for all $g \in B(S)$,*

$$\phi^{(q)}(g) \leq m^{-1} \log \sup \left\{ ||e^{mg} P^m f||_q : f \in B^+_q \right\}.$$

Proof. Let us recall that for $n \in \mathbb{N}$, $f \geq 0$,

$$K^{mn}_g f(x) = \mathbb{E}_x \left\{ f(X_{mn}) \left[\exp \left(\sum_{j=1}^{mn} g(X_j) \right) \right] \right\}.$$

For $i - 0, \ldots, m$, let

$$G_{n,i} = \sum_{j=0}^{n-1} g(X_{i+mj}).$$

Then

$$\sum_{j=1}^{mn} g(X_j) = \sum_{i=1}^{m} G_{n,i}$$

and

$$\begin{aligned}K^{mn}_g f(x) &= \mathbb{E}_x \left\{ f(X_{mn}) \left[\exp \left(m^{-1} \sum_{i=1}^{m} m G_{n,i} \right) \right] \right\} \\ &\leq m^{-1} \sum_{i=1}^{m} \mathbb{E}_x \left[f(X_{mn}) \exp(m G_{n,i}) \right].\end{aligned} \tag{7.42}$$

If $\mathcal{F}_i = \sigma(X_0, \ldots, X_i)$, we have by the Markov property,

$$\begin{aligned} \mathbb{E}_x \left[f(X_{mn}) \exp(mG_{n,i}) \right] &= \mathbb{E}_x \mathbb{E}_x \left[f(X_{mn}) \exp(mG_{n,i}) \mid \mathcal{F}_i \right] \\ &= \mathbb{E}_x \mathbb{E}_{X_i} \left[f(X_{mn-i}) \exp(mG_{n,0}) \right]. \end{aligned} \tag{7.43}$$

Next, for $y \in S$,

$$\begin{aligned} \mathbb{E}_y \left[f(X_{mn-i}) \exp(mG_{n,0}) \right] &= \mathbb{E}_y \mathbb{E}_y \left[f(X_{mn-i}) \exp(mG_{n,0}) \mid \mathcal{F}_{m(n-1)} \right] \\ &= \mathbb{E}_y \left[h_i(X_{m(n-1)}) \exp(mG_{n,0}) \right], \end{aligned}$$

where $h_i(z) = \mathbb{E}_z f(X_{m-i}) = P^{m-i} f(z)$,

$$\leq e^{m\|g\|} \mathbb{E}_y \left[h_i(X_{m(n-1)}) \exp(mG_{n-1,0}) \right]. \tag{7.44}$$

It will be convenient to rewrite the last expression in a slightly different way. For $h, f \in B(S)$, let

$$Q_h f(x) = e^{h(x)} P^m f(x) = e^{h(x)} \mathbb{E}_x f(X_m), \qquad x \in S.$$

Then it is easily proved by induction that

$$Q_h^n f = \mathbb{E}_\cdot \left[\left(\exp \sum_{j=0}^{n-1} h(X_{mj}) \right) f(X_{mn}) \right], \qquad n \in \mathbb{N}.$$

Taking $h = mg$, $f = h_i$, we have for $y \in S$,

$$\mathbb{E}_y \left[(\exp(mG_{n-1,0}))\, h_i(X_{m(n-1)}) \right] = Q_{mg}^{n-1} h_i(y). \tag{7.45}$$

By (7.43)–(7.45), we have, for $i = 1, \ldots, m$, $x \in S$,

$$\begin{aligned} \mathbb{E}_x \left[f(X_{mn}) \exp(mG_{n,i}) \right] &\leq e^{m\|g\|} \mathbb{E}_x Q_{mg}^{n-1} h_i(X_i) \\ &= e^{m\|g\|} P^i Q_{mg}^{n-1} h_i(x). \end{aligned} \tag{7.46}$$

Summarizing, from (7.42) and (7.46) we have

$$K_g^{mn} f \leq m^{-1} e^{m\|g\|} \sum_{i=1}^{m} P^i Q_{mg}^{n-1} h_i. \tag{7.47}$$

For the operators below acting on $L^q(\lambda)$, we will simplify the notation, writing

$$\| \cdot \| = \| \cdot \|_{q \to q}.$$

From (7.47),

$$\begin{aligned}
\|K_g^{mn} f\|_q &\le m^{-1} e^{m\|g\|} \sum_{i=1}^{m} \left\|P^i Q_{mg}^{n-1} h_i\right\|_q \\
&\le m^{-1} e^{m\|g\|} \sum_{i=1}^{m} \|P^i\| \left\|Q_{mg}^{n-1}\right\| \|h_i\|_q \\
&\le m^{-1} e^{m\|g\|} \sum_{i=1}^{m} \|P\|^i \left\|Q_{mg}\right\|^{n-1} \|P\|^{m-i} \|f\|_q \\
&= e^{m\|g\|} \|P\|^m \|Q_{mg}\|^{n-1} \|f\|_q.
\end{aligned}$$

Therefore

$$\begin{aligned}
\|K_g^{mn}\| &= \sup\left\{\|K_g^{mn} f\|_q : \|f\|_q \le 1\right\} \\
&\le e^{m\|g\|} \|P\|^m \|Q_{mg}\|^{n-1}.
\end{aligned} \tag{7.48}$$

By (7.48),

$$\begin{aligned}
\phi^{(q)}(g) = \lim_n (mn)^{-1} \log \|K_g^{mn}\| &\le m^{-1} \log \|Q_{mg}\| \\
&= m^{-1} \log \sup\left\{\|e^{mg} P^m f\|_q : f \in B_q^+\right\}. \qquad \square
\end{aligned}$$

In the proof of Theorem 7.17 we will use the following function. For $t \ge 0$, let $\varphi^* : \mathbb{R}^+ \to \mathbb{R}^+$ be defined by

$$\varphi^*(t) = \sup_{s \ge 0} \left[st - \varphi(s)\right].$$

Then it is easily proved that φ^* satisfies:

1. $\varphi^*(0) = 0$.
2. For all $t \ge 0$, $0 \le \varphi^*(t) < \infty$.
3. φ^* is increasing and continuous.
4. For all $s \ge 0$, $t \ge 0$,

$$st \le \varphi(s) + \varphi^*(t). \tag{7.49}$$

Proof of Theorem 7.17.

Claim 1: For $q \ge p$, if $0 \le g_j \in B(S)$, $g_j \downarrow 0$ pointwise, then

$$\lim_j \phi^{(q)}(g_j) \le (mq)^{-1} \log\left(C + \varphi^*(1)\right).$$

For, let $g \in B(S)$, $f \in B_q^+$. By (7.49),

$$(e^{mg} P^m f)^q \le \varphi\left((P^m f)^q\right) + \varphi^*(e^{qmg}),$$

$$\int (e^{mg} P^m f)^q \, d\lambda \le \int \varphi\left((P^m f)^q\right) d\lambda + \int \varphi^*(e^{qmg}) \, d\lambda.$$

Therefore, by Lemma 7.19,

$$\begin{aligned}
&\sup\left\{\|e^{mg}P^m f\|_q \colon f \in B_q^+\right\}\\
&\quad\le \sup\left\{\left[\int \varphi\left((P^m f)^q\right)\, d\lambda + \int \varphi^*(e^{qmg})\, d\lambda\right]^{1/q} : f \in B_q^+\right\}\\
&\quad\le \left(C + \int \varphi^*(e^{qmg})\, d\lambda\right)^{1/q}.
\end{aligned} \tag{7.50}$$

It follows from Lemma 7.20 and (7.50) that

$$\begin{aligned}
\lim_j \phi^{(q)}(g_j) &\le \lim_j\, (mq)^{-1} \log\left(C + \int \varphi^*(e^{qmg_j})\, d\lambda\right)\\
&= (mq)^{-1} \log\left(C + \varphi^*(1)\right),
\end{aligned}$$

since by dominated convergence,

$$\lim_j \int \varphi^*(e^{qmg_j})\, d\lambda = \int \varphi^*(1)\, d\lambda = \varphi^*(1).$$

Claim 1 is proved.

Claim 2: Let $r > 1$, $a > 1$, $M = M(r,a)$. If $0 \le g_j \in B(S)$ and $g_j \downarrow 0$ pointwise, then

$$\lim_j \phi_M(g_j) = 0.$$

For, let $q > \max\{p, (r-1)^{-1}r\}$. Then $q' = (q-1)^{-1}q < r$ and $M(q',a) \supset M(r,a)$. For $\nu \in M$, $g \in B(S)$,

$$\begin{aligned}
\mathbb{E}_\nu\left(\exp \sum_{k=1}^n g(X_k)\right) &= \int \nu(dx)\, \mathbb{E}_x\left(\exp \sum_{k=1}^n g(X_k)\right)\\
&= \int (K_g^n 1) h\, d\lambda, \quad \text{where } h = \frac{d\nu}{d\lambda},\\
&\le \|h\|_{q'} \|K_g^n 1\|_q\\
&\le \|h\|_{q'} \|K_g^n\|_{q\to q}.
\end{aligned}$$

Therefore

$$\begin{aligned}
\sup_{\nu\in M} \mathbb{E}_\nu\left(\exp \sum_{k=1}^n g(X_k)\right) &\le \sup\left\{\mathbb{E}_\nu\left(\exp \sum_{k=1}^n g(X_k)\right) : \nu \in M(q',a)\right\}\\
&\le a\|K_g^n\|_{q\to q},
\end{aligned}$$

$$
\begin{aligned}
\phi_M(g) &= \overline{\lim_n}\, n^{-1} \log \sup_{\nu \in M} \mathbb{E}_\nu \left(\exp \sum_{k=1}^{n} g(X_k) \right) \\
&\le \phi^{(q)}(g).
\end{aligned} \tag{7.51}
$$

From (7.51) and Claim 1,

$$
\lim_j \phi_M(g_j) \le \lim_j \phi^{(q)}(g_j) \le (mq)^{-1} \log\left(C + \varphi^*(1)\right).
$$

Letting $q \to \infty$, Claim 2 is proved. Now statement 1 of Theorem 7.17 follows from Claim 2 and Theorem 6.8 with $V = B(S)$.

To prove the first part of statement 2 of Theorem 7.17, since $M \subset \mathcal{P}(S, \lambda)$, if F is P-closed and $\lambda(F) = 1$, taking into account the fact that $\lambda P \ll \lambda$, we have $\nu P^n(F^c) = 0$ for all $\nu \in M$. Since $\lambda \in M$, by Proposition 4.17(1) it follows that $\phi^*_M = I_\lambda$.

Next, as in the proof of Proposition 5.4, we have $\{\mu\colon I_\lambda(\mu) = 0\} \neq \emptyset$, and if $I_\lambda(\mu) = 0$ then $\mu \ll \lambda$ and μ is invariant.

For the proof of statement 3, we show first that if π is invariant and $\pi \ll \lambda$, then π is ergodic and $\pi \equiv \lambda$. Let F be P-closed. If $\pi(F^c) > 0$ then $\lambda(F^c) > 0$. Since λ is ergodic, this implies $\lambda(F) = 0$, hence $\pi(F) = 0$. This shows that π is ergodic.

Next, if $\pi(F^c) = 0$ then $\lambda(F^c) = 0$: if $\lambda(F^c) > 0$ then $\lambda(F) = 0$, hence $\pi(F) = 0$, which is impossible. Using the fact that $\pi P = \pi$, if $A \in \mathcal{S}$ and $\pi(A) = 0$, then there exists a P-closed set F such that $A \subset F^c$ and $\pi(F^c) = 0$; this is proved as in Proposition B.2(3). It follows that $\lambda(A) \le \lambda(F^c) = 0$. This shows that $\lambda \ll \pi$, and consequently $\lambda \equiv \pi$.

Thus we have: if π, π' are invariant probability measures such that $\pi \ll \lambda$, $\pi' \ll \lambda$, then they are ergodic and $\pi \equiv \pi'$. Suppose $\pi \neq \pi'$, and let $\rho = (\pi - \pi')^+$, $\sigma = (\pi - \pi')^-$. Then $\rho \neq 0$, $\sigma \neq 0$, they are mutually singular and, as in the proof of Lemma 5.6, there exist disjoint P-closed sets B_ρ, B_σ such that

$$
\rho(B^c_\rho) = 0, \qquad \sigma(B^c_\sigma) = 0.
$$

Since $\pi \ge \rho$, we have $\pi(B_\rho) \ge \rho(B_\rho) > 0$ and therefore $\pi(B_\rho) = 1$. Similarly, $\pi'(B_\sigma) = 1$, and hence $\pi \perp \pi'$, which is impossible. Therefore $\pi = \pi'$. □

7.4 Notes

Theorem 7.1 unifies two statements in de Acosta (1990). Corollary 7.2 upgrades to the τ topology a result of Deuschel and Stroock (1989), where it was proved that if S is Polish then, under assumption (4.43), S is a uniformity set for the upper bound in the $C_b(S)$ topology with $C_b(S)$-tight rate function I.

Theorem 7.8, Corollary 7.9, and Corollary 7.10 are extensions of results in Wu (2000a), where condition (7.12) is introduced when S is Polish, P is Feller, and C is compact.

Theorems 7.11 and 7.12 are based on Deuschel and Stroock (1989, 5.1.12) and de Acosta (1990, Thm. 4). Corollary 7.14 covers the upper bound results in Donsker and Varadhan (1976) and Dupuis and Ellis (1997).

Theorem 7.15 was proved in Wu (2000a); our formulation is somewhat different.

Theorem 7.17 was proved in Wu (2000b, Thm. 5.1(a), upper bound part) by a different method for S Polish under the conditions $\lambda P \ll \lambda$ and P is λ-essentially irreducible; it is easily seen that this latter condition implies that λ is ergodic. The version we present here is partly based on Gao and Wang (2003) with some simplifications, in particular circumventing the use of Orlicz spaces and the Riesz–Thorin interpolation theorem.

8

The Large Deviation Principle for Empirical Measures

In this chapter we present several formulations of the large deviation principle for empirical measures. First, we study the large deviation principle for $\{\mathbb{P}_\nu[L_n \in \cdot]\}$ for an arbitrary $\nu \in \mathcal{P}(S)$. Next, we present conditions under which a set $M \subset \mathcal{P}(S)$ is a uniformity set for both the upper and lower bounds. It turns out that if $M \subset S$ is a uniformity set for the upper bound, then it is also a uniformity set for the lower bound: for, by Theorem 6.12, if M is a uniformity set for the upper bound then M is petite, so by Theorem 2.10 M is a uniformity set for the lower bound. Finally, we study the case when the large deviation principle holds for $\{\mathbb{P}_x[L_n \in \cdot]\}$ for every $x \in S$.

We consider first the general case when V satisfies V.1′–V.4. Then we study the case $V = B(S)$, where some simplifications are possible and stronger results hold.

8.1 Large Deviations in the V Topology

We start with the case $M = \{\nu\}$, $\nu \in \mathcal{P}(S)$. As we have seen in Theorem 2.10, under the assumption of irreducibility the lower bound in the V topology with rate function $(\Lambda \mid V)^*$, hence with rate function $\Lambda^* = I_\psi$ (Theorem 4.2), holds for $\{\mathbb{P}_\nu[L_n \in \cdot]\}$ for any $\nu \in \mathcal{P}(S)$. On the other hand, Theorem 6.8 gives a necessary and sufficient condition for the upper bound with rate function $(\phi_\nu \mid V)^*$. The present task is to combine Theorems 2.10 and 6.8 in order to obtain a large deviation principle for $\{\mathbb{P}_\nu[L_n \in \cdot]\}$. Of course, the crucial issue is the equality

$$(\phi_\nu \mid V)^* = I_\psi;$$

we note that in the context of Theorem 8.1 I_ψ is V-tight, so by Proposition 4.6 $(\Lambda \mid V)^* = \Lambda^* = I_\psi$.

For irreducible P, Theorem 8.1 gives necessary and sufficient conditions for the large deviation principle for $\{\mathbb{P}_\nu[L_n \in \cdot]\}$ in the V topology with V-tight rate function I_ψ. In particular, it is shown that this large deviation principle is equivalent to the statement: $\{\mathbb{P}_\nu[L_n \in \cdot]\}$ satisfies the upper bound in the V topology with a V-tight rate function J such that $\operatorname{dom} J \subset \mathcal{P}(S, \psi)$.

Theorem 8.1 *Assume that V satisfies V.1′–V.4. Let P be irreducible,* $\nu \in \mathcal{P}(S)$.

1. *The conditions* (8.1)–(8.5) *are equivalent:*

$$\text{If } 0 \le g_k \in V,\ \{g_k\} \text{ is decreasing, and } \int g_k\, d\psi \to 0, \text{ then } \phi_\nu(g_k) \to 0. \tag{8.1}$$

$$\phi_\nu \mid V \text{ is } V\text{-regular and } \operatorname{dom}(\phi_\nu \mid V)^* \subset \mathcal{P}(S, \psi). \tag{8.2}$$

$$\phi_\nu \mid V = \Lambda \mid V \quad \text{and} \quad I_\psi \text{ is } V\text{-tight}. \tag{8.3}$$

$$\{\mathbb{P}_\nu[L_n \in \cdot]\} \text{ satisfies the large deviation principle in the } V \text{ topology with } V\text{-tight rate function } I_\psi. \tag{8.4}$$

$$\{\mathbb{P}_\nu[L_n \in \cdot]\} \text{ satisfies the upper bound in the } V \text{ topology with a } V\text{-tight rate function } J \text{ such that } \operatorname{dom} J \subset \mathcal{P}(S, \psi). \tag{8.5}$$

2. *If any (hence all) of conditions* (8.1)–(8.5) *is satisfied, then P has a unique invariant probability measure* π *and* $\pi \equiv \psi$, *hence* $I_\psi = I_\pi$.

In the following lemma we return briefly to the general framework of Section 6.1.

Lemma 8.2 *Assume that V satisfies V.1–V.3 and* $\Gamma\colon V \to \mathbb{R}$ *satisfies G.1–G.4 of Theorem 6.1. Let* $\lambda \in \mathcal{P}(S)$. *Then the following conditions are equivalent:*

$$\text{If } 0 \le g_k \in V,\ \{g_k\} \text{ is decreasing, and } \int g_k\, d\lambda \to 0, \text{ then } \Gamma(g_k) \to 0. \tag{8.6}$$

$$\Gamma \text{ is } V\text{-regular and } \operatorname{dom} \Gamma^* \subset \mathcal{P}(S, \lambda). \tag{8.7}$$

Proof. (8.6) $\Longrightarrow$ (8.7): The fact that (8.6) implies that Γ is V-regular follows by dominated convergence. To prove that (8.6) implies the second statement in (8.7), we follow an argument given in the proof of (4.24) $\Longrightarrow$ (4.25) in Proposition 4.6. Let $\mu \in \mathcal{P}(S)$ and assume that $a = \Gamma^*(\mu) < \infty$. Let $f \ge 0$, $f \in C(V)$, where $C(V)$ is as in Appendix J, and assume that $\int f\, d\lambda = 0$. Then there exists a decreasing sequence $\{g_k\}$ such that $0 \le g_k \in V$ and $f = \lim_k g_k$, and therefore by dominated convergence $\int g_k\, d\lambda \to 0$. For all $t > 0$, we have by (8.6),

$$\begin{aligned} t \int f\, d\mu &= \lim_k \int (t g_k)\, d\mu \\ &= \lim_k \left\{ \left[\int (t g_k)\, d\mu - \Gamma(t g_k) \right] + \Gamma(t g_k) \right\} \le a. \end{aligned}$$

Since t is arbitrary, it follows that $\int f\,d\mu = 0$. By Proposition J.1(2), we have $\mu \ll \psi$.

(8.7) $\Longrightarrow$ (8.6): Again, we follow the argument for (4.24). Since Γ is V-regular, it is $\sigma(V, \mathcal{P}(S))$-lower semicontinuous, so by Proposition F.3 we have, for all $g \in V$,

$$\Gamma(g) = \sup\left\{\int g\,d\mu - \Gamma^*(\mu)\colon \mu \in \mathcal{P}(S)\right\}.$$

Assume now that $0 \le g_k \in V$, $\{g_k\}$ is decreasing, and $\int g_k\,d\lambda \to 0$. For $b \ge 0$,

$$\Gamma(g_k) \le \max\left\{\sup_{\mu \in L_b} \int g_k\,d\mu, \sup g_1 - b\right\},$$

where $L_b = \{\mu \in \mathcal{P}(S)\colon \Gamma^*(\mu) \le b\}$. By the second condition in (8.7), $L_b \subset \mathcal{P}(S, \lambda)$, and by the V-regularity of Γ, L_b is V-compact. Therefore by Proposition H.3,

$$\lim_k \sup\left\{\int g_k\,d\mu\colon \mu \in L_b\right\} = 0.$$

Since b is arbitrary, it follows that $\lim_k \Gamma(g_k) = 0$. □

Proof of Theorem 8.1. (8.1) $\Longleftrightarrow$ (8.2): This follows from Lemma 8.2.

(8.2) $\Longrightarrow$ (8.3): We show first that

$$(\phi_\nu \mid V)^* = (\Lambda \mid V)^*. \tag{8.8}$$

By Proposition 4.10, we have $(\phi_\nu \mid V)^* = I_\psi$. But

$$(\phi_\nu \mid V)^* \le (\Lambda \mid V)^* \le \Lambda^* = I_\psi$$

by Theorem 4.2, and (8.8) follows. Next, since $\phi_\nu \mid V$ is $\sigma(V, \mathcal{P}(S))$-lower semicontinuous by its V-regularity, $\Lambda \mid V$ is $\sigma(V, \mathcal{P}(S))$-lower semicontinuous by Theorem 2.13, and both functions are convex and proper, by Proposition F.3 (8.8) implies that $\phi_\nu \mid V = \Lambda \mid V$. Again by the V-regularity of $\phi_\nu \mid V$, $(\phi_\nu \mid V)^*$ is V-tight, and the second statement in (8.3) follows.

(8.3) $\Longrightarrow$ (8.2): If $\phi_\nu \mid V = \Lambda \mid V$, then by Theorem 2.13 $\phi_\nu \mid V$ is $\sigma(V, \mathcal{P}(S))$-lower semicontinuous. By Proposition 4.6, $(\Lambda \mid V)^* = I_\psi$. Therefore $(\phi_\nu \mid V)^* = I_\psi$, which implies that $(\phi_\nu \mid V)^*$ is V-tight and therefore $\phi_\nu \mid V$ is V-regular; we also have $\mathrm{dom}(\phi_\nu \mid V)^* \subset \mathcal{P}(S, \psi)$, proving (8.2).

(8.2) $\Longrightarrow$ (8.4): By Theorem 6.8, $\{\mathbb{P}_\nu[L_n \in \cdot]\}$ satisfies the upper bound in the V topology with V-tight rate function $(\phi_\nu \mid V)^*$. By Proposition 4.10, $(\phi_\nu \mid V)^* = I_\psi$. On the other hand, by Theorem 2.10 $\{\mathbb{P}_\nu[L_n \in \cdot]\}$ satisfies the lower bound in the V topology with rate function $\Lambda^* = I_\psi$ (Theorem 4.2).

(8.4) $\Longrightarrow$ (8.5): Obvious.

(8.5) $\Longrightarrow$ (8.1): Assume that $0 \leq g_k \in V$, $\{g_k\}$ is decreasing, and $\int g_k \, d\psi \to 0$. By Corollary E.4,

$$\phi_\nu(g_k) \leq \sup\left\{\int g_k \, d\mu - J(\mu) : \mu \in \mathcal{P}(S)\right\}.$$

Arguing as in the proof of (8.7) $\Longrightarrow$ (8.6) in Lemma 8.2, we obtain $\lim_k \phi_\nu(g_k) = 0$.

Finally, conclusion 2 follows from Proposition 5.4. □

In the next result we will study the case when a set $M \subset \mathcal{P}(S)$ is a uniformity set for both the upper and lower bounds in the V topology. Theorem 8.3 is an extension of Theorem 8.1, but for clarity we have stated them separately. Condition (6.21) is assumed in Theorem 8.3. This condition is satisfied in many situations; see the paragraph preceding Theorem 6.12. Note that if $M = \{\nu\}$, then (6.21) is unnecessary.

Theorem 8.3 *Assume that V satisfies V.1′–V.4. Let P be irreducible and assume that* (6.21) *holds. Let $M \subset \mathcal{P}(S)$.*

1. *The conditions* (8.9)–(8.13) *are equivalent:*

$$\text{If } 0 \leq g_k \in V,\ \{g_k\} \text{ is decreasing, } \int g_k \, d\psi \to 0, \text{ then } \phi_M(g_k) \to 0. \tag{8.9}$$

$$\phi_M \mid V \text{ is } V\text{-regular and } \operatorname{dom}(\phi_M \mid V)^* \subset \mathcal{P}(S, \psi). \tag{8.10}$$

$$\phi_M \mid V = \Lambda \mid V \quad \text{and} \quad I_\psi \text{ is } V\text{-tight}. \tag{8.11}$$

$$M \text{ is a uniformity set for both the upper and lower bounds in the } V \text{ topology with } V\text{-tight rate function } I_\psi. \tag{8.12}$$

$$M \text{ is a uniformity set for the upper bound in the } V \text{ topology with a } V\text{-tight rate function } J \text{ such that } \operatorname{dom} J \subset \mathcal{P}(S, \psi). \tag{8.13}$$

2. *As in Theorem 8.1.*

Proof. The proofs of the implications (8.9) $\Longleftrightarrow$ (8.10), (8.10) $\Longleftrightarrow$ (8.11), (8.12) $\Longrightarrow$ (8.13), (8.13) $\Longrightarrow$ (8.9) run exactly as in Theorem 8.1.

It remains to prove that (8.10) implies (8.12). The upper bound in (8.12) follows from Theorem 6.8 and Proposition 4.10. To prove the lower bound in

(8.12), by Theorems 2.10 and 4.2 it suffices to show that M satisfies: there exists a petite set C such that

$$\inf\{\nu(C) : \nu \in M\} > 0;$$

note that this condition is (2.1) with $h = 0$. But this condition does hold by Theorem 6.12, taking into account the upper bound and assumption (6.21). □

We consider now the situation when the large deviation principle in the V topology holds for $\{\mathbb{P}_x[L_n \in \cdot]\}$ for every $x \in S$. Here it turns out that $I_\psi = I$. This equality holds in a more general situation; see Remark 8.5.

Theorem 8.4 *Assume that V satisfies V.1′–V.4 and P is irreducible. Then the conditions* (8.14)–(8.19) *are equivalent:*

$$\text{If } 0 \le g_k \in V,\ \{g_k\} \text{ is decreasing, and } \int g_k\, d\psi \to 0, \text{ then } \tilde{\phi}(g_k) \to 0. \tag{8.14}$$

$$\text{If } 0 \le g_k \in V,\ \{g_k\} \text{ is decreasing, and } \int g_k\, d\psi \to 0, \quad \text{then } \phi_x(g_k) \to 0 \text{ for every } x \in S. \tag{8.15}$$

$$\text{For every } x \in S,\ \phi_x \mid V \text{ is } V\text{-regular and } \operatorname{dom}(\phi_x \mid V)^* \subset \mathcal{P}(S, \psi). \tag{8.16}$$

$$\phi_x \mid V = \Lambda \mid V \text{ for every } x \in S, \qquad I_\psi \text{ is } V\text{-tight and } I_\psi = I. \tag{8.17}$$

$$\text{For every } x \in S,\ \{\mathbb{P}_x[L_n \in \cdot]\} \text{ satisfies the large deviation principle in the } V \text{ topology with } V\text{-tight rate function } I \text{ and } I = I_\psi. \tag{8.18}$$

$$\text{There exists a } V\text{-tight function } J : \mathcal{P}(S) \to \overline{\mathbb{R}^+} \text{ such that } \operatorname{dom} J \subset \mathcal{P}(S, \psi) \text{ and for every } x \in S,\ \{\mathbb{P}_x[L_n \in \cdot]\} \text{ satisfies the upper bound in the } V \text{ topology with rate function } J. \tag{8.19}$$

Proof. (8.14) ⟹ (8.15): Obvious.

(8.15) ⟹ (8.16): Follows from Lemma 8.2.

(8.16) ⟹ (8.17): By Theorem 8.1, $\phi_x \mid V = \Lambda \mid V$ for every $x \in S$ and I_ψ is V-tight. Thus $\tilde{\phi} \mid V = \Lambda \mid V$, and by Theorem 4.1 and Proposition 4.6 $I = (\tilde{\phi} \mid V)^* = (\Lambda \mid V)^* = I_\psi$.

(8.17) ⟹ (8.16): As in Theorem 8.1 with $\nu = \delta_x$, $x \in S$.

(8.16) ⟹ (8.18): By Theorem 8.1, for every $x \in S$, $\{\mathbb{P}_x[L_n \in \cdot]\}$ satisfies the large deviation principle in the V topology with V-tight rate function I_ψ. But by (8.17) $I_\psi = I$.

(8.18) ⟹ (8.19): Obvious.

(8.19) $\Longrightarrow$ (8.14): As in the proof of Theorem 6.10, we have: for $g \in V$, $a \geq 0$, setting $L_a = \{\mu \in \mathcal{P}(S) \colon J(\mu) \leq a\}$,

$$\tilde{\phi}(g) \leq \max\left\{\sup_{\mu \in L_a} \int g\, d\mu, \|g\| - a\right\}.$$

Arguing as in the proof of (8.7) $\Longrightarrow$ (8.6) in Lemma 8.2, (8.14) follows. □

Remark 8.5 If in Theorem 8.3 any of assumptions (8.9)–(8.11) is assumed to hold for all members of a class $\mathcal{D} \subset S$ such that $\bigcup\{M \colon M \in \mathcal{D}\} = S$, then in (8.12) of Theorem 8.3 we have $I_\psi = I$. This follows from the proof of Theorem 8.4, for the equation $\phi_M \mid V = \Lambda \mid V$ for all $M \in \mathcal{D}$ implies $\tilde{\phi} \mid V = \Lambda \mid V$.

8.2 The Case $V = B(S)$

If $V = B(S)$, some simplifications and stronger results are possible. In particular, Theorem 8.3 can be reformulated as follows. Note that condition (6.21) holds automatically for $V = B(S)$.

Theorem 8.6 *Let P be irreducible, $M \subset \mathcal{P}(S)$.*

1. *The conditions* (8.20)–(8.24) *are equivalent:*

If $0 \leq g_k \in B(S)$, $\{g_k\}$ is decreasing, and $\int g_k\, d\psi \to 0$,
then $\phi_M(g_k) \to 0$. (8.20)

ϕ_M is $B(S)$-regular and if $0 \leq g \in B(S)$, $\int g\, d\psi = 0$,
then $\phi_M(g) = 0$. (8.21)

$\phi_M = \Lambda$ and I_ψ is τ-tight. (8.22)

M is a uniformity set for both the upper and lower bounds
in the τ topology with τ-tight rate function I_ψ. (8.23)

M is a uniformity set for the upper bound in the τ topology
with a τ-tight rate function J such that $\operatorname{dom} J \subset \mathcal{P}(S, \psi)$. (8.24)

2. *As in Theorem 8.1.*

Proof. To show that the result follows from Theorem 8.3 it suffices to prove that if (C) is the second condition in (8.21), then:

1. (C) implies $\operatorname{dom} \phi_M^* \subset \mathcal{P}(S, \psi)$.
2. If ϕ_M is $\sigma(B(S), \mathcal{P}(S))$-lower semicontinuous and $\operatorname{dom} \phi_M^* \subset \mathcal{P}(S, \psi)$, then (C) holds.

Part 1 follows from Proposition 4.11(1); part 2 from Proposition 4.11(2). □

Remark 8.7 The second condition in (8.21) can be replaced by any of the equivalent conditions given by Proposition 4.9.

Remark 8.8 The fact that conditions (8.20) and (8.21) of Theorem 8.6 are equivalent may be shown more directly as follows:

(8.20) $\Longrightarrow$ (8.21): The first statement in (8.21) is proved as in Lemma 8.2. The second statement follows obviously from (8.20).

(8.21) $\Longrightarrow$ (8.20): Let $0 \leq g_k \in B(S)$, $g_k \downarrow h$ pointwise, $\int g_k \, d\psi \to 0$. Then for all $t > 0$,

$$\int th \, d\psi = \lim_k \int tg_k \, d\psi = 0,$$

and therefore by (8.21) $\phi_M(th) = 0$. By the convexity of ϕ_M,

$$\begin{aligned} \phi_M(g_k) &= \phi_M \left\{ \frac{1}{2} \left[2(g_k - h) \right] + \frac{1}{2}(2h) \right\} \\ &\leq \frac{1}{2} \phi_M \left[2(g_k - h) \right] + \frac{1}{2} \phi_M(2h) \\ &= \frac{1}{2} \phi_M \left[2(g_k - h) \right] \longrightarrow 0. \end{aligned}$$

Also, (8.20) is equivalent to the condition: if $A_k \in \mathcal{S}$, $\{A_k\}$ is decreasing, and $\psi(A_k) \to 0$, then $\phi_M(t1_{A_k}) \to 0$ for all $t > 0$. This equivalence is proved as in Lemma 4.8.

Remark 8.9 It is of interest to study separately the equality $\phi_M = \Lambda$. This will be useful in Chapter 11, Section 11.6. The following conditions are equivalent:

$$\phi_M = \Lambda. \tag{8.25}$$

ϕ_M is $\sigma(B(S), \mathcal{P}(S))$-lower semicontinuous. (8.26a)

If $0 \leq g \in B(S)$ and $\int g \, d\psi = 0$, then $\phi_M(g) = 0$. (8.26b)

(8.25) $\Longrightarrow$ (8.26): Statement (8.26a) follows from Theorem 2.13. Next, for $g \in B(S)$,

$$\begin{aligned} \phi_M(g) = \Lambda(g) &= \sup \left\{ \int g \, d\mu - \Lambda^*(\mu) \colon \mu \in \mathcal{P}(S) \right\} \\ &= \sup \left\{ \int g \, d\mu - I(\mu) \colon \mu \ll \psi \right\} \end{aligned}$$

by Theorem 2.13, Proposition F.3, and Theorem 4.2. It follows that if $g \geq 0$ and $\int g \, d\psi = 0$, then $\phi_M(g) = 0$.

(8.26) $\Longrightarrow$ (8.25): By the condition (8.26b), Proposition 4.10, Proposition 4.11, and Theorem 4.2, we have $\phi_M^* = I_\psi = \Lambda^*$. The proof is completed as in the proof of (8.2) $\Longrightarrow$ (8.3) in Theorem 8.1.

Remark 8.10 If $M \subset \mathcal{P}(S,\psi)$, then in (8.21) of Theorem 8.6 the second condition automatically holds. For, by Proposition 4.9, it suffices to show that for all $\nu \ll \psi$, $A \in \mathcal{S}$ with $\psi(A) = 0$, we have $\nu P^n(A) = 0$ for all $n \in \mathbb{N}$. But this follows from the fact that $\psi P \ll \psi$ (Proposition B.2(1)).

We return in the case $V = B(S)$ to the situation studied in Theorem 8.4. Arguing as in the proof of Theorem 8.6, it is clear that the second condition in (8.16) of Theorem 8.4 can be replaced by the condition

$$\text{If } 0 \le g \in B(S) \text{ and } \int g\, d\psi = 0, \text{ then } \tilde{\phi}(g) = 0. \tag{8.27}$$

Theorem 8.4 with $V = B(S)$ and Theorem 5.8 are complemented by the following result.

Theorem 8.11 *The following conditions are equivalent:*

$$P \textit{ has an invariant ergodic probability measure } \pi. \tag{8.28a}$$

$$\textit{For every } x \in S,\ \phi_x \textit{ is } B(S)\textit{-regular.} \tag{8.28b}$$

$$\textit{If } 0 \le g \in B(S) \textit{ and } \int g\, d\pi = 0, \textit{ then } \tilde{\phi}(g) = 0. \tag{8.28c}$$

$$P \textit{ is irreducible.} \tag{8.29a}$$

$$\textit{For every } x \in S,\ \phi_x \textit{ is } B(S)\textit{-regular.} \tag{8.29b}$$

$$\textit{If } 0 \le g \in B(S) \textit{ and } \int g\, d\psi = 0, \textit{ then } \tilde{\phi}(g) = 0. \tag{8.29c}$$

$$\begin{gathered}\textit{For every } x \in S,\ \{\mathbb{P}_x[L_n \in \cdot]\} \textit{ satisfies the large deviation principle}\\ \textit{in the } \tau \textit{ topology with } \tau\textit{-tight rate function } I.\end{gathered} \tag{8.30}$$

$$\begin{gathered}\textit{There exists a } \tau\textit{-tight function } J : \mathcal{P}(S) \to \overline{\mathbb{R}^+} \textit{ such that}\\ \textit{for every } x \in S,\ \{\mathbb{P}_x[L_n \in \cdot]\} \textit{ satisfies the large deviation principle}\\ \textit{in the } \tau \textit{ topology with rate function } J.\end{gathered} \tag{8.31}$$

Proof. (8.28) $\Longrightarrow$ (8.29) By Proposition 4.16, (8.28c) implies: if $A \in \mathcal{S}$ and $\pi(A) = 0$, then for every $x \in S$, $P^n(x, A) \to 0$. Therefore condition 2(ii) of Lemma 5.7 holds, and by Lemma 5.7 it follows that P is irreducible. Moreover, π is a P-maximal irreducibility probability measure and (8.29c) holds.

(8.29) $\Longrightarrow$ (8.30): This follows from Theorem 8.4 together with the preceding remark on (8.27).

(8.30) $\Longrightarrow$ (8.31): Trivial.

(8.31) $\Longrightarrow$ (8.28): It was proved in Theorem 5.8 that (8.31) implies that P is irreducible, there exists a unique invariant probability measure π, and π is a P-maximal irreducibility probability measure. Therefore π is ergodic; in fact, $\pi(F) = 1$ for every P-closed set F (Proposition B.2). Also, $J = I_\pi = I$. Therefore (8.28b) and (8.28c) follow from the implication (8.23) $\Longrightarrow$ (8.21) in Theorem 8.6 with $M = \delta_x, x \in S$. □

From Theorems 7.8 and 8.6 we obtain the following sufficient condition for the large deviation principle.

Theorem 8.12 *Let P be irreducible, $M \subset \mathcal{P}(S)$. Assume:*

1. *For every $b > 0$, there exists a $(P\text{-}B(S))$-tight set $C \in \mathcal{S}$ such that*

$$\sup_{x \in C} \mathbb{E}_x e^{b\tau} < \infty, \quad \text{where } \tau = \tau_C. \tag{8.32a}$$

$$\sup_{\nu \in M} \mathbb{E}_\nu e^{b\tau} < \infty. \tag{8.32b}$$

2. *If $0 \le g \in B(S)$ and $\int g \, d\psi = 0$, then $\phi_M(g) = 0$.*

Then M is a uniformity set for both the upper and lower bounds in the $B(S)$ topology with $B(S)$-tight rate function I_ψ.

Corollary 8.13 *Assume:*

1. *S is Polish and for some $m \in \mathbb{N}$, P^m is strong Feller (Section B.4).*
2. *P is irreducible and $\psi(G) > 0$ for every nonempty open set G.*
3. *Let $M \subset \mathcal{P}(S)$. For every $b > 0$, there exists a compact set K such that*

$$\sup_{x \in K} \mathbb{E}_x e^{b\tau} < \infty, \quad \text{where } \tau = \tau_K. \tag{8.33a}$$

$$\sup_{\nu \in M} \mathbb{E}_\nu e^{b\tau} < \infty. \tag{8.33b}$$

Then M is a uniformity set for both the upper and lower bounds in the $B(S)$ topology with $B(S)$-tight rate function I_ψ and $I_\psi = I$.

Proof. From assumptions 1–2 and Remark 4.14(3), it follows that $I_\psi = I$. Also, since K is $(P\text{-}B(S))$-tight, assumption 1 of Theorem 8.12 holds.

We will prove now that assumption 2 of Theorem 8.12 is satisfied. By Proposition 4.9, it suffices to show: if $A \in \mathcal{S}$ and $\psi(A) = 0$, then $P^m(x, A) = 0$ for all $x \in S$. But this follows from the argument in Remark 4.14(3). Therefore Corollary 8.13 follows from Theorem 8.12. □

Remark 8.14 In the presence of assumption 1 of Corollary 8.13, assumption 2 can be stated in a different way. If

1. S is Polish and for some $m \in \mathbb{N}$, $P^m 1_K$ is lower semicontinuous for every compact set K,
2. $\sum_{n=1}^{\infty} P^n(x, G) > 0$ for every $x \in S$ and every nonempty open set G,

then

3. P is irreducible and $\psi(G) > 0$ for every nonempty open set G.

For, let $x_0 \in S$, $\mu = P^m(x_0, \cdot)$. We claim that μ is an irreducibility measure. Let $A \in \mathcal{S}$, $\mu(A) > 0$, and let K be a compact subset of A such that $\mu(K) > 0$. Let $G = \{x \in S\colon P^m(x, K) > 0\}$; then $G \neq \emptyset$ since $x_0 \in G$, and G is open. Let $y \in S$. Then

$$\sum_{n=1}^{\infty} P^{n+m}(y, A) \geq \sum_{n=1}^{\infty} P^{n+m}(y, K)$$
$$\geq \int_G \left[\sum_{n=1}^{\infty} P^n(y, \cdot)\right](dz) P^m 1_K(z) > 0.$$

This shows that P is irreducible. To prove the second claim in statement 3, we can assume that

$$\psi = \sum_{j=1}^{\infty} 2^{-j} \mu P^j$$

(see Section B.1). Therefore if U is open,

$$\psi(U) = \int \mu(dx) \left(\sum_{j=1}^{\infty} 2^{-j} P^j(x, U)\right) > 0.$$

Conversely, obviously statement 3 implies 2.

Corollary 8.15 *Let P be irreducible. Assume:*

1. *For every $b > 0$, there exists a (P-$B(S)$)-tight set $C \in \mathcal{S}$ such that*

$$\sup_{x \in C} \mathbb{E}_x e^{b\tau} < \infty, \quad \text{where } \tau = \tau_C. \tag{8.34a}$$

$$\text{For every } y \in S, \quad \mathbb{E}_y e^{b\tau} < \infty. \tag{8.34b}$$

2. *If $0 \leq g \in B(S)$ and $\int g\, d\psi = 0$, then $\tilde{\phi}(g) = 0$.*

Then for every $x \in S$, $\{\mathbb{P}_x[L_n \in \cdot]\}$ satisfies the large deviation principle in the $B(S)$ topology with $B(S)$-tight rate function I_ψ and $I_\psi = I$.

Proof. Except for the fact that $I_\psi = I$, the assertion follows from Theorem 8.12 with $M = \{\delta_x\}$, $x \in S$. The equality $I_\psi = I$ follows from assumption 2 and Theorem 4.12, taking into account Remark 4.13. □

In Theorem 8.19 we will show that under the assumption of irreducibility and only (8.32a) of Theorem 8.12, the large deviation principle for $\{\mathbb{P}_\nu[L_n \in \cdot]\}$ in the $B(S)$ topology with $B(S)$-tight rate function I_ψ holds for a broad class of initial distributions, which includes all petite probability measures (Definition B.4) and a set of initial states of full ψ measure. A class of uniformity sets is also presented.

In the following lemma we will consider a set $M \subset \mathcal{P}(S)$ satisfying the following condition, under irreducibility: there exist $m \in \mathbb{N}$, $h \in \mathcal{S}^+ \cap B(S)$ such that

$$\sum_{j=1}^{m} P^j \geq h \otimes \lambda \qquad \text{for all } \lambda \in M. \tag{8.35}$$

If $M = \{\lambda\}$, then (8.35) reduces to the definition of petite probability measure (Definition B.4). Condition (8.35) implies that $M \subset \mathcal{P}(S, \psi)$. For, $\psi P^j \ll \psi$ for $j \in \mathbb{N}$ (Proposition B.2(2)) and $\sum_{j=1}^{m} \psi P^j \geq (\int h \, d\psi)\lambda$ for all $\lambda \in M$.

Lemma 8.16 *Assume that P is irreducible. Suppose that C is petite, $b > 0$, and*

$$\sup_{x \in C} \mathbb{E}_x e^{b\tau} < \infty, \qquad \textit{where } \tau = \tau_C.$$

Then if $M \subset \mathcal{P}(S)$ satisfies (8.35),

$$\sup_{\lambda \in M} \mathbb{E}_\lambda e^{b\tau} < \infty.$$

Proof. Let $\tau(1) = \tau$, $\tau(2), \ldots,$ be the successive return times of C. Then for all $j \geq 1$, on $[\tau(j) < \infty]$,

$$\tau(j+1) = \tau(j) + \tau \circ \Theta^{\tau(j)},$$

where Θ is the shift on Ω and, as in the proof of Theorem 7.8,

$$\mathbb{P}_x\left[\tau(j) < \infty \text{ for all } j \geq 1\right] = 1 \qquad \text{for all } x \in C,$$

and for $j \in \mathbb{N}$,

$$\sup_{x \in C} \mathbb{E}_x\left(\exp b\tau(j)\right) \leq \left[\sup_{x \in C} \mathbb{E}_x\left(\exp b\tau\right)\right]^j < \infty.$$

Next, let $n \in \mathbb{N}$, $\alpha > 0$, $\mu \in \mathcal{P}(S)$ be such that

$$\sum_{i=1}^{n} P^i \geq \alpha(1_C \otimes \mu).$$

By irreducibility, there exists $k \in \mathbb{N}$ such that $\mu P^k h > 0$. Then if $q = k + n + m$, we have for some $c > 0$,

$$\begin{aligned}\sum_{j=1}^{q} P^j &\geq c\left(\sum_{i=1}^{n} P^i\right) P^k \left(\sum_{j=1}^{m} P^j\right) \\ &\geq c(\alpha \mathbf{1}_C \otimes \mu) P^k (h \otimes \lambda) \\ &= \beta(\mathbf{1}_C \otimes \lambda)\end{aligned} \tag{8.36}$$

for all $\lambda \in M$, where $\beta = c\alpha(\mu P^k h)$. Proceeding as in the proof of Lemma 2.2, (8.36) implies: for any measurable function $\Phi \colon \Omega \to \mathbb{R}^+$, for all $x \in S$, $\lambda \in M$,

$$\sum_{j=1}^{q} \mathbb{E}_x[\Phi \circ \Theta^j] \geq \beta \mathbf{1}_C(x) \mathbb{E}_\lambda \Phi.$$

Now let $\Phi = \exp(b\tau)$. Since $\tau(j) \geq j$ for $j \in \mathbb{N}$, we have $\tau \circ \Theta^j + j \leq \tau \circ \Theta^{\tau(j)} + \tau(j) = \tau(j+1)$, and for $x \in C$,

$$\begin{aligned}\sup_{\lambda \in M} \mathbb{E}_\lambda(\exp b\tau) &\leq \beta^{-1} \sum_{j=1}^{q} \mathbb{E}_x \left[\exp b(\tau \circ \Theta^j)\right] \\ &\leq \beta^{-1} \sum_{j=1}^{q} e^{-bj} \mathbb{E}_x \left(\exp b\tau(j+1)\right) \\ &\leq \beta^{-1} \sum_{j=1}^{q} e^{-bj} \left[\sup_{y \in C} \mathbb{E}_y(\exp b\tau)\right]^{j+1} < \infty.\end{aligned}$$

□

In the next proposition we will state a sufficient condition for the equalities $\phi_M(g) = \Lambda(g)$ and $\phi_x(g) = \Lambda(g)$ for $g \in B(S)$. We recall (Chapter 1) that

$$T_n(g) = \sum_{j=1}^{n} g(X_j).$$

For $C \in \mathcal{S}$, $\mu \in \mathcal{P}(S)$, $g \in B(S)$, let

$$\begin{aligned}\alpha(\mu, g, C) &= \overline{\lim_n}\, n^{-1} \log \mathbb{E}_\mu \left[(\exp T_n(g))\, \mathbf{1}_C(X_n)\right] \\ &= \overline{\lim_n}\, n^{-1} \log \mu K_g^n \mathbf{1}_C, \\ \beta(\mu, g, C) &= \overline{\lim_n}\, n^{-1} \log \mathbb{E}_\mu \left[(\exp T_n(g))\, \mathbf{1}[\tau \geq n]\right],\end{aligned}$$

where $\tau = \tau_C$,

$$\gamma(g, C) = \overline{\lim_n}\, n^{-1} \log \sup_{x \in C} \mathbb{E}_x \left[(\exp T_n(g))\, \mathbf{1}[\tau \geq n]\right].$$

Lemma 8.17 *For all $C \in \mathcal{S}$, $\mu \in \mathcal{P}(S)$, $g \in B(S)$,*

$$\phi_\mu(g) \le \max\{\alpha(\mu, g, C), \beta(\mu, g, C), \gamma(g, C)\}.$$

For the proof, see de Acosta and Ney (2014), Lemma 3.2 (note that in the notation of de Acosta and Ney, 2014, what we call here T_n is denoted S_n).

For $C \in \mathcal{S}$, $\tau = \tau_C$, let

$$\begin{aligned} \lambda(C) &= \sup\left\{\lambda \ge 0\colon \sup_{x\in C} \mathbb{E}_x e^{\lambda\tau} < \infty\right\}, \\ \bar{\lambda} &= \sup\{\lambda(C)\colon C \text{ is petite}\}. \end{aligned} \tag{8.37}$$

Regarding the condition $\bar{\lambda} = \infty$, which appears below, see Remark 11.28(1).

Proposition 8.18 *Assume that P is irreducible and $\bar{\lambda} = \infty$. Then:*

1. *For every $M \subset \mathcal{P}(S)$ satisfying* (8.35), *$g \in B(S)$,*

$$\phi_M(g) = \Lambda(g).$$

2. *For every $g \in B(S)$, there exists a ψ-null set $N(g) \in \mathcal{S}$ such that for $x \notin N(g)$,*

$$\phi_x(g) = \Lambda(g).$$

Proof.

1. We first prove: if ν is petite, $g \in B(S)$, then $\phi_\nu(g) \le \Lambda(g)$. For, let $b > \|g\| - \Lambda(g)$, and let C be a petite set such that

$$\sup_{x\in C} \mathbb{E}_x e^{b\tau} < \infty, \quad \text{where } \tau = \tau_C.$$

Then

$$\begin{aligned} \mathbb{E}_x\left[(\exp T_n(g))\, \mathbf{1}[\tau \ge n]\right] &\le e^{n\|g\|} e^{-bn} \mathbb{E}_x e^{b\tau}, \\ \gamma(g, C) &\le \|g\| - b < \Lambda(g), \end{aligned}$$

and since ν is petite, by Lemma 8.16 we have similarly,

$$\beta(\nu, g, C) < \Lambda(g).$$

Since $(\mathbf{1}_C, \nu)$ is a petite pair, $\alpha(\nu, g, C) = \Lambda(g)$, so by Lemma 8.17 $\phi_\nu(g) \le \Lambda(g)$.

2. For any $\mu \in \mathcal{P}(S)$, $g \in B(S)$, we have $\phi_M(g) \le \phi_\mu(g)$. For, by irreducibility there exists $k \in \mathbb{N}$ such that $c = \mu P^k h > 0$. Setting $q = k + m$, we have for $\lambda \in M$,

$$\begin{aligned} \lambda K_g^n 1 &\le c^{-1} \sum_{j=1}^{q} \mu P^j K_g^n 1 \\ &\le c^{-1} q \max_{1\le j\le q} \mu P^j K_g^n 1, \end{aligned}$$

and therefore

$$\begin{aligned}\phi_M(g) &= \overline{\lim_n}\, n^{-1} \log \sup_{\lambda\in M} \lambda K_g^n 1 \\ &\le \max_{1\le j\le q} \overline{\lim_n}\, n^{-1} \log \mu P^j K_g^n 1 \\ &= \max_{1\le j\le q} \phi_{\mu P^j}(g) \le \phi_\mu(g),\end{aligned}$$

as in part 2 of the proof of Theorem 4.2.

3. Since $\phi_M \ge \Lambda$ by part 2 of the proof of Theorem 4.2, by 1 and 2 above conclusion 1 is proved.
4. Again by the proof of Theorem 4.2, we have: for all $x \in S$, $\phi_x(g) \ge \Lambda(g)$. Next, we prove

$$\phi_x(g) \le \Lambda(g) \quad \text{a.s. } [\psi]. \tag{8.38}$$

Let $a > \Lambda(g)$,

$$F_n(a) = \left\{x \in S\colon n^{-1} \log K_g^n 1(x) > a\right\}.$$

Then for any small probability measure λ,

$$\lambda(F_n(a)) \le e^{-na} \lambda K_g^n 1,$$

and since by conclusion 1 with $M = \{\lambda\}$,

$$\overline{\lim_n}\, n^{-1} \log \lambda K_g^n 1 = \phi_\lambda(g) = \Lambda(g),$$

we have

$$\sum_{n=1}^{\infty} \lambda(F_n(a)) < \infty.$$

By the Borel–Cantelli lemma, for all $a > \Lambda(g)$,

$$\lambda\left(\overline{\lim_n}\, F_n(a)\right) = 0.$$

This implies that $\lambda(N) = 0$, where

$$N = \{x \in S\colon \phi_x(g) > \Lambda(g)\}.$$

For a fixed λ, the probability measure λP^k, $k \in \mathbb{N}$, is also small. Therefore $\rho(N) = 0$, where $\rho = \sum_{k=1}^{\infty} 2^{-k} \lambda P^k$. Since $\rho \equiv \psi$ (Section B.1), (8.38) is proved, hence also conclusion 2. □

Let $\tilde{\lambda} = \sup\{\lambda(C)\colon C \text{ is } (P\text{-}B(S))\text{-tight}\}$. Assumption (8.32a) of Theorem 8.12 may be rephrased in the form: $\tilde{\lambda} = \infty$. The quantity $\tilde{\lambda}$ is related to $\bar{\lambda}$ as follows: by Remark 7.4, $\tilde{\lambda} \le \bar{\lambda}$.

Theorem 8.19 *Assume that P is irreducible and $\tilde{\lambda} = \infty$. Then:*

1. *Every set $M \subset \mathcal{P}(S)$ satisfying (8.35) is a uniformity set for both the upper and lower bounds in the $B(S)$ topology with $B(S)$-tight rate function I_ψ.*
2. *There exists a ψ-null set $N \in \mathcal{S}$ such that if $x \notin N$ then $\{\mathbb{P}_x[L_n \in \cdot]\}$ satisfies the large deviation principle in the $B(S)$ topology with $B(S)$-tight rate function I_ψ.*

Proof. For $b > 0$, let C_b be a $(P$-$B(S))$-tight set such that

$$\sup_{x \in C_b} \mathbb{E}_x e^{b\tau_b} < \infty,$$

where $\tau_b = \tau_{C_b}$. By Remark 7.4, C_b is petite. Therefore by Lemma 8.16,

$$\sup_{\nu \in M} \mathbb{E}_\nu e^{b\tau_b} < \infty.$$

Also $M \subset \mathcal{P}(S, \psi)$, and therefore statement 1 follows from Remark 8.10 and Theorem 8.12.

If ν is a small probability measure, then so is $\nu_k \triangleq \nu P^k$ for $k \in \mathbb{N}$, and therefore by Lemma 8.16,

$$\mathbb{E}_{\nu_k} e^{b\tau_b} < \infty \qquad \text{for all } b \in \mathbb{N}.$$

Let $M_b = \{y \in S : \mathbb{E}_y e^{b\tau_b} = \infty\}$. Then for all $k \in \mathbb{N}$, $b \in \mathbb{N}$, $\nu_k(M_b) = 0$, and therefore

$$\rho(M_b) = 0 \qquad \text{for all } b \in \mathbb{N},$$

where

$$\rho = \sum_{k=1}^{\infty} 2^{-k} \nu_k.$$

But $\rho \equiv \psi$ (Section B.1). Therefore if

$$M = \bigcup_{b \in \mathbb{N}} M_b,$$

we have $\psi(M) = 0$, and if $y \notin M$ then

$$\mathbb{E}_y e^{b\tau_b} < \infty \qquad \text{for all } b \in \mathbb{N}.$$

By Theorem 7.8, if $y \notin M$ then:

1. If $0 \le g_k \in B(S)$ and $g_k \downarrow 0$ pointwise, then $\phi_y(g_k) \to 0$.
2. $\{\mathbb{P}_y[L_n \in \cdot]\}$ satisfies the upper bound in the $B(S)$ topology with $B(S)$-tight rate function ϕ_y^*.

Also, since $\phi_y \geq \Lambda$ for all $y \in S$ (part 2 of the proof of Theorem 4.2), we have by statement 1,

$$\lim_k \Lambda(g_k) \leq 0. \tag{8.39}$$

Recalling that $\mathcal{S}$ is countably generated, let C be a countable family such that $\sigma(C) = \mathcal{S}$, and let $\widehat{C}$ be the algebra generated by C; then $\widehat{C}$ is countable. Let

$$W_0 = \left\{ \sum_{i=1}^{k} r_i \mathbf{1}_{A_i} : k \in \mathbb{N}, r_i \in \mathbb{Q}, A_i \in \widehat{C}, i = 1, \ldots, k \right\};$$

then W_0 is countable. Let

$$N = \bigcup \{N(g) \colon g \in W_0\},$$

where $N(g)$ is as in Proposition 8.18(2). Then $\psi(N) = 0$ and if $x \notin N$, then $\phi_x(g) = \Lambda(g)$ for $g \in W_0$. For fixed $y \notin M \cup N$, let

$$\mathcal{H}_y = \left\{ g \in B(S) \colon \phi_y(g) = \Lambda(g) \right\}.$$

Taking into account item 1 above, (8.39), Corollary 2.19, and Corollary 6.3, we have: if $g_n \in \mathcal{H}_y$, $g \in B(S)$, and $g_n \uparrow g$ (respectively, $g_n \downarrow g$) pointwise, then $g \in \mathcal{H}_y$. Since $W_0 \subset \mathcal{H}_y$, clearly

$$W = \left\{ \sum_{i=1}^{k} a_i \mathbf{1}_{A_i} : k \in \mathbb{N}, a_i \in \mathbb{R}, A_i \in \widehat{C}, i = 1, \ldots, k \right\}$$

is also contained in $\mathcal{H}_y$. Since W satisfies V.1–V.3, by Proposition I.1 it follows that $\mathcal{H}_y = B(S)$. Since $\phi_y(g) = \Lambda(g)$ for all $g \in B(S)$, it follows that

$$\phi_y^* = \Lambda^* = I_\psi$$

by Theorem 4.2. By item 2 above and Theorem 2.10, statement 2 is proved. □

From Theorems 7.17 and 8.6 we obtain another sufficient condition for the large deviation principle in the τ topology.

Theorem 8.20 *Let $\lambda \in \mathcal{P}(S)$. Suppose:*

1. *Assumptions* (7.34) *and* (7.35) *of Theorem 7.17 hold.*
2. *$\lambda P \ll \lambda$, λ is ergodic and if $B \in \mathcal{S}$, $\lambda(B) = 1$, then*

$$\sum_{n=1}^{\infty} P^n(x, B) > 0 \qquad \text{for all } x \in S.$$

Then for every $r > 1$, $a > 1$, $M(r, a)$ is a uniformity set for both the upper and lower bounds in the τ topology with τ-tight rate function I_λ.

Remark 8.21 If P is irreducible and $\lambda = \psi$, then Condition 2 is trivially satisfied.

Proof of Theorem 8.20. By Theorems 7.17(1)–(2), $M(r, a)$ is a uniformity set for the upper bound in the τ topology with τ-tight rate function I_λ. By Theorem 7.17(3), there exists an ergodic invariant probability measure π and $\pi \equiv \lambda$. Clearly π satisfies Condition 2(ii) of Lemma 5.7. Therefore by Lemma 5.7, P is irreducible and π, hence λ, is a P-maximal irreducibility probability measure. The conclusion follows now from the implication (8.24) $\Longrightarrow$ (8.23) in Theorem 8.6.

Alternatively, since $M(r, a) \subset \mathcal{P}(S, \psi)$ and the fact that ϕ_M is $B(S)$-regular was proved in Theorem 7.17, the result follows from Remark 8.10 and the implication (8.21) $\Longrightarrow$ (8.23) in Theorem 8.6. □

8.3 Notes

A result closely related to Theorem 8.11, assuming that S is Polish and $J = I$, appears in Wu (2000a, Prop. 2.5). The lower bound is not proved there; instead, de Acosta (1988) and Jain (1990) are referred to for the proof.

Corollary 8.13 is a variant of Theorem 1.2 in Wu (2000a). Theorem 8.20 is a variant for general state space of Theorem 5.1(a) in Wu (2000b), which is proved for Polish state space. In both cases, the proofs presented here follow different lines.

Lemma 8.16 and Proposition 8.18 are based on de Acosta and Ney (2014).

9

The Case When S Is Countable and P Is Matrix Irreducible

In this chapter we study the situation when S is countable, $\mathcal{S}$ is the power set of S, and P is matrix irreducible: for all $x, y \in S$,

$$\sum_{n=1}^{\infty} P^n(x, y) > 0.$$

Equivalently, P is irreducible and counting measure is a P-maximal irreducibility measure. Therefore, if ψ is a P-maximal irreducibility probability measure, we have $\psi(x) > 0$ for all $x \in S$ and, by Theorems 4.1 and 4.2, the various rate functions coincide:

I. $\phi^* = I = I_\psi = \Lambda^*$.
II. For all $M \subset \mathcal{P}(S)$, $\phi^*_M = I$.

$\mathcal{P}(S)$ will be endowed with the τ topology; note that in the present situation $B(S)$ is the space of bounded real-valued functions on S. If S is endowed with the discrete topology, then $C_b(S) = B(S)$ and the weak and τ topologies on $\mathcal{P}(S)$ coincide.

We will present some results which, though essentially contained in previous chapters, take a simpler or more transparent form in the present context.

9.1 A Weak Large Deviation Principle

The title refers to the situation when the lower bound holds for all measurable sets but the upper bound is required to hold only for measurable sets with compact closure, both with the same rate function (see, e.g., Definition 2.17 of Rassoul-Agha and Seppäläinen, 2015).

Theorem 9.1

1. *For every measurable set $B \subset \mathcal{P}(S)$ such that its τ-closure is compact,*
$$\overline{\lim_n}\, n^{-1} \log \sup_{x \in S} \mathbb{P}_x[L_n \in B] \le -\inf\{I(\mu) \colon \mu \in \mathrm{cl}_\tau(B)\}. \tag{9.1}$$
2. *Assume that P is matrix irreducible. Then for every set $M \subset \mathcal{P}(S)$ satisfying* (2.1) *and every measurable set $B \subset \mathcal{P}(S)$,*
$$\underline{\lim_n}\, n^{-1} \log \inf_{\nu \in M} \mathbb{P}_\nu[L_n \in B] \ge -\inf\{I(\mu) \colon \mu \in \mathrm{int}_\tau(B)\}. \tag{9.2}$$

Proof.

1. (9.1) follows from Theorem 3.2.1 and II.
2. (9.2) follows from Theorem 2.10 and I. □

9.2 Upper Bounds

We will obtain now upper bounds for general measurable sets; irreducibility is not assumed. We will consider two cases:

1. the upper bound holds pointwise over S with respect to the initial state;
2. the upper bound holds uniformly over S with respect to the initial state.

Theorem 9.2

1. *The following conditions are equivalent:*

   For every $b > 0$, there exists a finite set $F \subset S$ such that for all $x \in S$,
   $$\mathbb{E}_x e^{b\tau} < \infty, \quad \text{where } \tau = \tau_F. \tag{9.3}$$
   $$\tilde{\phi} \text{ is } B(S)\text{-regular.} \tag{9.4}$$
   For every $x \in S$, $\{\mathbb{P}_x[L_n \in \cdot]\}$ satisfies the upper bound with
   $$B(S)\text{-tight rate function } I. \tag{9.5}$$
   There exists $J \colon \mathcal{P}(S) \to \overline{\mathbb{R}^+}$, $B(S)$-tight, such that for every $x \in S$,
   $$\{\mathbb{P}_x[L_n \in \cdot]\} \text{ satisfies the upper bound with rate function } J. \tag{9.6}$$

2. *The following conditions are equivalent:*

   For every $b > 0$, there exists a finite set $F \subset S$ such that
   $$\sup_{x \in S} \mathbb{E}_x e^{b\tau} < \infty. \tag{9.7}$$
   $$\phi \text{ is } B(S)\text{-regular.} \tag{9.8}$$

S is a uniformity set for the upper bound with
$B(S)$-tight rate function I. (9.9)

There exists $J\colon \mathcal{P}(S) \to \overline{\mathbb{R}^+}$, $B(S)$-tight, such that S is uniformity set
for the upper bound with rate function J. (9.10)

Proof.

1. The implication (9.3) $\Longrightarrow$ (9.4) is a simpler version of Theorem 7.8.

   Let $0 \le g_k \in B(S)$, $g_k \downarrow 0$ pointwise, and let $b = \|g_1\|$. Let F be the set associated with b in (9.3). Let $\tau(1) = \tau_F$ and for $j \ge 1$, inductively, $\tau(j+1) = \inf\{n \ge 1 + \tau(j)\colon X_n \in F\}$. For all $x \in S$, $f \in B(S)$, $f \ge 0$,

$$\mathbb{E}_x \exp\left(\sum_{j=1}^{n} f(X_j)\right) \le \mathbb{E}_x \exp\left(\sum_{j=1}^{\tau(n)} f(X_j)\right). \tag{9.11}$$

   Using the strong Markov property, it is easily proved that

$$\mathbb{E}_x \exp\left(\sum_{j=1}^{\tau(n)} f(X_j)\right) \le \left[\sup_{y\in F} \mathbb{E}_y \exp\left(\sum_{j=1}^{\tau} f(X_j)\right)\right] \times \mathbb{E}_x \exp\left(\sum_{j=1}^{\tau(n-1)} f(X_j)\right),$$

   and iterating,

$$\mathbb{E}_x \exp\left(\sum_{j=1}^{\tau(n)} f(X_j)\right) \le \left[\sup_{y\in F} \mathbb{E}_y \exp\left(\sum_{j=1}^{\tau} f(X_j)\right)\right]^{n-1} \times \mathbb{E}_x \exp\left(\sum_{j=1}^{\tau} f(X_j)\right). \tag{9.12}$$

   Also, for all $x \in S$,

$$\mathbb{E}_x \exp\left(\sum_{j=1}^{\tau} f(X_j)\right) \le \mathbb{E}_x \exp(\|f\|\tau). \tag{9.13}$$

   Setting $f = g_k$ we have, from (9.3) and (9.11)–(9.13),

$$\begin{aligned} \phi_x(g_k) &= \overline{\lim_n}\, n^{-1} \log \mathbb{E}_x \exp\left(\sum_{j=1}^{n} g_k(X_j)\right) \\ &\le \log \sup_{y\in F} \mathbb{E}_y \exp\left(\sum_{j=1}^{\tau} g_k(X_j)\right). \end{aligned} \tag{9.14}$$

Next, for $m \in \mathbb{N}$,

$$\mathbb{E}_y \exp\left(\sum_{j=1}^{\tau} g_k(X_j)\right) \le \mathbb{E}_y \exp\left(\sum_{j=1}^{m} g_k(X_j)\right) + \mathbb{E}_y\left(e^{b\tau}\mathbf{1}[\tau > m]\right). \quad (9.15)$$

For fixed m, by dominated convergence,

$$\lim_k \sup_{y \in F} \mathbb{E}_y \exp\left(\sum_{j=1}^{m} g_k(X_j)\right) = 1. \quad (9.16)$$

From (9.14)–(9.16) we obtain for each $m \in \mathbb{N}$,

$$\begin{aligned}\overline{\lim_k}\, \tilde{\phi}(g_k) &= \overline{\lim_k} \sup_{x \in S} \phi_x(g_k) \\ &\le \log\left[1 + \sup_{y \in F} \mathbb{E}_y\left(e^{b\tau}\mathbf{1}[\tau > m]\right)\right].\end{aligned}$$

Letting $m \to \infty$, by dominated convergence we have

$$\overline{\lim_k}\, \tilde{\phi}(g_k) = 0,$$

proving (9.4).

(9.4) $\Longrightarrow$ (9.5): Follows from Theorem 6.10.

(9.5) $\Longrightarrow$ (9.6): Obvious.

(9.6) $\Longrightarrow$ (9.3): Let $\{F_k\}$ be an increasing sequence of finite sets such that $\bigcup_k F_k = S$, and let $0 < a < 1$,

$$M_k = \left\{\mu \in \mathcal{P}(S) \colon \mu(F_k^c) \ge a\right\}.$$

For $b > 0$, let

$$L_b = \{\mu \in \mathcal{P}(S) \colon J(\mu) \le b + 1\}.$$

Since L_b is $B(S)$-compact, M_k is τ-closed and $M_k \downarrow \emptyset$, there exists $h \in \mathbb{N}$ such that

$$M_h \cap L_b = \emptyset,$$

which implies

$$\inf\{J(\mu): \mu \in M_h\} \geq b + 1.$$

For $x \in S$ and all sufficiently large n,

$$\begin{aligned}\mathbb{P}_x[\tau_{F_h} > n] &= \mathbb{P}_x[X_j \in F_h^c,\ j = 1, \dots, n] \\ &\leq \mathbb{P}_x[L_n \in M_h],\end{aligned}$$

and therefore

$$\begin{aligned}\overline{\lim_n}\, n^{-1} \log \mathbb{P}_x[\tau_{F_h} > n] &= \overline{\lim_n}\, n^{-1} \log \mathbb{P}_x[L_n \in M_h] \\ &\leq -\inf\{J(\mu): \mu \in M_h\} \\ &\leq -(b + 1).\end{aligned}$$

But this implies $\mathbb{E}_x(\exp b\tau_{F_h}) < \infty$, proving (9.3).

2. To prove (9.7) $\Longrightarrow$ (9.8), let $\{g_k\}$ be as in part 1, and let F be the set associated with b as in (9.7). From (9.11)–(9.13) we have, setting $f = g_k$,

$$\sup_{x \in S} \mathbb{E}_x \exp\left(\sum_{j=1}^{n} g_k(X_j)\right) \leq \left[\sup_{y \in F} \mathbb{E}_y \exp\left(\sum_{j=1}^{\tau} g_k(X_j)\right)\right]^{n-1} \sup_{x \in S} \mathbb{E}_x e^{b\tau}, \quad (9.17)$$

and from (9.7) and (9.17),

$$\begin{aligned}\phi(g_k) &= \overline{\lim_n}\, n^{-1} \log \sup_{x \in S} \mathbb{E}_x \exp\left(\sum_{j=1}^{n} g_k(X_j)\right) \\ &\leq \log \sup_{y \in F} \mathbb{E}_y \exp\left(\sum_{j=1}^{\tau} g_k(X_j)\right).\end{aligned}$$

Proceeding now as in 1, we have

$$\lim_k \phi(g_k) = 0,$$

proving (9.8).

(9.8) $\Longrightarrow$ (9.9): Follows from Theorem 6.8 with $M = S$, $V = B(S)$, and I.

(9.9) $\Longrightarrow$ (9.10): Obvious.

(9.10) $\Longrightarrow$ (9.7): As in the proof of (9.6) $\Longrightarrow$ (9.3),

$$\overline{\lim_n}\, n^{-1} \log \sup_{x \in S} \mathbb{P}_x[\tau_{F_h} > n] \leq -(b + 1),$$

and (9.7) follows. □

9.3 The Large Deviation Principle

We will discuss now the large deviation principle under the assumption of matrix irreducibility. For clarity, we will consider separately the cases when the large deviation principle holds pointwise or uniformly with respect to the initial state.

Theorem 9.3 *Assume that P is matrix irreducible.*

1. *The following conditions are equivalent:*

   For every $b > 0$, there exists a finite set $F \subset S$ such that for all $x \in F$,
   $$\mathbb{E}_x e^{b\tau} < \infty. \tag{9.18}$$
   $$\tilde{\phi} \textit{ is } B(S)\textit{-regular.} \tag{9.19}$$
   For every $x \in S$, $\{\mathbb{P}_x[L_n \in \cdot]\}$ satisfies the large deviation principle
   $$\textit{with } B(S)\textit{-tight rate function } I. \tag{9.20}$$
   There exists $J\colon \mathcal{P}(S) \to \overline{\mathbb{R}^+}$, $B(S)$-tight, such that for every $x \in S$,
   $$\{\mathbb{P}_x[L_n \in \cdot]\} \textit{ satisfies the upper bound with rate function } J. \tag{9.21}$$

2. *If any, hence all, of the equivalent conditions above is satisfied, then P has a unique invariant probability measure π, and $\pi(x) > 0$ for all $x \in S$.*

Proof.

1. In view of Theorems 9.1.2 and 9.2.1, the only fact that remains to be proved is that (9.18) implies (9.3). We proceed to prove it. In what follows, $\Theta\colon S^{\mathbb{N}_0} \to S^{\mathbb{N}_0}$ is the usual shift.

   Let $y \in S$. Let $x \in F$, and let $k \in \mathbb{N}$ be such that $a = P^k(x, y) > 0$. Then for all $z \in S$,
   $$P^{k+1}(x, z) \geq P^k(x, y)P(y, z),$$
   which implies: for all $f\colon S \to \mathbb{R}^+$,
   $$\sum_{z\in S} P^{k+1}(x, z)f(z) \geq a \sum_{z\in S} P(y, z)f(z). \tag{9.22}$$
   Let $\Phi\colon S^{\mathbb{N}_0} \to \overline{\mathbb{R}^+}$ be measurable, and let $f(z) = \mathbb{E}_z\Phi$. Observing that for $h \in \mathbb{N}$,
   $$\mathbb{E}_x(\Phi \circ \Theta^h) = \sum_{z\in S} P^h(x, z)\mathbb{E}_z\Phi$$
   by the Markov property, we obtain from (9.22),
   $$\mathbb{E}_x(\Phi \circ \Theta^{k+1}) \geq a\mathbb{E}_y(\Phi \circ \Theta). \tag{9.23}$$

Let $\sigma = \inf\{n \geq 0 : X_n \in F\}$, $\Phi = \exp(b(1+\sigma))$. Taking into account that $\tau = 1 + \sigma \circ \Theta$, we have from (9.23),

$$\begin{aligned} \mathbb{E}_x \exp\left(b(\tau \circ \Theta^k)\right) &= \mathbb{E}_x \exp\left(b(1 + \sigma \circ \Theta^{k+1})\right) \\ &= \mathbb{E}_x(\Phi \circ \Theta^{k+1}) \\ &\geq a\mathbb{E}_y(\Phi \circ \Theta) = a\mathbb{E}_y e^{b\tau}. \end{aligned} \tag{9.24}$$

For $j \in \mathbb{N}$, let $\tau(j)$ be as in the proof of Theorem 9.2. Since obviously $\tau(k) \geq k$, we have

$$\begin{aligned} \tau \circ \Theta^k + k &= \inf\{j > k : X_j \in F\} \\ &\leq \inf\left\{j > \tau(k) : X_j \in F\right\} = \tau(k+1). \end{aligned} \tag{9.25}$$

Finally, by (9.24), (9.25), and the strong Markov property,

$$\begin{aligned} \mathbb{E}_y e^{b\tau} &\leq a^{-1}\mathbb{E}_x \exp\left(b(\tau \circ \Theta^k)\right) \\ &\leq a^{-1}\mathbb{E}_x \exp\left(b\tau(k+1)\right) \\ &\leq a^{-1}\left(\sup_{z \in F} \mathbb{E}_z e^{b\tau}\right)^{k+1} < \infty. \end{aligned}$$

2. The existence and uniqueness of π follows from Theorem 8.1.2. Since π is a P-maximal irreducibility probability measure and P is matrix irreducible, we have $\pi(x) > 0$ for all $x \in S$. □

Remark 9.4 The conditions in Theorem 9.3.1 are equivalent to the following condition, formally weaker than (9.19):

(C) For every $x \in S$, ϕ_x is $B(S)$-regular.

For, if (C) holds then by Theorem 6.8, for every $x \in S$, $\{\mathbb{P}_x[L_n \in \cdot]\}$ satisfies the upper bound with $B(S)$-tight rate function ϕ_x^*. But, since P is matrix irreducible, we have $\phi_x^* = I$.

Theorem 9.5 *Assume that P is matrix irreducible.*

1. *The following conditions are equivalent:*

For every $b > 0$, there exists a finite set $F \subset S$ such that

$$\sup_{x \in S} \mathbb{E}_x e^{b\tau} < \infty. \tag{9.26}$$

$$\phi \text{ is } B(S)\text{-regular.} \tag{9.27}$$

$$S \text{ is a uniformity set for both the upper and lower bounds with} \\ \mathcal{B}(S)\text{-tight rate function } I. \tag{9.28}$$

$$\text{There exists } J\colon \mathcal{P}(S) \to \overline{\mathbb{R}^+},\ \mathcal{B}(S)\text{-tight, such that } S \text{ is a uniformity set} \\ \text{for the upper bound with rate function } J. \tag{9.29}$$

2. *The statement about π in Theorem 9.3.2 also holds here.*

Proof. In view of Theorem 9.2.2, only the lower bound part of (9.28) requires proof. By Theorem 6.12, the fact that S is a uniformity set for the upper bound in the τ topology with a τ-tight rate function implies that S is petite. But then the lower bound follows from Theorem 9.1.2. □

Remark 9.6 If S is finite and P is matrix irreducible, then for all $g \in B(S)$,

$$\phi(g) = \Lambda(g), \tag{9.30}$$

and their common value equals $\log \rho(g)$, where $\rho(g)$ is the Perron–Frobenius eigenvalue of the matrix irreducible transform matrix

$$K_g(x, y) = P(x, y)e^{g(y)}, \qquad x, y \in S.$$

This is a general fact about finite matrix irreducible nonnegative matrices (see, e.g., Seneta, 1981). In our present context we can recover (9.30) from the fact that $\phi^* = \Lambda^*$ and the duality theorem for convex functions, since both ϕ and Λ are finite convex functions on the finite-dimensional vector space $B(S)$.

9.4 Notes

Theorems 9.1.1 and 9.1.2, the latter statement only in the case $M = \{\delta_x\}$, $x \in S$, are proved in Theorem 13.5 of Rassoul-Agha and Seppäläinen (2015) by different methods.

Of course, the condition that $\tilde{\phi}$ or ϕ be $B(S)$-regular in Theorems 9.2, 9.3, or 9.5 can be implemented by appealing to the results in Chapter 7. Here is a generalization of a condition given in Theorem 13.7 of Rassoul-Agha and Seppäläinen (2015), in which the upper bound is proved by other methods. Assume that there exist $\lambda \in \mathcal{P}(S)$, $k \in \mathbb{N}$ and a nondecreasing convex function $\varphi\colon \mathbb{R}^+ \to \mathbb{R}^+$ such that $\varphi(0) = 0$ and $\lim_{t\to\infty} t^{-1}\varphi(t) = \infty$, satisfying:

$$\text{For all } x \in S,\ P^k(x, \cdot) \ll \lambda. \tag{9.31}$$

$$\sup_{x\in S} \sum_{y\in A(\lambda)} \varphi\left(\frac{P^k(x, y)}{\lambda(y)}\right)\lambda(y) < \infty, \tag{9.32}$$

where $A(\lambda) = \{y \in S: \lambda(y) > 0\}$. Then by Theorem 7.1 and Remark 7.6, ϕ is $B(S)$-regular.

The condition given in Rassoul-Agha and Seppäläinen (2015) is (9.31) and (9.32) in the case $\varphi(t) = t^r$, $t \in \mathbb{R}^+$, for some $r > 1$. Under this assumption, the large deviation principle for $\{\mathbb{P}_x[L_n \in \cdot]\}$ is extended to an arbitrary Markov matrix in Corollary 3.6 and Theorem 3.7 of Rassoul-Agha and Seppäläinen (2015). This result previously appeared in Jiang and Wu (2005).

Rassoul-Agha and Seppäläinen (2015) presents a detailed discussion of the case when S is finite.

10

Examples

In this chapter we present several examples which show boundaries of the results developed in the previous chapters.

The first two examples show that, even if S is finite, the irreducibility assumption allows for a variety of different behaviors. In the first one, $S = \{1, 2, 3\}$ and for each $i \in S$, $\{\mathbb{P}_i[L_n \in \cdot]\}$ satisfies the large deviation principle with rate function J_i, but J_1, J_2, J_3 are all different. In the second example, $|S| = 4$ and for a certain $i \in S$, $\{\mathbb{P}_i[L_n \in \cdot]\}$ satisfies the large deviation principle with a rate function J which is not convex.

These examples are in contrast to the case of matrix irreducibility; this stricter condition precludes the existence of such situations. For, if S is countable, P is matrix irreducible and for some $\nu \in \mathcal{P}(S)$, $\{\mathbb{P}_\nu[L_n \in \cdot]\}$ satisfies the large deviation principle with rate function J, then by Theorem 5.1 and Theorems 4.1–4.2, we have $J = I$.

The third example is actually a counterexample to the large deviation principle. Here $S = \mathbb{N}_0$, P is matrix irreducible and uniformly ergodic, but $\{\mathbb{P}_\nu[L_n \in \cdot]\}$ does not satisfy the large deviation principle.

10.1 Different Rate Functions

In our first example, $S = \{1, 2, 3\}$ and we identify $\mathcal{P}(S)$ with

$$\Delta_3 = \{(p_1, p_2, p_3) \colon 0 \le p_i \le 1,\ i = 1, 2, 3,\ p_1 + p_2 + p_3 = 1\}$$

via $\mathcal{P}(S) \ni \mu \equiv (\mu(1), \mu(2), \mu(3)) \in \Delta_3$.

Let

$$P = \begin{bmatrix} 1/2 & 1/2 & 0 \\ 0 & 1/2 & 1/2 \\ 0 & 0 & 1 \end{bmatrix}. \tag{10.1}$$

Then P is irreducible with P-maximal irreducibility probability measure $\psi = \delta_3$.

Proposition 10.1 *Let P be given by* (10.1).

1. (i) $\{\mathbb{P}_1[L_n \in \cdot]\}$ *satisfies the large deviation principle with rate function* $J_1 \colon \Delta_3 \to \mathbb{R}^+$ *given by*

$$J_1(p_1, p_2, p_3) = (p_1 + p_2) \log 2.$$

   (ii) $J_1 = I = \phi_1^*$.
2. (i) $\{\mathbb{P}_2[L_n \in \cdot]\}$ *satisfies the large deviation principle with rate function* $J_2 \colon \Delta_3 \to \overline{\mathbb{R}^+}$ *given by*

$$J_2(p_1, p_2, p_3) = \begin{cases} p_2 \log 2 & \text{if } p_1 = 0, \\ \infty & \text{if } p_1 > 0. \end{cases}$$

   (ii) $J_2 = \phi_2^*$.
3. (i) $\{\mathbb{P}_3[L_n \in \cdot]\}$ *satisfies the large deviation principle with rate function*

$$J_3(p_1, p_2, p_3) = \begin{cases} 0 & \text{if } p_3 = 1, \\ \infty & \text{otherwise.} \end{cases}$$

   (ii) $J_3 = I_\psi = \phi_3^*$.

We will need the following simple lemma:

Lemma 10.2 *Let T be a topological space. Let $f \colon T \to \mathbb{R}^+$ be a bounded lower semicontinuous function. Let $\{E_n\}$ be a sequence of finite sets in T such that:*

1. *For every nonempty open set G, $E_n \cap G \neq \emptyset$ for all sufficiently large n.*
2. $|E_n|^{1/n} \to 1$.

Let $u_n \colon E_n \to \mathbb{R}^+$ and suppose $\{u_n\}_{n \in \mathbb{N}}$ satisfies, for some $0 < a < b < \infty$, $a \leq u_n \leq b$ for all n. Then

$$\lim_n \left[\sum_{x \in E_n} u_n(x) \, (f(x))^n \right]^{1/n} = \sup_{x \in T} f(x).$$

Proof.

1. Let $s = \sup_{x \in T} f(x)$. Then

$$\sum_{x \in E_n} u_n(x) \, (f(x))^n \leq b |E_n| s^n,$$

and therefore by 2,

$$\overline{\lim_n} \left[\sum_{x \in E_n} u_n(x) \, (f(x))^n \right]^{1/n} \leq s.$$

2. Given $\epsilon > 0$, let $G = \{x \in T : f(x) > s - \epsilon\}$, a nonempty open set. Therefore by 1, for all sufficiently large n, $E_n \cap G \neq \emptyset$, and it follows that

$$\sum_{x \in E_n} u_n(x) \, (f(x))^n \geq a(s - \epsilon)^n.$$

Hence

$$\underline{\lim_n} \left[\sum_{x \in E_n} u_n(x) \, (f(x))^n \right]^{1/n} \geq s - \epsilon.$$

But ϵ is arbitrary. □

Proof of Proposition 10.1. 1 (i): Let $h : \Delta_3 \to \mathbb{R}$ be continuous, and let

$$\Phi_1(h) = \sup_{(p,q) \in T_1} \left[h\,(p, q, 1 - (p + q)) - (p + q) \log 2 \right],$$

where

$$T_1 = \left\{ (p, q) \in \mathbb{R}^2 : 0 \leq p \leq 1, \ 0 \leq q \leq 1, \ p + q \leq 1 \right\}.$$

Then

$$\lim_n n^{-1} \log \mathbb{E}_1 \exp\,(nh(L_n)) = \Phi_1(h). \tag{10.2}$$

For, let $\tau = \inf\{n \geq 1 : X_n = 2\}$, $\sigma = \inf\{n \geq 1 : X_n = 3\}$. Then $\mathbb{P}_1[\tau < \sigma] = 1$ and

$$\begin{aligned} \mathbb{E}_1 \exp\,(nh(L_n)) &= \sum_{k=1}^{n-1} \mathbb{E}_1 \left\{ \exp\,(nh(L_n)) \, \mathbf{1}[\tau = k] \right\} \\ &\quad + \mathbb{E}_1 \left\{ \exp\,(nh(L_n)) \, \mathbf{1}[\tau \geq n] \right\} \\ &= \sum_{k=1}^{n-2} \mathbb{E}_1 \left\{ \exp\,(nh(L_n)) \, \mathbf{1}[\tau = k] \mathbf{1}[\sigma \leq n - 1] \right\} \\ &\quad + \sum_{k=1}^{n-1} \mathbb{E}_1 \left\{ \exp\,(nh(L_n)) \, \mathbf{1}[\tau = k] \mathbf{1}[\sigma \geq n] \right\} \\ &\quad + \mathbb{E}_1 \left\{ \exp\,(nh(L_n)) \, \mathbf{1}[\tau \geq n] \right\}. \end{aligned} \tag{10.3}$$

If $A, B \in \mathcal{S}^{\mathbb{N}_0}$ and $\mathbb{P}_1[A \backslash B] = 0$, we write

$$A \subset B \quad [\mathbb{P}_1].$$

By the form of P, we have for $1 \le k < j \le n-1$,

$$[\tau = k, \sigma = j] \subset \left[L_n = \left(\frac{k}{n}, \frac{j-k}{n}, \frac{n-j}{n}\right)\right] \quad [\mathbb{P}_1].$$

Then

$$\begin{aligned}
&\mathbb{E}_1 \{\exp(nh(L_n))\, \mathbf{1}[\tau = k]\mathbf{1}[\sigma = j]\} \\
&= \exp\left\{nh\left(\frac{k}{n}, \frac{j-k}{n}, \frac{n-j}{n}\right)\right\} \mathbb{P}_1[\tau = k, \sigma = j] \\
&= \exp\left\{nh\left(\frac{k}{n}, \frac{j-k}{n}, \frac{n-j}{n}\right)\right\}\left(\frac{1}{2}\right)^j \qquad (10.4)\\
&= \exp\left(n\left\{h\left[\frac{k}{n}, \frac{j-k}{n}, 1-\left(\frac{k}{n}+\frac{j-k}{n}\right)\right] - \left(\frac{k}{n}+\frac{j-k}{n}\right)\log 2\right\}\right) \\
&= \left[f_1\left(\frac{k}{n}, \frac{j-k}{n}\right)\right]^n,
\end{aligned}$$

where $f_1(p,q) = \exp\{h[p,q,1-(p+q)] - (p+q)\log 2\}$, $(p,q) \in T_1$. Similarly, for $1 \le k \le n-1$,

$$\mathbb{E}_1 \{\exp(nh(L_n))\, \mathbf{1}[\tau = k]\mathbf{1}[\sigma \ge n]\} = \left[f_1\left(\frac{k}{n}, \frac{n-k}{n}\right)\right]^n, \tag{10.5}$$

$$\mathbb{E}_1 \{\exp(nh(L_n))\, \mathbf{1}[\tau \ge n]\} = (f_1(1,0))^n. \tag{10.6}$$

By (10.3)–(10.6),

$$\begin{aligned}
\mathbb{E}_1 \exp(nh(L_n)) = &\sum_{k=1}^{n-2}\sum_{j=k+1}^{n-1}\left[f_1\left(\frac{k}{n}, \frac{j-k}{n}\right)\right]^n \\
&+ \sum_{k=1}^{n-1}\left[f_1\left(\frac{k}{n}, \frac{n-k}{n}\right)\right]^n + [f_1(1,0)]^n.
\end{aligned}$$

Let

$$\begin{aligned}
E_{n,1} &= \left\{\left(\frac{k}{n}, \frac{l}{n}\right) : 1 \le k \le n-2,\ 1 \le l \le n-2,\ k+l \le n-1\right\} \\
E_{n,2} &= \left\{\left(\frac{k}{n}, \frac{l}{n}\right) : 1 \le k \le n-1,\ 1 \le l \le n-1,\ k+l = n\right\} \\
E_{n,3} &= \{(1,0)\},
\end{aligned}$$

and $E_n = E_{n,1} \cup E_{n,2} \cup E_{n,3}$. Then by (10.3)–(10.6),

$$\mathbb{E}_1 \exp(nh(L_n)) = \sum_{x \in E_n} (f_1(x))^n,$$

and by Lemma 10.2,

$$\lim n^{-1} \log \mathbb{E}_1 \exp(nh(L_n)) = \Phi_1(h),$$

proving (10.2).

By Theorem 4.4.2 of Dembo and Zeitouni (1998), it follows that $\{\mathbb{P}_1[L_n \in \cdot]\}$ satisfies the large deviation principle with rate function

$$J_1(p_1, p_2, p_3) = \sup_{h \in C(\Delta_3)} \left[h(p_1, p_2, p_3) - \Phi_1(h)\right].$$

We proceed now to calculate this supremum. We have, by the definition of $\Phi_1(h)$,

$$h(p_1, p_2, p_3) - (p_1 + p_2) \log 2 \le \Phi_1(h)$$

for all $(p_1, p_2, p_3) \in \Delta_3$ and $h \in C(\Delta_3)$, and therefore

$$J_1(p_1, p_2, p_3) \le (p_1 + p_2) \log 2.$$

On the other hand, if $h_0(p_1, p_2, p_3) = (p_1 + p_2) \log 2$, then

$$h_0(p_1, p_2, p_3) - \Phi_1(h_0) = (p_1 + p_2) \log 2,$$

and it follows that

$$J_1(p_1, p_2, p_3) = (p_1 + p_2) \log 2.$$

1 (ii): Since J_1 is convex, by Theorem 5.1.3 $\phi_1^* = J_1$. This can also be shown as follows: let $g\colon S \to \mathbb{R}$ be given by $g(i) = \xi_i$, $i = 1, 2, 3$. Then

$$\begin{aligned}
\phi_1(g) &= \overline{\lim_n}\, n^{-1} \log \mathbb{E}_1 \exp\left(\sum_{j=0}^{n-1} g(X_j)\right) \\
&= \lim_n n^{-1} \log \mathbb{E}_1 \exp(nh(L_n)),
\end{aligned}$$

where $h(p_1, p_2, p_3) = \sum_{i=1}^{3} \xi_i p_i$. We have

$$\begin{aligned}
\Phi_1(h) &= \sup_{(p,q)\in T} \left[\xi_1 p + \xi_2 q + \xi_3 (1 - (p+q)) - (p+q) \log 2\right] \\
&= \max\{\xi_1 - \log 2, \xi_2 - \log 2, \xi_3\} \\
&= \varphi_1(\xi_1, \xi_2, \xi_3),
\end{aligned}$$

say, and by (10.2) $\phi_1(g) = \varphi_1(\xi_1, \xi_2, \xi_3)$. Now by a direct calculation one can show that $\phi_1^* = J_1$. The other equality in 1 (ii) will be shown later.

2 (i): The proof is similar to that of 1 (i) but simpler. Let $h\colon \Delta_3 \to \mathbb{R}$ be continuous, and let

$$\Phi_2(h) = \sup_{q \in T_2} \left[h(0, q, 1 - q) - q \log 2\right],$$

where $T_2 = [0, 1]$. Then

$$\lim_n n^{-1} \log \mathbb{E}_2 \exp\left(nh(L_n)\right) = \Phi_2(h). \tag{10.7}$$

This is proved like (10.2); we omit the details.

Applying again Theorem 4.4.2 of Dembo and Zeitouni (1998) it follows that $\{\mathbb{P}_2[L_n \in \cdot]\}$ satisfies the large deviation principle with rate function

$$J_2(p_1, p_2, p_3) = \sup_{h \in C(\Delta_3)} \left[h(p_1, p_2, p_3) - \Phi_2(h)\right]$$

By the definition of $\Phi_2(h)$, if $(p_1, p_2, p_3) \in \Delta_3$ and $p_1 = 0$, we have

$$h(0, p_2, p_3) - p_2 \log 2 \le \Phi_2(h),$$

and therefore

$$J_2(0, p_2, p_3) \le p_2 \log 2. \tag{10.8}$$

On the other hand, if $h_0(p_1, p_2, p_3) = p_2 \log 2$, then

$$h_0(p_1, p_2, p_3) - \Phi_2(h_0) = p_2 \log 2,$$

which implies

$$J_2(p_1, p_2, p_3) \ge p_2 \log 2 \tag{10.9}$$

for $(p_1, p_2, p_3) \in \Delta_3$.

Now let $h_t(p_1, p_2, p_3) = tp_1$, $t > 0$. Then $\Phi_2(h_t) = 0$ and $J_2(p_1, p_2, p_3) \ge tp_1$ for all $t > 0$, which implies that if $p_1 > 0$,

$$J_2(p_1, p_2, p_3) = \infty. \tag{10.10}$$

From (10.8)–(10.10), we have

$$J_2(p_1, p_2, p_3) = \begin{cases} p_2 \log 2 & \text{if } p_1 = 0, \\ \infty & \text{if } p_1 > 0, \end{cases}$$

proving 2 (i). Since J_2 is convex, $\phi_2^* = J_2$ follows from Theorem 5.1.3. Also, arguing as in 1 (ii), we can show that for $g\colon S \to \mathbb{R}$,

$$\phi_2(g) = \varphi_2(\xi_1, \xi_2, \xi_3),$$

where $\varphi_2(\xi_1, \xi_2, \xi_3) = \max\{\xi_2 - \log 2, \xi_3\}$. Now an elementary calculation shows again that $\phi_2^* = J_2$.

3 (i)–3 (ii): Since $P^2 \geq \beta 1 \otimes \delta_3$ for some $\beta > 0$, as is easily seen, $(1, \delta_3)$ is a small pair, hence a K_g-small pair for any $g \in B(S)$. Let $g(i) = \xi_i$, $i = 1, 2, 3$. Then

$$K_g = \begin{bmatrix} 1/2 e^{\xi_1} & 1/2 e^{\xi_2} & 0 \\ 0 & 1/2 e^{\xi_2} & 1/2 e^{\xi_3} \\ 0 & 0 & e^{\xi_3} \end{bmatrix},$$

and by a simple computation,

$$\Lambda(g) = \overline{\lim_n}\, n^{-1} \log \delta_3 K_g^n 1 = \xi_3$$

and

$$\Lambda^*(p_1, p_2, p_3) = \begin{cases} 0 & \text{if } p_3 = 1, \\ \infty & \text{if } p_3 < 1. \end{cases}$$

But also $\phi_3(g) = \Lambda(g)$. Therefore by Theorem 8.1, $\{\mathbb{P}_3[L_n \in \cdot]\}$ satisfies the large deviation principle with rate function

$$\phi_3^* = \Lambda^* = I_\psi.$$

Finally, since S is finite, for $g \in B(S)$,

$$\begin{aligned} \phi(g) &= \overline{\lim_n}\, n^{-1} \log \sup_{x \in S} K_g^n 1(x) \\ &= \sup_{x \in S} \overline{\lim_n}\, n^{-1} \log K_g^n 1(x) \\ &= \sup_{x \in S} \{\phi_1(g), \phi_2(g), \phi_3(g)\} = \phi_1(g), \end{aligned}$$

recalling the expressions for $\phi_i(g)$, $i = 1, 2, 3$, and therefore $\phi_1^* = \phi^* = I$, completing the proof of 1 (ii). □

10.2 A Nonconvex Rate Function

Let

$$P = \begin{bmatrix} 0 & 1/2 & 1/2 & 0 \\ 0 & 1/2 & 0 & 1/2 \\ 0 & 0 & 1/2 & 1/2 \\ 0 & 0 & 0 & 1 \end{bmatrix}. \tag{10.11}$$

Then P is irreducible with P-maximal irreducibility probability measure $\psi = \delta_4$.

Proposition 10.3 *Let P be given by* (10.11). *Then*

1. $\{\mathbb{P}_1[L_n \in \cdot]\}$ *satisfies the large deviation principle with rate function* $J\colon \Delta_4 \to \mathbb{R}_+$ *given by*

$$J(p_1, p_2, p_3, p_4) = \begin{cases} \infty & \text{if } p_1 > 0 \text{ or } p_2 \wedge p_3 > 0, \\ (1 - p_4)\log 2 & \text{otherwise.} \end{cases} \tag{10.12}$$

2. *J is not convex.*

Proof 1. It is similar to that of Proposition 10.1. We will present a sketch, indicating the main points.

For $h \in C(\Delta_4)$, let

$$\Phi(h) = \max\left\{ \sup_{p\in[0,1]} \left[h(0, p, 0, 1 - p) - p\log 2\right], \right.$$
$$\left. \sup_{p\in[0,1]} \left[h(0, 0, p, 1 - p) - p\log 2\right] \right\}.$$

It will be convenient to introduce $L_n' = 1/n \sum_{j=1}^{n} \delta_{X_j}$. Then

$$\begin{aligned} \lim_n n^{-1} \log \mathbb{E}_1 \exp\left(nh(L_n)\right) &= \lim_n n^{-1} \log \mathbb{E}_1 \exp\left(nh(L_n')\right) \\ &= \Phi(h). \end{aligned} \tag{10.13}$$

The first equality follows from the uniform continuity of h and the fact that $|L_n - L_n'| \le 2/n$. We will prove now the second equality.

Let $\tau = \inf\{n \ge 1 : X_n = 4\}$. We have

$$\begin{aligned} \mathbb{E}_1 \left\{\exp\left(nh(L_n')\right) \mathbf{1}[X_1 = 2]\right\} &= \sum_{k=2}^{n} \mathbb{E}_1 \left\{\exp\left(nh(L_n')\right) \mathbf{1}[X_1 = 2]\mathbf{1}[\tau = k]\right\} \\ &\quad + \mathbb{E}_1 \left\{\exp\left(nh(L_n')\right) \mathbf{1}[X_1 = 2]\mathbf{1}[\tau > n]\right\} \\ &= \sum_{k=2}^{n} \exp\left(nh\left(0, \frac{k-1}{n}, 0, \frac{n-k+1}{n}\right)\right)\left(\frac{1}{2}\right)^k \\ &\quad + \exp\left(nh(0, 1, 0, 0)\right)\left(\frac{1}{2}\right)^n. \end{aligned}$$

As in Proposition 10.1,

$$\begin{aligned} &\lim_n n^{-1} \log \mathbb{E}_1 \exp\left(nh(L_n')\right) \mathbf{1}[X_1 = 2] \\ &\qquad = \sup_{p\in[0,1]} \left[h(0, p, 0, 1 - p) - p\log 2\right]. \end{aligned} \tag{10.14}$$

Similarly,

$$\lim_n n^{-1} \log \mathbb{E}_1 \exp (nh(L'_n))\, 1[X_1 = 3] \\ = \sup_{p\in[0,1]} [h(0, 0, p, 1 - p) - p \log 2] . \tag{10.15}$$

Now (10.13) follows from (10.14) and (10.15).

By Theorem 4.4.2 of Dembo and Zeitouni (1998), it follows that $\{\mathbb{P}_1[L_n \in \cdot]\}$ satisfies the large deviation principle with rate function

$$J(p_1, p_2, p_3, p_4) = \sup_{h\in C(\Delta_4)} [h(p_1, p_2, p_3, p_4) - \Phi(h)] .$$

To obtain the expression (10.12) for J, let $h_t(p_1, p_2, p_3, p_4) = tp_1$, $t > 0$. Then

$$J(p_1, p_2, p_3, p_4) \geq h_t(p_1) - \Phi(h_t) \\ = tp_1,$$

and therefore $J(p_1, p_2, p_3, p_4) = \infty$ if $p_1 > 0$. Similarly, $J(p_1, p_2, p_3, p_4) = \infty$ if $p_2 \wedge p_3 > 0$. The fact that

$$J(0, 0, 1 - p_4, p_4) = J(0, 1 - p_4, 0, p_4) = (1 - p_4) \log 2$$

is proved similarly to Proposition 10.1.

2. Let $x = (0, 1, 0, 0)$, $y = (0, 0, 1, 0)$. Then $J(x) = J(y) = \log 2$ but $J((1/2)x + (1/2)y) = J(0, 1/2, 1/2, 0) = \infty$. □

10.3 A Counterexample

We present a counterexample to the large deviation principle in a case where S is countable and P is matrix irreducible and uniformly ergodic.

Let $S = \mathbb{N}_0$, $0 < \rho < 1$, and define P as follows: for $i \in S$,

$$P(i, i + 1) = \rho, \qquad P(i, 0) = 1 - \rho, \\ \text{or} \qquad P(i, \cdot) = \rho\delta_{i+1} + (1 - \rho)\delta_0. \tag{10.16}$$

Clearly P is matrix irreducible. Also, S is small, and therefore P is uniformly ergodic (see Proposition B.9).

Proposition 10.4 *Let P be given by* (10.16), *$\nu\in\mathcal{P}(S)$, $\nu\neq\delta_0$. Then $\{\mathbb{P}_\nu[L_n \in \cdot]\}$ does not satisfy the large deviation principle in the τ topology.*

Proof. Suppose $\{\mathbb{P}_\nu[L_n \in \cdot]\}$ satisfies the large deviation principle in the τ topology with rate function J. Then, as argued in the beginning of this chapter, we must have $J = I$.

We will reach a contradiction by exhibiting a closed set $C \subset \mathcal{P}(S)$ for which the large deviation upper bound with rate function I does not hold.

For $m \geq 1, l \geq 1$, let

$$\mu_{m,l} = m^{-1} \sum_{j=0}^{m-1} \delta_{j+l},$$

$C = \{\mu_{m,l} \colon m \geq 1, l \geq 1\}$. Also, for $k \in \mathbb{N}$, $l \geq 1$, let

$$f_{k,l}(j) = \begin{cases} k & j = l \\ 1 & j \neq l. \end{cases}$$

Let $\mathcal{U} = \{u \colon S \to \mathbb{R} \colon \sup u < \infty, \inf u > 0\}$. Then for $\mu \in \mathcal{P}(S)$,

$$\begin{aligned} I(\mu) &= \sup_{u \in \mathcal{U}} \left[\sum_{j \in S} \mu(j) \log \left(\frac{u}{Pu} \right)(j) \right] \\ &= \sup_{u \in \mathcal{U}} \left[\sum_{j=0}^{\infty} \mu(j) \log \left(\frac{u(j)}{(1-\rho)u(0) + \rho u(j+1)} \right) \right] \qquad (10.17) \\ &\geq \sum_{j=0}^{\infty} \mu(j) g_{k,l}(j), \end{aligned}$$

where

$$g_{k,l}(j) = \log \left(\frac{f_{k,l}(j)}{(1-\rho) f_{k,l}(0) + \rho f_{k,l}(j+1)} \right).$$

Now

$$g_{k,l}(j) = \begin{cases} \log \left\{ [(1-\rho) + \rho k]^{-1} \right\} & j = l-1 \\ \log k & j = l \\ 0 & \text{otherwise,} \end{cases}$$

so from (10.17) we have

$$I(\mu) \geq \mu(l-1) \log [(1-\rho) + \rho k]^{-1} + \mu(l) \log k.$$

Since $\mu_{m,l}(l-1) = 0$, $\mu_{m,l}(l) = m^{-1}$, we have $I(\mu_{m,l}) \geq m^{-1} \log k$. Therefore $I(\mu_{m,l}) = \infty$,

$$\inf_{\mu \in C} I(\mu) = \infty,$$

and it follows that

$$\overline{\lim_n}\, n^{-1} \log \mathbb{P}_\nu[L_n \in C] = -\infty. \tag{10.18}$$

On the other hand, let $l \geq 1$ be such that $\nu(l) > 0$. Then

$$\begin{aligned}
\mathbb{P}_\nu[L_n \in C] &\geq \mathbb{P}_\nu\,[X_0 = l,\ X_1 = l+1, \ldots, X_{n-1} = l+(n-1)] \\
&\geq \nu(l)\mathbb{P}_l\,[X_1 = l+1, \ldots, X_{n-1} = l+(n-1)] \\
&= \nu(l)\rho^{n-1},
\end{aligned}$$

$$\underline{\lim_n}\, n^{-1} \log \mathbb{P}_\nu[L_n \in C] \geq \log \rho,$$

contradicting (10.18). □

Remark 10.5 The weak large deviation principle for $\{\mathbb{P}_\nu[L_n \in \cdot]\}$ in the τ topology with rate function I does hold by Theorem 9.1.

10.4 Notes

Propositions 10.1 and 10.3 are stated in Dinwoodie (1993). The quoted result from Dembo and Zeitouni (1998) is due to W. Bryc. Proposition 10.4 is proved in Baxter et al. (1991).

11

Large Deviations for Vector-Valued Additive Functionals

Let E be a separable Banach space, $\mathcal{E}$ its Borel σ-algebra, and let $f : S \to E$ be a measurable function. In this chapter we study the large deviations for $\{\mathbb{P}_\nu[n^{-1}S_n(f) \in \cdot]\}$.

In Section 11.1 we extend the lower bound with rate function Λ_f^*, proved in Theorem 2.1 for bounded f, to an arbitrary measurable function.

In Section 11.2 the rate function Λ_f^* is identified in terms of I_ψ. Theorem 11.13 extends to an arbitrary measurable function Theorem 2.17 and Remark 2.18, in which Λ_f^* was identified in terms of $\Lambda^* = I_\psi$ when f is bounded.

In Section 11.3, Theorem 11.16, we obtain the upper bound for

$$\left\{\sup_{\mu \in M} \mathbb{P}_\mu \left[n^{-1}S_n(f) \in \cdot\right]\right\},$$

where $M \subset \mathcal{P}(S)$, with rate function $\phi_{f,M}^*$. The tightness and integrability assumptions can be relaxed if E is finite-dimensional; see Remark 11.18.

In Section 11.4, Theorem 11.21, we give a necessary and sufficient condition for the large deviation principle under the assumptions of Theorems 11.1 and 11.16.

In Section 11.5 we study the zero set of Λ_f^*; as it was pointed out in the paragraph following Proposition 1.12, under suitable conditions (Proposition 1.12 and, more generally, Theorem 11.21) the equality $(\Lambda_f^*)^{-1}(0) = \{\pi(f)\}$ ensures the exponential decay of $\{\mathbb{P}_\nu[n^{-1}S_n(f) \in B]\}$, where $B \in \mathcal{E}$ and $\pi(f) \notin \bar{B}$.

In Section 11.6, we give sufficient conditions for the large deviation principle. In particular we state a condition on P which, together with 2(i) and (11.35) of Theorem 11.16 implies the large deviation principle for $\{\mathbb{P}_\nu[n^{-1} S_n(f) \in \cdot]\}$ for all petite ν and all $\nu = \delta_x$, where x belongs to a set in $\mathcal{S}$ of full ψ measure.

Finally, in Section 11.7 we discuss the relationship between large deviations for empirical measures and large deviations for additive functionals.

11.1 Lower Bounds: The General Case

We recall some definitions in Chapter 1. Let $F(S)$ be the space of real-valued measurable functions on S. For $g \in F(S)$, $x \in S$, $A \in \mathcal{S}$, let

$$K_g(x, A) = \int_A e^{g(y)} P(x, dy).$$

If P is irreducible, then so is K_g and its convergence parameter $R(K_g)$ exists (Appendix C). We define

$$K_{f,\xi}(x, A) = K_{\langle f,\xi\rangle}(x, A) = \int_A e^{\langle f(y),\xi\rangle} P(x, dy);$$
$$\Lambda_f(\xi) = -\log R(K_{f,\xi}), \qquad \xi \in E^*;$$
$$\Lambda_f^*(u) = \sup\left\{\langle u, \xi\rangle - \Lambda_f(\xi) : \xi \in E^*\right\}, \qquad u \in E.$$

The main result of this section is

Theorem 11.1 *Let P be irreducible and let $f : S \to E$ be measurable. Then for every set $M \subset \mathcal{P}(S)$ satisfying* (2.1) *and* $\lim_k \sup_{\mu\in M} \mu(\{x \in S : \|f(x)\| > k\}) = 0$ *and every open set $G \subset E$,*

$$\varliminf_n n^{-1} \log \inf_{\mu\in M} \mathbb{P}_\mu\left[n^{-1} S_n(f) \in G\right] \geq -\inf_{u\in G} \Lambda_f^*(u).$$

In particular, the conclusion holds if $M \subset S$, M is petite and $\sup_{x\in M} \|f(x)\| < \infty$.

A plausible strategy to prove this result is as follows. We express $\mathbb{P}_\mu[n^{-1} S_n(f) \in G]$ as iterated integrals involving the kernel P and minorize them by the corresponding iterated integrals involving P truncated at sets of boundedness of f. Then we apply a suitable version of Theorem 2.1 for sub-Markov kernels — that is, kernels P on $(S, \mathcal{S})$ satisfying $P(x, S) \leq 1$ for all $x \in S$ — to obtain a lower bound; the rate function will involve a parameter indexing the truncation. Finally, we obtain the correct lower bound for $\{\mathbb{P}_\mu[n^{-1} S_n(f) \in \cdot]\}$ by taking limits and removing the truncation parameter.

The actual steps in the proof of Theorem 11.1 are as follows:

1. We show that in fact Theorem 2.1, suitably reformulated, holds for an irreducible sub-Markov kernel P.

2. We show that the strategy described above works and Theorem 11.1 is true, under the assumption that P truncated at sets of boundedness of f remains irreducible.
3. We show how to remove the additional assumption in step 2.

In order to reformulate Theorem 2.1 for sub-Markov kernels, we introduce the following notation: $\mathcal{P}'(S)$ is the class of measures μ on $\mathcal{S}$ such that $\mu(S) \leq 1$, and for $\mu \in \mathcal{P}'(S)$, P a sub-Markov kernel on $(S, \mathcal{S})$, $n \in \mathbb{N}$, $F\colon S^{n+1} \to \mathbb{R}^+$ measurable, we define

$$\mathbb{E}_\mu^{(n)} F(X_0, \ldots, X_n) = \int \mu(dx_0) \int P(x_0, dx_1) \int \cdots \int P(x_{n-1}, dx_n) F(x_0, \ldots, x_n). \quad (11.1)$$

The functions $\Lambda_f\colon E^* \to \mathbb{R}$, $\Lambda_f^*\colon E \to \mathbb{R}$ are defined as in the case when P is Markov.

The version of Theorem 2.1 for sub-Markov kernels is

Proposition 11.2 *Let P be an irreducible sub-Markov kernel, and let $f\colon S \to E$ be a bounded measurable function. Then for every set $M \subset \mathcal{P}'(S)$ satisfying* (2.1) *and every open set $G \subset E$,*

$$\underline{\lim}_n\, n^{-1} \log \inf_{\mu \in M} \mathbb{E}_\mu^{(n-1)} \mathbf{1}_G\left(n^{-1} S_n(f)\right) \geq -\inf_{u \in G} \Lambda_f^*(u).$$

For the proof of Proposition 11.2 we need the following versions of certain technical tools used in the proof of Theorem 2.1.

Lemma 11.3 *Let $\mathbb{E}_\mu^{(n)} F(X_0, \ldots, X_n)$ be defined by* (11.1). *Then:*

1. *Markov property: for $k, n \in \mathbb{N}$ and measurable functions $F_0\colon S^{k+1} \to \mathbb{R}^+$, $F_1\colon S^n \to \mathbb{R}^+$,*

$$\mathbb{E}_\mu^{(k+n)} F_0(X_0, \ldots, X_k) F_1(X_{k+1}, \ldots, X_{k+n}) = \mathbb{E}_\mu^{(k)} F_0(X_0, \ldots, X_k) \mathbb{E}_{X_k}^{(n)} F_1(X_1, \ldots, X_n).$$

2. *Assume that P satisfies* (2.3). *Then for $n \in \mathbb{N}$, $x \in S$, $\Phi\colon S^n \to \mathbb{R}^+$ measurable,*

$$\mathbb{E}_x^{(n)} \Phi(X_1, \ldots, X_n) \geq \alpha \mathbf{1}_C(x) \mathbb{E}_\nu^{(n-1)} \Phi(X_0, \ldots, X_{n-1}).$$

3. *For* $0 < t < 1$, *let* P_t *be as in Appendix D and let* $\mathbb{E}_{t,\mu}^{(n)}$ *be as in* (11.1) *with* P_t *instead of* P. *Then the following version of Lemma 2.7 holds:*

$$\mathbb{E}_{t,\mu}^{(n-1)} 1_{G_\epsilon}\left(n^{-1}S_n(f)\right) \leq \mathbb{E}_{\mu}^{(n-1)} 1_{G}\left(n^{-1}S_n(f)\right) + r_n(t,\epsilon).$$

Proof. We will prove part 1. The proof of part 2 is a straightforward modification of the proof of (2.5) and the proof of part 3 can be obtained by adapting the proofs of Lemmas D.1 and 2.7.

We have

$$\begin{aligned}
&\mathbb{E}_{\mu}^{(k+n)} F_0(X_0,\ldots,X_k)F_1(X_{k+1},\ldots,X_{k+n})\\
&\quad = \int \mu(dx_0)\int P(x_0,dx_1)\int\cdots\int P(x_{k-1},dx_k)F_0(x_0,\ldots,x_k)\\
&\quad\quad \times \int P(x_k,dx_{k+1})\int\cdots\int P(x_{k+n-1},dx_{k+n})F_1(x_{k+1},\ldots,x_{k+n})\\
&\quad = \int \mu(dx_0)\int P(x_0,dx_1)\int\cdots\int P(x_{k-1},dx_k)F_0(x_0,\ldots,x_k)\\
&\quad\quad \times \mathbb{E}_{x_k}^{(n)} F_1(X_1,\ldots,X_n)\\
&\quad = \mathbb{E}_{\mu}^{(k)} F_0(X_0,\ldots,X_k)\mathbb{E}_{X_k}^{(n)} F_1(X_1,\ldots,X_n).
\end{aligned}$$

□

Using Lemma 11.3, Proposition 11.2 can be proved by retracing the proof of Theorem 2.1 with some straightforward changes. We omit the details.

In Lemma 11.5 we will need the following continuity property of convex conjugation, proved in de Acosta (1988, Thm. B.3).

Lemma 11.4 *Let* E *be a separable Banach space and let* h_k, $k \in \mathbb{N}$, h *be functions from* E^* *into* $\overline{\mathbb{R}}$, *such that:*

1. *For* $k \in \mathbb{N}$, h_k *is convex, proper (defined in Appendix F) and* $\sigma(E^*, E)$*-lower semicontinuous.*
2. $h_k \uparrow h$ *pointwise and* h *is proper.*
3. *There exist constants* $a \in \mathbb{R}$, $b > 0$ *such that for all* $\xi \in E^*$, $h_1(\xi) \geq a - b\|\xi\|$.

Then, for every $u \in E$, *there exists a sequence* $\{u_k\} \subset E$ *such that* $u_k \to u$ *and*

$$\overline{\lim_k}\, h_k^*(u_k) \leq h^*(u).$$

For $f\colon S \to E$ measurable, $k \in \mathbb{N}$, let

$$B_k = \{u \in E\colon \|u\| \leq k\}, \qquad C_k = f^{-1}(B_k).$$

Let P_k be the sub-Markov kernel defined by

$$P_k(x, A) = P(x, A \cap C_k), \qquad x \in S,\ A \in \mathcal{S}.$$

Lemma 11.5 *Let P, f and M be as in Theorem 11.1. Assume that for each $k \in \mathbb{N}$, P_k is irreducible. Then the conclusion of Theorem 11.1 holds.*

Proof. Let $\mu \in \mathcal{P}(S)$ and let $f_k = f\mathbf{1}_{C_k}$ ($k \in \mathbb{N}$). For any open set $G \subset E$,

$$\begin{aligned}
&\mathbb{P}_\mu\left[n^{-1}S_n(f) \in G\right] \\
&= \int \mu(dx_0) \int P(x_0, dx_1) \int \cdots \int P(x_{n-2}, dx_{n-1}) \mathbf{1}_G\left(n^{-1}\sum_{j=0}^{n-1} f(x_j)\right) \\
&\geq \int \mu(dx_0) \int P(x_0, dx_1) \int \cdots \\
&\qquad \int P(x_{n-2}, dx_{n-1}) \prod_{j=0}^{n-1} \mathbf{1}_{C_k}(x_j) \mathbf{1}_G\left(n^{-1}\sum_{j=0}^{n-1} f_k(x_j)\right) \\
&= \int \mu_k(dx_0) \int P_k(x_0, dx_1) \int \cdots \int P_k(x_{n-2}, dx_{n-1}) \mathbf{1}_G\left(n^{-1}\sum_{j=0}^{n-1} f_k(x_j)\right),
\end{aligned}$$

where $\mu_k(A) = \mu(A \cap C_k)$,

$$= \mathbb{E}_{k,\mu_k}^{(n-1)} \mathbf{1}_G\left(n^{-1}S_n(f_k)\right),$$

where $\{\mathbb{E}_{k,\rho}^{(m)}\}$ are as in (11.1) for the measure ρ and the kernel P_k.

Let $M_k = \{\mu_k : \mu \in M\}$. Then for all sufficiently large $k \in \mathbb{N}$, M_k satisfies (2.1). For: since for $i = 0, \ldots, h$,

$$\begin{aligned}
\mu P^i(D) - \mu_k P^i(D) &= \int P^i(x, D) \mathbf{1}_{C_k^c}(x) \mu(dx) \\
&\leq \mu(C_k^c),
\end{aligned}$$

we have

$$\sum_{i=0}^{h} \mu_k P^i(D) \geq -(h+1)\mu(C_k^c) + \sum_{i=0}^{h} \mu P^i(D),$$

and the claim follows from the second condition on M.

Therefore by Proposition 11.2, for all $u \in G$ and all sufficiently large $k \in \mathbb{N}$,

$$\begin{aligned}
&\varliminf_n n^{-1} \log \inf_{\mu \in M} \mathbb{P}_\mu\left[n^{-1}S_n(f) \in G\right] \\
&\quad \geq \varliminf_n n^{-1} \log \inf_{\mu \in M} \mathbb{E}_{k,\mu_k}^{(n-1)} \mathbf{1}_G\left(n^{-1}S_n(f_k)\right) \geq -\Lambda_{k,f_k}^*(u),
\end{aligned} \tag{11.2}$$

where

$$\Lambda^*_{k,f_k}(u) = \sup\left\{\langle u, \xi\rangle - \Lambda_{k,f_k}(\xi) \colon \xi \in E^*\right\},$$
$$\Lambda_{k,f_k}(\xi) = -\log R(K_{k,f_k,\xi}),$$

and

$$K_{k,f_k,\xi}(x, A) = \int_A e^{\langle f_k(y),\xi\rangle} P_k(x, dy).$$

For all $x \in S$, $A \in \mathcal{S}$,

$$K_{k,f_k,\xi}(x, A) = \int_{C_k \cap A} e^{\langle f(y),\xi\rangle} P(x, dy) \uparrow K_{f,\xi}(x, A).$$

Therefore by Corollary C.4, for all $\xi \in E^*$,

$$\Lambda_{k,f_k}(\xi) = -\log R(K_{k,f_k,\xi}) \uparrow -\log R(K_{f,\xi}) = \Lambda_f(\xi). \tag{11.3}$$

Let $h_k = \Lambda_{k,f_k}$, $h = \Lambda_f$. Then $\{h_k\}$, h satisfy the assumptions of Lemma 11.4:

1. For $k \in \mathbb{N}$, h_k is convex and, since f_k is bounded, real-valued (see de Acosta and Ney, 2014, Lemma 2.5). Also, by Theorem 2.17, h_k is $\sigma(E^*, E)$-lower semicontinuous.
2. $h_k \uparrow h$ pointwise by (11.3). Also, for all $\xi \in E^*$, $h(\xi) > -\infty$ (Appendix C) and $h(0) \leq 0$, so h is proper.
3. By Lemma 2.5 of de Acosta and Ney (2014), for $\xi \in E^*$,

$$|h_1(\xi) - h_1(0)| \leq \sup_{x\in S} |\langle f_1(x), \xi\rangle| \leq ||\xi||,$$

and therefore

$$h_1(\xi) \geq h_1(0) - ||\xi||.$$

Let $u \in G$. Then by Lemma 11.4 there exists a sequence $\{u_k\} \subset G$ such that $u_k \to u$ and

$$\overline{\lim_k}\, \Lambda^*_{k,f_k}(u_k) \leq \Lambda^*_f(u).$$

Using now (11.2), the proof is finished. □

Having established steps 1 and 2 described above in Proposition 11.2 and Lemma 11.5, we proceed to step 3.

In general, the truncated kernel P_k defined above may not be irreducible. In order to overcome this difficulty, we introduce a "smoothing" of P which suitably approximates P and has the desired irreducibility property. This is the kernel P_σ on $(S \times E, \mathcal{S} \otimes \mathcal{E})$ defined by

$$P_\sigma((x,u),A) = (P(x,\cdot) \otimes \gamma_\sigma)(A), \qquad (x,u) \in S \times E,\ A \in \mathcal{S} \otimes \mathcal{E},$$

where γ_σ is a Gaussian measure on E with full support and "small standard deviation" $\sigma > 0$. We extend $f \colon S \to E$ to $g \colon S \times E \to E$ as follows:

$$g(x,u) = f(x) + u, \qquad x \in S,\ u \in E.$$

We will show that P_σ is irreducible and P_σ truncated at sets of boundedness of g is also irreducible. The fact that γ_σ has full support is crucial here, so for completeness we prove the following lemma. If Z is as in Lemma 11.6, we define $\gamma_\sigma = \mathcal{L}(\sigma Z)$.

Lemma 11.6 *There exists an E-valued centered Gaussian vector Z such that for every $\sigma > 0$ and every open set $G \subset E$,*

$$\mathbb{P}[\sigma Z \in G] > 0.$$

Proof. Let $\{Y_n \colon n \geq 1\}$ be an independent sequence of $N(0,1)$ random variables, let $\{a_n \colon n \geq 1\}$ be a positive sequence such that for all $\epsilon > 0$,

$$\prod_{n=1}^{\infty} \mathbb{P}[|Y_1| \leq \epsilon a_n] > 0,$$

let $\{b_n \colon n \geq 1\}$ be a positive sequence such that $\sum_n a_n b_n \leq 1$, and let $\{u_n \colon n \geq 1\}$ be a dense subset of the unit ball of E. Define

$$Z = \sum_{n=1}^{\infty} b_n Y_n u_n;$$

the series converges in $L^1(E)$ since

$$\sum_{n=1}^{\infty} \mathbb{E}\|b_n Y_n u_n\| \leq (\mathbb{E}|Y_1|) \sum_{n=1}^{\infty} b_n < \infty,$$

and Z is a centered Gaussian vector. Given $u \in E$, $\epsilon > 0$, choose $n_0 \in \mathbb{N}$, $\rho > 0$ such that $\|u - \rho u_{n_0}\| < \epsilon/2$. Then

$$
\begin{aligned}
0 &< \prod_{n \neq n_0} \mathbb{P}\left[|Y_n| \leq \left(\frac{\epsilon}{4}\right) a_n\right] \cdot \mathbb{P}\left[|b_{n_0} Y_{n_0} - \rho| \leq \left(\frac{\epsilon}{4}\right)\right] \\
&= \mathbb{P}\left\{\bigcap_{n \neq n_0} \left[|Y_n| \leq \left(\frac{\epsilon}{4}\right) a_n\right] \cap \left[|b_{n_0} Y_{n_0} - \rho| \leq \left(\frac{\epsilon}{4}\right)\right]\right\} \\
&\leq \mathbb{P}\left\{\left\|\sum_{n \neq n_0} b_n Y_n u_n + (b_{n_0} Y_{n_0} - \rho) u_{n_0}\right\| \leq \left(\frac{\epsilon}{2}\right)\right\} \\
&\leq \mathbb{P}[\|Z - u\| < \epsilon].
\end{aligned}
$$

Since obviously the same argument applies to σZ, the result is proved. □

For $k \in \mathbb{N}$, let $C_k = g^{-1}(B_k)$. Let $P_{\sigma,k}$ be the sub-Markov kernel on $(S \times E, \mathcal{S} \otimes \mathcal{E})$ defined by

$$P_{\sigma,k}((x,u), A) = P_\sigma((x,u), A \cap C_k).$$

Lemma 11.7 *Let φ be an irreducibility measure for P. Then:*

1. *P_σ is irreducible with irreducibility measure $\varphi \otimes \gamma_\sigma$.*
2. *Let*

$$\rho_k(A) = (\varphi \otimes \gamma_\sigma)(A \cap C_k), \qquad A \in \mathcal{S} \otimes \mathcal{E}.$$

Then $\rho_k \neq 0$ and $P_{\sigma,k}$ is irreducible with irreducibility measure ρ_k.

Proof.

1. It is easily verified from the definition of P_σ that for all $n \geq 1$, $(x,u) \in S \times E$, $A \in \mathcal{S} \otimes \mathcal{E}$,

$$P_\sigma^n((x,u), A) = (P^n(x,\cdot) \otimes \gamma_\sigma)(A). \tag{11.4}$$

Assume that $(\varphi \otimes \gamma_\sigma)(A) > 0$. For $y \in S$, let $A_y = \{u \in E \colon (y,u) \in A\}$, $h(y) = \gamma_\sigma(A_y)$. By Fubini's theorem,

$$\int_S h(y)\varphi(dy) = (\varphi \otimes \gamma_\sigma)(A) > 0.$$

By the irreducibility of P, given $x \in S$ there exists $n \in \mathbb{N}$ such that

$$\int h(y) P^n(x, dy) > 0. \tag{11.5}$$

By (11.4), (11.5), and Fubini's theorem, for any $u \in E$,

$$P_\sigma^n((x,u), A) = \int h(y) P^n(x, dy) > 0.$$

This shows that P_σ is irreducible with irreducibility measure $\varphi \otimes \gamma_\sigma$.

2. As in step 1, $\rho_k(S \times E) = (\varphi \otimes \gamma_\sigma)(C_k) = \int \gamma_\sigma((C_k)_y)\varphi(dy)$. But $(C_k)_y = \{u \in E\colon g(y,u) \in B_k\} = B_k - f(y)$. By Lemma 11.6, $\gamma_\sigma(B_k - f(y)) > 0$ for all $y \in S$ and therefore $\rho_k \neq 0$.

   Assume that $\rho_k(A) > 0$. By step 1, given $x \in S$ there exists $n \in \mathbb{N}$ such that for all $u \in E$,

$$P^n_\sigma((x,u), A \cap C_k) > 0. \tag{11.6}$$

   We have

$$\begin{aligned}
&P^n_{\sigma,k}((x,u),A)\\
&= \int P_{\sigma,k}((x,u), d(x_1,u_1)) \int \cdots\\
&\qquad \int P_{\sigma,k}((x_{n-1},u_{n-1}), d(x_n,u_n))\, \mathbf{1}_A(x_n,u_n)\\
&= \int P(x_0,dx_1)\gamma_\sigma(du_1) \int \cdots \int P(x_{n-1},dx_n)\gamma_\sigma(du_n)\\
&\qquad \times \prod_{j=1}^{n} \mathbf{1}_{C_k}(x_j,u_j)\mathbf{1}_A(x_n,u_n)\\
&= \int P(x_0,dx_1) \int \cdots \int P(x_{n-2},dx_{n-1})\\
&\qquad \times q(x_1,\ldots,x_{n-1}) \int P(x_{n-1},dx_n)\gamma_\sigma(du_n)\mathbf{1}_{A\cap C_k}(x_n,u_n),
\end{aligned} \tag{11.7}$$

where $q(x_1,\ldots,x_{n-1}) = \prod_{j=1}^{n-1} \gamma_\sigma(B_k - f(x_j))$, since

$$\mathbf{1}_{C_k}(x_j,u_j) = \mathbf{1}_{B_k - f(x_j)}(u_j).$$

On the other hand, as in (11.7),

$$P^n_\sigma((x,u), A\cap C_k) = \int P(x_0,dx_1) \int \cdots \int P(x_{n-1},dx_n)\gamma_\sigma(du_n)\mathbf{1}_{A\cap C_k}(x_n,u_n).$$

By (11.6), (11.7), and the fact that $q(x_1,\ldots,x_{n-1}) > 0$ for all $(x_1,\ldots,x_{n-1}) \in S^{n-1}$ by Lemma 11.6, we have

$$P^n_{\sigma,k}((x,u),A) > 0.$$

This shows that $P_{\sigma,k}$ is irreducible with irreducibility measure ρ_k. □

The next step toward the proof of Theorem 11.1 is to apply Lemma 11.5 to P_σ and g. To prepare the ground for this, we introduce the transform kernel

$$H_{\sigma,\xi}((x,u),A) = \int_A [\exp\langle g(y,w),\xi\rangle]\, P_\sigma((x,u), d(y,w)),$$

$(x,u) \in S \times E$, $A \in \mathcal{S} \otimes \mathcal{E}$, $\xi \in E^*$, and

$$\Lambda^\sigma(\xi) = -\log R(H_{\sigma,\xi});$$

note that since P_σ is irreducible, so is $H_{\sigma,\xi}$ and $R(H_{\sigma,\xi})$ exists.

We will need the following lemma relating Λ^σ to Λ_f.

Lemma 11.8 *For all $\sigma > 0$, $\xi \in E^*$,*

$$\Lambda^\sigma(\xi) = \Lambda_f(\xi) + \sigma^2 V(\xi),$$

where

$$V(\xi) = \frac{1}{2}\int \langle u, \xi\rangle^2 \gamma_1(du).$$

Proof. For $\xi \in E^*$, let

$$\gamma_{\sigma,\xi}(B) = \int_B (\exp\langle u, \xi\rangle)\, \gamma_\sigma(du), \qquad B \in \mathcal{E}.$$

Then $\hat{\gamma}_\sigma(\xi) \triangleq \gamma_{\sigma,\xi}(E) = \exp\{\sigma^2 V(\xi)\}$ by an elementary fact about Gaussian random variables.

It is easily verified that for all $n \in \mathbb{N}$, $(x,u) \in S \times E$, $A \in \mathcal{S} \otimes \mathcal{E}$,

$$H^n_{\sigma,\xi}((x,u), A) = (\hat{\gamma}_\sigma(\xi))^{n-1} \left(K^n_{f,\xi}(x,\cdot) \otimes \gamma_{\sigma,\xi}\right)(A). \qquad (11.8)$$

Let (s, ν) be a $K_{f,\xi}$-small pair: for some $m \in \mathbb{N}$,

$$K^m_{f,\xi} \geq s \otimes \nu.$$

Then it is easily verified using (11.8) that for all $(x,u) \in S \times E$, $A \in \mathcal{S} \otimes \mathcal{E}$,

$$H^m_{\sigma,\xi}((x,u), A) \geq (\hat{\gamma}_\sigma(\xi))^{m-1}\, s(x)(\nu \otimes \gamma_{\sigma,\xi})(A).$$

$R(H_{\sigma,\xi})$ is the radius of convergence of the power series with coefficients

$$(\nu \otimes \gamma_{\sigma,\xi}) H^n_{\sigma,\xi} s = (\nu K^n_{f,\xi} s)\, (\hat{\gamma}_\sigma(\xi))^n \qquad (11.9)$$

(see Appendix C). It follows from (11.9) that

$$R(H_{\sigma,\xi}) = R(K_{f,\xi})\, (\hat{\gamma}_\sigma(\xi))^{-1},$$

which implies the conclusion. □

We will also need to relate condition (2.1) for P to the same condition for P_σ.

Lemma 11.9

1. *Suppose D is P-petite. Then $D \times E$ is P_σ-petite.*
2. *Suppose $M \subset \mathcal{P}(S)$ satisfies (2.1) for P. Then $M_\sigma \triangleq \{\mu \otimes \gamma_\sigma : \mu \in M\}$ satisfies (2.1) for P_σ.*
3. *Suppose $\lim_k \sup_{\mu \in M} \mu(\{x \in S : \|f(x)\| > k\}) = 0$. Then*

$$\lim_k \sup_{\mu \in M} (\mu \otimes \gamma_\sigma)(\{(x, u) \in S \times E : \|g(x, u)\| > k\}) = 0.$$

Proof.

1. Suppose that for some $m \in \mathbb{N}$, $\nu \in \mathcal{P}(S)$, $\alpha > 0$,

$$\sum_{j=1}^{m} P^j \geq \alpha \mathbf{1}_D \otimes \nu.$$

Then for any $(x, u) \in S \times E$, $A \in \mathcal{S} \otimes \mathcal{E}$,

$$\begin{aligned} \sum_{j=1}^{m} P_\sigma^j ((x, u), A) &= \sum_{j=1}^{m} \left(P^j(x, \cdot) \otimes \gamma_\sigma\right)(A) \\ &\geq \alpha \mathbf{1}_D(x)(\nu \otimes \gamma_\sigma)(A) \\ &= \alpha \mathbf{1}_{D \times E}(x, u)(\nu \otimes \gamma_\sigma)(A). \end{aligned}$$

2. Suppose that for some $h \in \mathbb{N}$, D petite,

$$a = \inf \left\{ \sum_{i=0}^{h} \mu P^i(D) : \mu \in M \right\} > 0.$$

We have

$$\begin{aligned} (\mu \otimes \gamma_\sigma) P_\sigma^i (D \times E) &= \int (\mu \otimes \gamma_\sigma)\, d(x, u) P_\sigma^i ((x, u), D \times E) \\ &= \int (\mu \otimes \gamma_\sigma)\, d(x, u) \left(P^i(x, \cdot) \otimes \gamma_\sigma\right)(D \times E) \\ &= \int \mu(dx) P^i(x, D), \end{aligned}$$

and therefore

$$\inf \left\{ \sum_{i=0}^{h} \lambda P_\sigma^i (D \times E) : \lambda \in M_\sigma \right\} = a.$$

3. By Fubini's theorem, for $\mu \in M$,

$$\begin{aligned}(\mu \otimes \gamma_\sigma)(\{(x,u) \in S \times E \colon \|g(x,u)\| > k\}) \\ &= \int \gamma_\sigma(du)\mu(\{x \in S \colon \|f(x) + u\| > k\}) \\ &\leq \int \gamma_\sigma(du) h_k(u),\end{aligned}$$

where $h_k(u) = \sup_{\nu \in M} \nu(\{x \in S \colon \|f(x)\| > k - \|u\|\})$. By the assumption, $h_k(u) \to 0$ for all $u \in E$. Therefore by dominated convergence,

$$\sup_{\mu \in M} (\mu \otimes \gamma_\sigma)(\{(x,u) \in S \times E \colon \|g(x,u)\| > k\}) \leq \int \gamma_\sigma(du) h_k(u) \longrightarrow 0.$$

□

Proof of Theorem 11.1. Let $\lambda \in \mathcal{P}(S \times E)$ and let $(\Omega' = (S \times E)^{\mathbb{N}_0}, \mathbb{P}^\sigma_\lambda, \{X_j, U_j\}_{j \geq 0})$ be the canonical Markov chain with transition kernel P_σ and initial distribution λ; $\{X_j\}_{j\geq 0}$ (respectively, $\{U_j\}_{j \geq 0}$) are the coordinate functions on $S^{\mathbb{N}_0}$ (respectively, on $E^{\mathbb{N}_0}$). If $\lambda = \mu \otimes \gamma_\sigma$, $\mu \in \mathcal{P}(S)$, it is not difficult to verify that

$$\mathbb{P}^\sigma_\lambda = \mathbb{P}_\mu \otimes \gamma_\sigma^{\mathbb{N}}. \tag{11.10}$$

Let $G \subset E$ be an open set, and for $\epsilon > 0$ let $G_\epsilon = \{u \in E \colon d(u, G^c) > \epsilon\}$. Since $g(X_j, U_j) = f(X_j) + U_j$, the following set-theoretic relationship holds (in $(S \times E)^{\mathbb{N}_0}$):

$$\left[n^{-1} S_n(g) \in G_\epsilon\right] \subset \left[n^{-1} S_n(f) \in G\right] \cup \left[\left\| n^{-1} \sum_{j=0}^{n-1} U_j \right\| > \epsilon\right]. \tag{11.11}$$

Let $\{Z^\sigma_j\}_{j \geq 0}$ be independent E-valued random vectors with $\mathcal{L}(Z^\sigma_j) = \gamma_\sigma$, defined on a probability space $(\Omega'', \mathcal{A}, \mathbb{P})$. Then if $\lambda = \mu \otimes \gamma_\sigma$, we have by (11.10) and (11.11),

$$\mathbb{P}^\sigma_\lambda \left[n^{-1} S_n(g) \in G_\epsilon\right] \leq \mathbb{P}_\mu \left[n^{-1} S_n(f) \in G\right] + \mathbb{P}\left[\left\| n^{-1} \sum_{j=0}^{n-1} Z^\sigma_j \right\| > \epsilon\right]. \tag{11.12}$$

Let $u \in G$, and let ϵ be such that $u \in G_\epsilon$. Then by Lemmas 11.5, 11.7, 11.9, and (11.12), for $M \subset \mathcal{P}(S)$ as in the statement of the theorem,

$$\begin{aligned}-(\Lambda^\sigma)^*(u) &\leq \underline{\lim}_n\, n^{-1} \log \inf_{\lambda \in M_\sigma} \mathbb{P}^\sigma_\lambda \left[n^{-1} S_n(g) \in G_\epsilon\right] \\ &\leq \max\left\{ \underline{\lim}_n\, n^{-1} \log \inf_{\mu \in M} \mathbb{P}_\mu \left[n^{-1} S_n(f) \in G\right],\ \ell(\epsilon, \sigma)\right\},\end{aligned} \tag{11.13}$$

where

$$\ell(\epsilon, \sigma) = \overline{\lim_n}\, n^{-1} \log \mathbb{P}\left[\left\| n^{-1} \sum_{j=0}^{n-1} Z_j^{\sigma} \right\| > \epsilon\right].$$

By the large deviation theorem for Gaussian measures (see, e.g., Deuschel and Stroock, 1989, Chap. 3),

$$\ell(\epsilon, \sigma) \le -\frac{1}{\sigma^2} \inf \{V^*(u) \colon \|u\| \ge \epsilon\},$$

where $V(\xi)$ is as in Lemma 11.8, $\xi \in E^*$. Since V^* is tight, $\inf\{V^*(u) \colon \|u\| \ge \epsilon\} > 0$, and therefore

$$\lim_{\sigma \to 0} \ell(\epsilon, \sigma) = -\infty. \tag{11.14}$$

On the other hand, by Lemma 11.8,

$$\begin{aligned} \lim_{\sigma \to 0} (\Lambda^{\sigma})^*(u) &= \lim_{\sigma \to 0} \sup_{\xi} \left[\langle u, \xi\rangle - \left(\Lambda_f(\xi) + \sigma^2 V(\xi)\right)\right] \\ &= \sup_{\sigma} \sup_{\xi} \left[\langle u, \xi\rangle - \left(\Lambda_f(\xi) + \sigma^2 V(\xi)\right)\right] \\ &= \sup_{\xi} \sup_{\sigma} \left[\langle u, \xi\rangle - \left(\Lambda_f(\xi) + \sigma^2 V(\xi)\right)\right] \\ &= \sup_{\xi} \left[\langle u, \xi\rangle - \Lambda_f(\xi)\right] = \Lambda_f^*(u). \end{aligned} \tag{11.15}$$

Letting $\sigma \to 0$ in (11.13), by (11.14) and (11.15) Theorem 11.1 is proved. □

Remark 11.10 Let P, f be as in Theorem 11.1, and recall the notation $T_n(f) = \sum_{j=1}^{n} f(X_j)$. Then for every set $M \subset \mathcal{P}(S)$ satisfying (2.1) and every open set $G \subset E$,

$$\underline{\lim_n}\, n^{-1} \log \inf_{\mu \in M} \mathbb{P}_{\mu}\left[n^{-1} T_n(f) \in G\right] \ge -\inf_{u \in G} \Lambda_f^*(u). \tag{11.16}$$

In particular, the conclusion holds if $M \subset S$ and M is petite.

Note that the second condition on M in Theorem 11.1 is not needed to prove (11.16). The claim will follow from the following steps:

1. Proposition 11.2 holds also for $\{T_n(f)\}$ instead of $\{S_n(f)\}$. Let $u \in G$, and let $\epsilon > 0$ be such that $u \in G_{\epsilon}$. Since for $n \in \mathbb{N}$,

$$\left\|(n+1)^{-1} S_{n+1}(f) - n^{-1} T_n(f)\right\| \le (n+1)^{-1} 2\|f\|,$$

if $(n+1) > 2\|f\|\epsilon^{-1}$ we have, setting $S_n(f) = S_n$, $T_n(f) = T_n$,

$$\left[(n+1)^{-1} S_{n+1} \in G_{\epsilon}\right] \subset [n^{-1} T_n \in G],$$

and therefore if M is as in Proposition 11.2,

$$\underline{\lim}_n n^{-1} \log \inf_{\mu \in M} \mathbb{E}_\mu^{(n)} \mathbf{1}_G(n^{-1}T_n \in G)$$
$$\geq \underline{\lim}_n (n+1)^{-1} \inf_{\mu \in M} \mathbb{E}_\mu^{(n)} \mathbf{1}_{G_\epsilon}\left((n+1)^{-1} S_{n+1}\right) \geq -\Lambda_f^*(u).$$

2. Lemma 11.5 holds for $\{T_n(f)\}$ instead of $\{S_n(f)\}$ with $M \subset \mathcal{P}(S)$ satisfying (2.1). For, referring to the proof of Lemma 11.5, since $T_n(f)$ does not involve X_0, it is not necessary to truncate μ at C_k, and therefore (11.2) holds for $T_n(f)$ (respectively, $T_n(f_k)$; respectively, M) instead of $S_n(f)$ (respectively, $S_n(f_k)$; respectively, M_k) by invoking step 1.
3. The proof of (11.16) is completed as that of Theorem 11.1, on the basis of the version of Lemma 11.5 for $\{T_n(f)\}$, given by step 2.

11.2 Identification of the Rate Function Λ_f^*

Let $F \colon S \to \mathbb{R}$ be a measurable function such that $\inf F > 0$,

$$B_F(S) = \left\{h \colon S \to \mathbb{R} \colon h \text{ is measurable and } \sup\left\{\frac{|h(x)|}{F(x)} \colon x \in S\right\} < \infty\right\},$$
$$\mathcal{M}_F(S) = \left\{\mu \in \mathcal{M}(S) \colon \int F \, d|\mu| < \infty\right\},$$

where $\mathcal{M}(S)$ is the space of finite signed measures on $(S, \mathcal{S})$ and for $\mu \in \mathcal{M}(S)$, $|\mu|$ is its total variation measure; also, let

$$\mathcal{P}_F(S) = \mathcal{M}_F(S) \cap \mathcal{P}(S).$$

Clearly, if $\mu \in \mathcal{M}_F(S)$ and $h \in B_F(S)$, then h is $|\mu|$-integrable.

The following result is a slight extension of Theorem 4.2, (4.9).

Proposition 11.11 *Let P be irreducible. For $\mu \in \mathcal{M}_F(S)$, let*

$$\Lambda_{(F)}^*(\mu) = \sup\left\{\int h \, d\mu - \Lambda(h) \colon h \in B_F(S)\right\}.$$

Then:

$$\text{For } \mu \notin \mathcal{P}_F(S), \ \Lambda_{(F)}^*(\mu) = \infty. \tag{11.17}$$

$$\text{For } \mu \in \mathcal{P}_F(S), \ \Lambda_{(F)}^*(\mu) = I_\psi(\mu). \tag{11.18}$$

Proof.

1. Assume that $\mu \in \mathcal{M}_F(S)$ and $c = \Lambda^*_{(F)}(\mu) < \infty$. We will show: if $0 \leq g \in B(S)$, then

$$\int g \, d\mu \geq 0. \tag{11.19}$$

For, let $t < 0$. Since $B(S) \subset B_F(S)$,

$$\begin{aligned} t \int g \, d\mu = \int (tg) \, d\mu &\leq \Lambda(tg) + c \\ &\leq \Lambda(0) + c. \end{aligned}$$

If $\int g \, d\mu < 0$, letting $t \to -\infty$ we obtain a contradiction. Therefore (11.19) holds.

Similarly, for all $t \in \mathbb{R}$,

$$\begin{aligned} t\mu(S) = \int t \, d\mu &\leq \Lambda(t) + c \\ &= \Lambda(0) + t + c, \\ t(\mu(S) - 1) &\leq \Lambda(0) + c, \end{aligned}$$

which implies

$$\mu(S) = 1. \tag{11.20}$$

Now (11.17) follows from (11.19) and (11.20).

2. Arguing as in the proof of (4.14) in Theorem 4.2, if $\mu \in \mathcal{P}_F(S)$ and $\mu \not\ll \psi$, then $\Lambda^*_{(F)}(\mu) = \infty = I_\psi(\mu)$. Next, as in the proof of Theorem 4.2, by Theorem 4.1, we have

$$I(\mu) = \phi^*(\mu) \leq \Lambda^*(\mu) \leq \Lambda^*_{(F)}(\mu).$$

Therefore to finish the proof we need to show: if $\mu \in \mathcal{P}_F(S)$ and $\mu \ll \psi$, then $I(\mu) \geq \Lambda^*_{(F)}(\mu)$.

By (4.17), since for $h \in B_F(S)$ we have $\int |h| \, d\mu < \infty$, it follows that $I(\mu) \geq \int h \, d\mu - \Lambda(h)$. Therefore

$$I(\mu) \geq \sup \left\{ \int h \, d\mu - \Lambda(h) \colon h \in B_F(S) \right\} = \Lambda^*_{(F)}(\mu). \qquad \square$$

Next, we prove the following consequence of Theorem 2.13. Let $\mathcal{M}_F(S, \psi) = \{\mu \in \mathcal{M}_F(S) \colon |\mu| \ll \psi\}$ and $\mathcal{P}_F(S, \psi) = \mathcal{M}_F(S, \psi) \cap \mathcal{P}(S)$.

Proposition 11.12 *Let P be irreducible. The map $\Lambda \colon B_F(S) \to \overline{\mathbb{R}}$ is $\sigma(B_F(S), \mathcal{M}_F(S, \psi))$-lower semicontinuous.*

Proof. We will prove: if D is a directed set and $\{h_\delta\} \subset B_F(S)$ for $\delta \in D$, $h \in B_F(S)$, and for all $\lambda \in \mathcal{P}_F(S, \psi)$,

$$\lim_\delta \int h_\delta \, d\lambda = \int h \, d\lambda,$$

then

$$\varliminf_\delta \Lambda(h_\delta) \geq \Lambda(h).$$

Let $g_\delta = h_\delta / F$, $g = h/F$. Then $g_\delta, g \in B(S)$ and for $\mu \in \mathcal{P}(S, \psi)$,

$$\int g_\delta \, d\mu = \int h_\delta (F^{-1} d\mu) \longrightarrow \int h (F^{-1} d\mu) = \int g \, d\mu,$$

since

$$\left(\int F^{-1} \, d\mu \right)^{-1} F^{-1} d\mu \in \mathcal{P}_F(S, \psi).$$

Therefore by Theorem 2.13,

$$\varliminf_\delta \Lambda(h_\delta) = \varliminf_\delta \Lambda(g_\delta F) \geq \Lambda(gF) = \Lambda(h).$$

We have shown: $\Lambda \colon B_F(S) \to \overline{\mathbb{R}}$ is $\sigma(B_F(S), \mathcal{P}_F(S, \psi))$-lower semicontinuous. But clearly this is equivalent to the stated conclusion. □

Λ_f^* is identified in terms of I_ψ in statements 2 and 3 of the following result.

Theorem 11.13 *Let P be irreducible and let $f \colon S \to E$ be a measurable function, where E is a separable Banach space.*

1. *$\Lambda_f \colon E^* \to \overline{\mathbb{R}}$ is $\sigma(E^*, E)$-lower semicontinuous.*
2. *For $u \in E$, let*

$$\widetilde{\Lambda_f}(u) = \begin{cases} \inf \left\{ I_\psi(\mu) \colon \mu \in \Phi_f^{-1}(u), \int \|f\| \, d\mu < \infty \right\} & \text{if } u \in \Phi_f(\mathcal{P}_F(S)) \\ \infty & \text{otherwise,} \end{cases}$$

   where $F = \|f\| + 1$ and for $\mu \in \mathcal{P}_F(S)$, $\Phi_f(\mu) = \int f \, d\mu$. Then $\Lambda_f^* = \widetilde{\Lambda_f}^{**}$.
3. *If for each $a \geq 0$, $\{\mu \in \mathcal{P}_F(S) \colon I_\psi(\mu) \leq a\}$ is $\sigma(\mathcal{P}_F(S), B_F(S))$-compact, then $\Lambda_f^* = \widetilde{\Lambda_f}$.*

The proof is essentially the same as that of Theorem 2.17.

Proof.

1. By Proposition 11.11, for $\xi \in E^*$,

$$\begin{aligned}\widetilde{\Lambda_f}^*(\xi) &= \sup\left\{\langle u,\xi\rangle - \widetilde{\Lambda}_f(u)\colon u\in E\right\}\\ &= \sup\Big\{\langle u,\xi\rangle - \inf\left\{\Lambda^*_{(F)}(\mu)\colon \mu\in\Phi_f^{-1}(u),\right.\\ &\qquad\qquad \left.\mu\in\mathcal{P}_F(S)\right\}\colon u\in\Phi_f\left(\mathcal{P}_F(S)\right)\Big\}\\ &= \sup\left\{\sup\left\{\left\langle\int f\,d\mu,\xi\right\rangle - \Lambda^*_{(F)}(\mu)\colon \mu\in\Phi_f^{-1}(u),\right.\right.\\ &\qquad\qquad \left.\left.\mu\in\mathcal{P}_F(S)\right\}\colon u\in\Phi_f\left(\mathcal{P}_F(S)\right)\right\} \qquad (11.21)\\ &= \sup\left\{\int\langle f,\xi\rangle\,d\mu - \Lambda^*_{(F)}(\mu)\colon \mu\in\mathcal{P}_F(S)\right\}\\ &= \sup\left\{\int\langle f,\xi\rangle\,d\mu - \Lambda^*_{(F)}(\mu)\colon \mu\in\mathcal{M}_F(S)\right\}\\ &= \Lambda\left(\langle f,\xi\rangle\right) = \Lambda_f(\xi);\end{aligned}$$

in the last step we have used the fact that $\langle f,\xi\rangle\in B_F(S)$, Proposition 11.12, and Proposition F.1, applied to $X = B_F(S)$, $Y = \mathcal{M}_F(S)$, and $\Lambda\colon B_F(S)\to\overline{\mathbb{R}}$. Since $\widetilde{\Lambda_f}^*$ is $\sigma(E^*,E)$-lower semicontinuous, this proves part 1.

2. It follows that $\Lambda_f^* = \widetilde{\Lambda_f}^{**}$.
3. To prove that $\Lambda_f^* = \widetilde{\Lambda}_f$, we must show that $\widetilde{\Lambda_f}^{**} = \widetilde{\Lambda}_f$. To prove this, by Proposition F.1 we must show that $\widetilde{\Lambda}_f\colon E\to\mathbb{R}$ is convex, proper, and $\sigma(E,E^*)$-lower semicontinuous. From its definition, it is easily seen that $\widetilde{\Lambda}_f$ is convex, and by (11.21) it is proper. Let $a\geq 0$ and

$$\begin{aligned}L_a &= \left\{\mu\in\mathcal{P}_F(S)\colon I_\psi(\mu)\leq a\right\},\\ M_a &= \left\{u\in E\colon \widetilde{\Lambda}_f(u)\leq a\right\}.\end{aligned}$$

We will prove that the map Φ_f is $\sigma(\mathcal{P}_F(S),B_F(S))/\sigma(E,E^*)$-continuous and $\Phi_f(L_a) = M_a$. Since L_a is $\sigma(\mathcal{P}_F(S),B_F(S))$-compact, it will follow that M_a is $\sigma(E,E^*)$-compact and therefore $\widetilde{\Lambda}_f$ is $\sigma(E,E^*)$-lower semicontinuous.

To prove the first assertion, assume that $\{\mu_\alpha\}_{\alpha\in D}$ is a net in $\mathcal{P}_F(S)$, $\mu\in\mathcal{P}_F(S)$, and $\int h\,d\mu_\alpha\to\int h\,d\mu$ for all $h\in B_F(S)$. Then for all $\xi\in E^*$, $\langle f,\xi\rangle\in B_F(S)$ and therefore

$$\left\langle\Phi_f(\mu_\alpha),\xi\right\rangle = \int\langle f,\xi\rangle\,d\mu_\alpha \longrightarrow \int\langle f,\xi\rangle\,d\mu = \left\langle\Phi_f(\mu),\xi\right\rangle,$$

proving the continuity of Φ_f. The proof is finished as in Theorem 2.17.3. □

11.3 Upper Bounds

The main result of this section is Theorem 11.16. In order to handle the issues involving compactness in that theorem – exponential tightness and the tightness of the rate function – we start by proving Lemma 11.14. This lemma is then used together with an integrability result, Theorem 11.15, to obtain the upper bound in Theorem 11.16.

We need several definitions. First, for a convex and symmetric set $A \subset E$, its Minkowski functional q_A is defined by

$$q_A(u) = \inf\{\lambda > 0 \colon u \in \lambda A\}, \qquad u \in E$$

(with the customary convention: $\inf \emptyset = +\infty$). Then $q_A(u) < \infty$ if and only if $u \in \bigcup_n (nA)$, and q_A is subadditive and positively homogeneous. Also, $q_{cA} = c^{-1} q_A$ for $c > 0$.

Next, let $g \in F(S)$ and assume

$$\sup_{x \in S} K_g(x, S) < \infty.$$

Then K_g acts as a bounded linear operator on $B(S)$, endowed with the supremum norm: if $h \in B(S)$, then

$$\begin{aligned} \sup_{x \in S} \left| K_g h(x) \right| &= \sup_{x \in S} \left| \int e^{g(y)} h(y) P(x, dy) \right| \\ &\le \|h\| \sup_{x \in S} K_g 1(x). \end{aligned}$$

We have: for $n \in \mathbb{N}$,

$$\left\| K_g^n \right\| = \left\| K_g^n 1 \right\| = \sup_{x \in S} K_g^n 1(x).$$

Let

$$\begin{aligned} \rho(g) &= \overline{\lim_n}\, n^{-1} \log \sup_{x \in S} K_g^n 1(x) \\ &= \overline{\lim_n}\, n^{-1} \log \left\| K_g^n \right\|; \end{aligned}$$

note that for $g \in B(S)$, $\rho(g) = \phi(g)$. Then by the spectral radius formula (see, e.g., Dunford and Schwartz, 1958), $\rho(g) = \log r(K_g)$, where $r(K_g)$ is the spectral radius of K_g.

Let $f \colon S \to E$ be measurable and assume for every $\xi \in E^*$,

$$d(\xi) = \sup_{x \in S} K_{f,\xi}(x, S) < \infty.$$

Let $\rho_f(\xi) = \rho(\langle f, \xi\rangle)$; then, as above, $\rho_f(\xi)$ is the logarithm of the spectral radius of $K_{f,\xi}$. For $u \in E$, we define

$$\rho_f^*(u) = \sup\left\{\langle u, \xi\rangle - \rho_f(\xi) \colon \xi \in E^*\right\}.$$

Note that since $\rho_f(0) = 0$, we have $\rho_f^*(u) \geq 0$.

Lemma 11.14 *Let $f \colon S \to E$ be measurable. Assume: that there exist $m \in \mathbb{N}$ and a compact, convex, symmetric set $K \subset E$ such that*

$$\kappa = \sup_{x\in S} \int \exp\left(q_K\left(f(y)\right)\right) P^m(x, dy) < \infty.$$

Then:

1. *For every $a > 0$, there exists $b > 0$ such that*

$$\overline{\lim_n}\, n^{-1} \log \sup_{x\in S} \mathbb{P}_x\left[n^{-1}\sum_{j=m}^{n-1} f(X_j) \in (bK)^c\right] \leq -a.$$

2. *If, moreover, $d(\xi) < \infty$ for every $\xi \in E^*$, then for every $a \geq 0$,*

$$L_a = \left\{u \in E \colon \rho_f^*(u) \leq a\right\}$$

is compact.

Proof. Let $n \in \mathbb{N}$, $p = [n/m] + 1$. Let $H = (2m)K$. Setting $h = 2q_H \circ f$, we have by the convexity of the exponential function

$$\begin{aligned}\exp\left(\sum_{j=m}^{n} h(X_j)\right) &\leq \exp\left(\sum_{j=m}^{pm} h(X_j)\right)\\ &\leq \exp\left(\sum_{i=0}^{m-1}\sum_{j=1}^{p} h(X_{i+mj})\right)\\ &\leq m^{-1}\sum_{i=0}^{m-1}\exp\left(\sum_{j=1}^{p} mh(X_{i+mj})\right).\end{aligned} \tag{11.22}$$

Setting $g = mh$, by the Markov property, for $x \in S$,

$$\begin{aligned}\mathbb{E}_x\left[\exp\left(\sum_{j=1}^{p} g(X_{mj})\right)\right] &= \mathbb{E}_x\left[\exp\left(\sum_{j=1}^{p-1} g(X_{mj})\right)\right]\mathbb{E}_{X_{m(p-1)}}\left[\exp g(X_m)\right]\\ &\leq \mathbb{E}_x\left[\exp\left(\sum_{j=1}^{p-1} g(X_{mj})\right)\right]\sup_{y\in S}\mathbb{E}_y\left[\exp g(X_m)\right].\end{aligned}$$

Iterating, we obtain

$$\mathbb{E}_x\left[\exp\left(\sum_{j=1}^{p} g(X_{mj})\right)\right] \leq \mathbb{E}_x\left[\exp g(X_m)\right]\left\{\sup_{y\in S}\mathbb{E}_y\left[\exp g(X_m)\right]\right\}^{p-1}$$
$$\leq \left\{\sup_{y\in S}\mathbb{E}_y\left[\exp g(X_m)\right]\right\}^{p}. \tag{11.23}$$

Similarly, for $i = 1, \ldots, m-1$,

$$\mathbb{E}_x\left[\exp\left(\sum_{j=1}^{p} g(X_{i+mj})\right)\right] \leq \mathbb{E}_x\left[\exp g(X_{i+m})\right]\left\{\sup_{y\in S}\mathbb{E}_y\left[\exp g(X_m)\right]\right\}^{p-1}. \tag{11.24}$$

But

$$\begin{aligned}\mathbb{E}_x\left[\exp g(X_{i+m})\right] &= \int P^{i+m}(x,dz)\,(\exp g(z)) \\ &= \int P^{i}(x,dy)\int P^{m}(y,dz)\,(\exp g(z)) \\ &\leq \sup_{y\in S}\int P^{m}(y,dz)\,(\exp g(z)).\end{aligned} \tag{11.25}$$

Therefore from (11.22)–(11.25) we have: for all $x \in S$, $n \geq m$,

$$\begin{aligned}\mathbb{E}_x\left[\exp\left(\sum_{j=m}^{n} 2q_H\left(f(X_j)\right)\right)\right] &\leq \left\{\sup_{y\in S}\int P^{m}(y,dz)\exp\left(2mq_H\left(f(z)\right)\right)\right\}^{p} \\ &\leq \kappa^{(n/m)+1} \leq \kappa^{2n/m},\end{aligned} \tag{11.26}$$

since $(2m)q_H = q_K$.

Next, for $b > 0$, by (11.26),

$$\begin{aligned}\mathbb{P}_x\left[n^{-1}\sum_{j=m}^{n-1} f(X_j) \in (bK)^c\right] &= \mathbb{P}_x\left[q_K\left(\sum_{j=m}^{n-1} f(X_j)\right) > nb\right] \\ &\leq \exp\left[-n(2m)^{-1}b\right]\mathbb{E}_x\left[\exp(2m)^{-1}q_K\left(\sum_{j=m}^{n-1} f(X_j)\right)\right] \\ &\leq \exp\left\{-n\left[(2m)^{-1}b - 2(m^{-1})\log\kappa\right]\right\},\end{aligned}$$

and part 1 follows by choosing $b > 2ma + 4\log\kappa$.

To prove part 2, let

$$\begin{aligned}H^0 &= \{\xi \in E^* : \langle u,\xi\rangle \leq 1 \text{ for all } u \in H\}, \\ H^{00} &= \left\{u \in E : \langle u,\xi\rangle \leq 1 \text{ for all } \xi \in H^0\right\}.\end{aligned}$$

By the properties of H and the bipolar theorem (see, e.g., Schaefer, 1966, p. 126), we have $H^{00} = H$.

For $\xi \in H^0$, $u \in E$, we have $\langle u, \xi \rangle \le q_H(u)$, so

$$\left\langle \sum_{j=m}^{n} f(X_j), \xi \right\rangle \le q_H\left(\sum_{j=m}^{n} f(X_j) \right),$$

and for $x \in S$, with T_n as in Remark 11.10,

$$\begin{aligned} K_{f,\xi}^n 1(x) &= \mathbb{E}_x \left[\exp T_n \left(\langle f, \xi \rangle \right) \right] \\ &\le \mathbb{E}_x \left[\exp T_{m-1} \left(\langle f, \xi \rangle \right) \right] \left(\exp \sum_{j=m}^{n} q_H \left(f(X_j) \right) \right) \\ &\le \left[\mathbb{E}_x \left(\exp 2T_{m-1} \left(\langle f, \xi \rangle \right) \right) \right]^{1/2} \left[\mathbb{E}_x \left(\exp 2 \sum_{j=m}^{n} q_H \left(f(X_j) \right) \right) \right]^{1/2} \\ &\le d(2\xi)^{(m-1)/2} (\kappa^{2n/m})^{1/2}. \end{aligned}$$

Therefore for $\xi \in H^0$,

$$\rho_f(\xi) = \overline{\lim_n}\, n^{-1} \log \sup_{x \in S} K_{f,\xi}^n 1(x) \le \gamma,$$

where $\gamma = \kappa^{1/m}$. It follows that for $u \in L_a$, $\xi \in H^0$,

$$\langle u, \xi \rangle \le \rho_f(\xi) + \rho_f^*(u) \le \gamma + a.$$

Thus for $u \in L_a$, $(\gamma + a)^{-1} u \in H^{00}$, and therefore

$$L_a \subset (\gamma + a) H^{00} = (\gamma + a) H.$$

Since $(\gamma + a)H$ is compact, this proves part 2. □

Theorem 11.15 *Let E be a separable Banach space and let $\{\mu_\alpha : \alpha \in A\}$ be a family of probability measures on E such that:*

1. *$\{\mu_\alpha : \alpha \in A\}$ is tight.*
2. *For every $t > 0$,*

$$c_t = \sup_{\alpha \in A} \int \exp(t \|\cdot\|) \, d\mu_\alpha < \infty.$$

Then there exists a compact, convex, symmetric set $K \subset E$ such that

$$\sup_{\alpha \in A} \int \exp(q_K) \, d\mu_\alpha < \infty.$$

Proof. We prove first: if $\tau_\alpha(t) = \mu_\alpha(\{u\colon \|u\| > t\})$, then

$$\lim_{t\to\infty} \sup_{\alpha\in A} (\tau_\alpha(t))^{1/t} = 0. \tag{11.27}$$

For, given $\epsilon > 0$, choose $a > 0$ so that $e^{-a} < \epsilon$. Then for all $t > 0$, $\alpha \in A$,

$$\tau_\alpha(t) \le e^{-at} \int \exp(a\|\cdot\|)\, d\mu_\alpha \le e^{-at} c_a,$$

and it follows that

$$\sup_{\alpha\in A} (\tau_\alpha(t))^{1/t} \le e^{-a} c_a^{1/t},$$

$$\overline{\lim_{t\to\infty}} \sup_{\alpha\in A} (\tau_\alpha(t))^{1/t} \le e^{-a} < \epsilon,$$

proving (11.27).

We will prove: for every $\beta \in (0, 1)$, there exist $c > 0$ and a compact, convex, symmetric set $K \subset E$ such that

$$\mu_\alpha(\{x\colon q_K(x) > t\}) \le c\beta^t \qquad \text{for } \alpha \in A,\ t \ge 1. \tag{11.28}$$

This statement implies the conclusion.

Let $t_{\alpha,n} = \inf\{t > 0\colon \tau_\alpha(t) < \beta^n\}$. Then

$$\tau_\alpha(t_{\alpha,n}) \le \beta^n \tag{11.29}$$

and

$$\tau_\alpha\left(\frac{t_{\alpha,n}}{2}\right) \ge \beta^n. \tag{11.30}$$

By assumption 1, there exists a compact set K_n such that for all $\alpha \in A$, $\mu_\alpha(K_n^c) < \beta^n$; by Mazur's theorem (Conway, 1985, p. 180), we may assume that K_n is convex and symmetric. We may also assume $K_n \subset K_{n+1}$ for $n \in \mathbb{N}$. Now let

$$A_n = \bigcup_{\alpha\in A} n^{-1}(K_n \cap B(t_{\alpha,n})),$$

where $B(r) = \{u \in E\colon \|u\| \le r\}$ for $r \ge 0$,

$$K = \text{ closed convex symmetric hull of } \bigcup_n A_n.$$

We claim that K is compact. To prove this, we first show that

$$d_n = n^{-1} u_n \longrightarrow 0 \qquad \text{as } n \longrightarrow \infty, \tag{11.31}$$

where $u_n = \sup_{\alpha\in A} t_{\alpha,n}$.

Given $\epsilon > 0$, let $n_0 \in \mathbb{N}$ be such that $n_0^{-1/2} < \epsilon/2$ and for all $\alpha \in A$, all $t > n_0^{1/2}$,

$$(\log \beta) \Big/ \left(t^{-1} \log \tau_\alpha(t)\right) < \epsilon/2; \tag{11.32}$$

this is possible by (11.27). We claim now that for $n > n_0$, we have $d_n \le \epsilon$. For, if $t_{\alpha,n}/2 > n^{1/2}$, then by (11.30) and (11.32),

$$\frac{t_{\alpha,n}}{2n} \le (\log \beta) \Big/ \left[\left(\frac{2}{t_{\alpha,n}}\right) \log \tau_\alpha \left(\frac{t_{\alpha,n}}{2}\right)\right] < \epsilon/2,$$
$$\frac{t_{\alpha,n}}{n} < \epsilon.$$

On the other hand, if $t_{\alpha,n}/2 \le n^{1/2}$, then

$$\frac{t_{\alpha,n}}{n} \le \frac{2}{n^{1/2}} < \frac{2}{n_0^{1/2}} < \epsilon.$$

Thus for $n > n_0$ we have $d_n \le \epsilon$, proving (11.31). It follows that $(\bigcup_n A_n)$ is totally bounded. For, let n_0 be as above for a given $\epsilon > 0$. Then

$$\bigcup_n A_n = \left(\bigcup_{n \le n_0} A_n\right) \cup \left(\bigcup_{n > n_0} A_n\right)$$
$$\subset K_{n_0} \cup B(\epsilon),$$

proving the total boundedness of $(\bigcup_n A_n)$. By Mazur's theorem (Conway, 1985, p. 180), it follows that K is compact.

We will show now that (11.28) holds. Given $t \ge 1$, let $n = [t]$. Then for all $\alpha \in A$, by (11.29),

$$\begin{aligned}
\mu_\alpha\left(\{x \colon q_K(x) > t\}\right) &= \mu_\alpha\left((tK)^c\right) \\
&\le \mu_\alpha\left((nK)^c\right) \\
&\le \mu_\alpha\left((K_n \cap B(t_{\alpha,n}))^c\right) \\
&\le \mu_\alpha(K_n^c) + \tau_\alpha(t_{\alpha,n}) \le \beta^n + \beta^n = 2\beta^n,
\end{aligned}$$

and since $\beta^{n+1} < \beta^t$, we have

$$\mu_\alpha\left(\{x \colon q_K(x) > t\}\right) \le (2\beta^{-1})\beta^t. \qquad \square$$

To state Theorem 11.16, we extend some definitions in Chapters 1 and 3. For $g \in F(S)$, $M \subset \mathcal{P}(S)$, we define

$$\phi_M(g) = \overline{\lim_n}\, n^{-1} \log \sup_{\mu \in M} \mathbb{E}_\mu\left(\exp S_n(g)\right),$$

and for $f: S \to E$ measurable, $\xi \in E^*$,

$$\begin{aligned}\phi_{f,M}(\xi) &= \phi_M(\langle f, \xi\rangle) \\ &= \overline{\lim_n}\, n^{-1} \log \sup_{\mu \in M} \mathbb{E}_\mu \left(\exp \langle S_n(f), \xi\rangle\right),\end{aligned}$$

and

$$\phi^*_{f,M}(u) = \sup\left\{\langle u, \xi\rangle - \phi_{f,M}(\xi) \colon \xi \in E^*\right\}, \qquad u \in E.$$

Since $\phi_{f,M}(0) = 0$, we have $\phi^*_{f,M}(u) \geq 0$.

Theorem 11.16 *Let E be a separable Banach space, and let $f: S \to E$ be measurable. Let $M \subset \mathcal{P}(S)$. Then:*

1. *For every $\sigma(E, E^*)$-compact set $F \subset E$,*

$$\overline{\lim_n}\, n^{-1} \log \sup_{\mu \in M} \mathbb{P}_\mu \left[n^{-1} S_n(f) \in F\right] \leq -\inf\left\{\phi^*_{f,M}(u) \colon u \in F\right\}. \tag{11.33}$$

2. *Assume:*

   (i) *For some $m \in \mathbb{N}$, $\{P^m(x, \cdot) \circ f^{-1} : x \in S\}$ is tight.*
   (ii) *For all $r > 0$,*

$$c_r = \sup_{\mu \in M} \int \exp(r\|f(y)\|)\, \mu(dy) < \infty. \tag{11.34}$$

$$d_r = \sup_{x \in S} \int \exp(r\|f(y)\|)\, P(x, dy) < \infty. \tag{11.35}$$

   Then

   (a) (11.33) *holds for every closed set $F \subset E$;*
   (b) $\phi^*_{f,M} \geq \rho^*_f$ *and* $\phi^*_{f,M}$ *is tight.*

Proof.

1. The argument is essentially the same as in Theorem 3.1.1. Let $a = \inf\{\phi^*_{f,M}(u) \colon u \in F\}$, and assume that $a < \infty$. Let $\epsilon > 0$ and for $\xi \in E^*$ such that $\phi_{f,M}(\xi) < \infty$, let

$$H(\xi) = \left\{u \in E \colon \langle u, \xi\rangle - \phi_{f,M}(\xi) > a - \epsilon\right\}.$$

   Then

$$F \subset \left\{u \in E \colon \phi^*_{f,M}(u) > a - \epsilon\right\} = \bigcup \left\{H(\xi) \colon \xi \in E^*, \phi_{f,M}(\xi) < \infty\right\}.$$

   Since F is $\sigma(E, E^*)$-compact, and $H(\xi)$ is $\sigma(E, E^*)$-open, there exist $\xi_1, \ldots,$

$\xi_k \in E^*$ with $\phi_{f,M}(\xi_i) < \infty$ such that $F \subset \bigcup_{i=1}^k H(\xi_i)$. Therefore for $\mu \in M$, setting $S_n(f) = S_n$,

$$\begin{aligned}
\mathbb{P}_\mu[n^{-1}S_n \in F] &\le \sum_{i=1}^{k} \mathbb{P}_\mu\left[S_n \in nH(\xi_i)\right] \\
&= \sum_{i=1}^{k} \mathbb{P}_\mu\left[\langle S_n, \xi_i\rangle > n\left(\phi_{f,M}(\xi_i) + a - \epsilon\right)\right] \\
&\le \sum_{i=1}^{k} \exp\left[-n\left(\phi_{f,M}(\xi_i) + a - \epsilon\right)\right] \mathbb{E}_\mu\left(\exp\langle S_n, \xi_i\rangle\right).
\end{aligned}$$

Let

$$h_n(\xi_i) = n^{-1} \log \sup_{\mu\in M} \mathbb{E}_\mu\left(\exp\langle S_n, \xi_i\rangle\right) - \phi_{f,M}(\xi_i).$$

Then

$$\sup_{\mu\in M} \mathbb{P}_\mu[n^{-1}S_n \in F] \le k e^{(-a+\epsilon)n} \exp\left(n \sup_i h_n(\xi_i)\right),$$

and therefore

$$\overline{\lim_n}\, n^{-1} \log \sup_{\mu\in M} \mathbb{P}_\mu[n^{-1}S_n \in F] \le -a + \epsilon.$$

But ϵ is arbitrary. This proves the statement for $a < \infty$. If $a = \infty$, one proceeds in a similar way.

2. (i) For $\delta > 0$, let $F^{[\delta} = \{u \in E \colon d(u, F) \le \delta\}$. Then for any $\mu \in M$, $n > m$,

$$\begin{aligned}
\mathbb{P}_\mu\left[n^{-1}S_n(f) \in F\right] &\le \mathbb{P}_\mu\left[\left\| n^{-1} \sum_{j=0}^{m-1} f(X_j)\right\| > \delta\right] \\
&\quad + \mathbb{P}_\mu\left[n^{-1} \sum_{j=m}^{n-1} f(X_j) \in F^{[\delta}\right]. \qquad (11.36)
\end{aligned}$$

We proceed to estimate the two terms on the right-hand side of (11.36). By the Markov property, for $r > 0$,

$$\begin{aligned}
\mathbb{E}_\mu\left(\exp\left\|\sum_{j=0}^{m-1} rf(X_j)\right\|\right) &\le \mathbb{E}_\mu\left(\exp\left\|\sum_{j=0}^{m-2} rf(X_j)\right\|\right) \\
&\quad \times \mathbb{E}_{X_{m-2}}\left(\exp\|rf(X_1)\|\right) \\
&\le \mathbb{E}_\mu\left(\exp\left\|\sum_{j=0}^{m-2} rf(X_j)\right\|\right) d_r,
\end{aligned}$$

and, iterating,

$$\le \mathbb{E}_\mu (\exp r \|f(X_0)\|)\, d_r^{m-1}.$$

Therefore

$$\sup_{\mu\in M} \mathbb{E}_\mu \left(\exp \left\| \sum_{j=0}^{m-1} r f(X_j) \right\| \right) \le c_r d_r^{m-1}.$$

It follows that

$$\sup_{\mu\in M} \mathbb{P}_\mu \left[\left\| n^{-1} \sum_{j=0}^{m-1} f(X_j) \right\| > \delta \right] \le e^{-nr\delta} c_r d_r^{m-1},$$

$$\overline{\lim_n}\, n^{-1} \log \sup_{\mu\in M} \mathbb{P}_\mu \left[\left\| n^{-1} \sum_{j=0}^{m-1} f(X_j) \right\| > \delta \right] \le -r\delta.$$

But r is arbitrary. Therefore for all $\delta > 0$,

$$\overline{\lim_n}\, n^{-1} \log \sup_{\mu\in M} \mathbb{P}_\mu \left[\left\| n^{-1} \sum_{j=0}^{m-1} f(X_j) \right\| > \delta \right] = -\infty. \tag{11.37}$$

Next, by Theorem 11.15, assumption 2(i) and (11.35) imply that there exists a compact, convex, symmetric set $K \subset E$ such that the assumption of Lemma 11.14(1) is satisfied; note that (11.35) easily implies that

$$\sup_{x\in S} \int \exp(r\|f(y)\|)\, P^m(x, dy) \le d_r < \infty.$$

Given $a > 0$, let $b > 0$ be as in Lemma 11.14 and set $L - bK$. Since

$$\mathbb{P}_\mu \left[n^{-1} \sum_{j=m}^{n-1} f(X_j) \in F^{[\delta} \right] \le \mathbb{P}_\mu \left[n^{-1} \sum_{j=m}^{n-1} f(X_j) \in F^{[\delta} \cap L \right] + \mathbb{P}_\mu \left[n^{-1} \sum_{j=m}^{n-1} f(X_j) \in L^c \right],$$

by (11.36) we have

$$\mathbb{P}_\mu \left[n^{-1} S_n(f) \in F \right] \le 3 \max \left\{ \mathbb{P}_\mu \left[\left\| n^{-1} \sum_{j=0}^{m-1} f(X_j) \right\| > \delta \right], \mathbb{P}_\mu \left[n^{-1} \sum_{j=m}^{n-1} f(X_j) \in F^{[\delta} \cap L \right], \mathbb{P}_\mu \left[n^{-1} \sum_{j=m}^{n-1} f(X_j) \in L^c \right] \right\},$$

and by (11.33), Lemma 11.14, and (11.37),

$$\begin{aligned}\overline{\lim_n}\, n^{-1} \log \sup_{\mu\in M} \mathbb{P}_\mu \left[n^{-1}S_n(f) \in F\right] &\le \max\left\{-\inf\left\{\phi^*_{f,M}(u) \colon u \in F^{[\delta} \cap L\right\}, -a\right\}\\ &\le \max\left\{-\inf\left\{\phi^*_{f,M}(u) \colon u \in F^{[\delta]}\right\}, -a\right\}.\end{aligned}$$

But a is arbitrary. Therefore

$$\overline{\lim_n}\, n^{-1} \log \sup_{\mu\in M} \mathbb{P}_\mu \left[n^{-1}S_n(f) \in F\right] \le -\inf\left\{\phi^*_{f,M}(u) \colon u \in F^{[\delta]}\right\}. \quad (11.38)$$

Assuming (b), we complete the proof of (a) as follows. Since

$$\bigcap_{\delta>0} F^{[\delta} = F,$$

by the tightness of $\phi^*_{f,M}$ and Lemma 4.16 of Dembo and Zeitouni (1998) we have

$$\inf\left\{\phi^*_{f,M}(u) \colon u \in F^{[\delta]}\right\} \uparrow \inf\left\{\phi^*_{f,M}(u) \colon u \in F\right\},$$

and therefore (a) follows from (11.38).

(i) We will show that $\phi_{f,M} \le \rho_f$. This implies the first assertion in (b) and the second one then follows from Lemma 11.14(2). Let $\xi \in E^*$. Then (11.34) implies that

$$\sup_{\mu\in M} \int (\exp\langle f(y), \xi\rangle)\, \mu(dy) < \infty,$$

and (11.35) implies that $d(\xi) < \infty$. Let $g = \langle f, \xi\rangle$. Then, using the notation in Remark 11.10, for $x \in S$,

$$\begin{aligned}\mathbb{E}_x (\exp S_n(g)) &= e^{g(x)} \mathbb{E}_x (\exp T_{n-1}(g))\\ &= e^{g(x)} K_g^{n-1} 1(x),\end{aligned}$$

and therefore

$$\sup_{\mu\in M} \mathbb{E}_\mu (\exp S_n(g)) \le \left(\sup_{x\in S} K_g^{n-1} 1(x)\right) \sup_{\mu\in M} \int e^{g(y)} \mu(dy),$$

and $\phi_{f,M}(\xi) = \phi_M(g) \le \rho(g) = \rho_f(\xi)$. □

Remark 11.17

1. If f is bounded and 2(i) is satisfied, then (a) of Theorem 11.16 holds with $M = S$ and rate function $\phi^*_{f,S} = \rho^*_f$.
2. If 2(i) and (11.35) are satisfied, then for all $x \in S$, (a) of Theorem 11.16 holds for $M = \{\delta_x\}$ with rate function $\phi^*_{f,x} \ge \rho^*_f$.

3. Let T_n be as in Remark 11.10. If 2(i) and (11.35) are satisfied, then the upper bound for $\{\sup_{x\in S} \mathbb{P}_x[n^{-1}T_n(f) \in \cdot]\}$ holds with tight rate function ρ_f^*. For, if $x \in S$ and $B \in \mathcal{E}$, then

$$\mathbb{P}_x\left[n^{-1}T_n(f) \in B\right] = \int P(x, dy)\mathbb{P}_y\left[n^{-1}S_n(f) \in B\right],$$

and therefore

$$\sup_{x\in S} \mathbb{P}_x\left[n^{-1}T_n(f) \in \cdot\right] = \sup_{\mu\in M} \mathbb{P}_\mu\left[n^{-1}S_n(f) \in \cdot\right],$$

where $M = \{P(x, \cdot)\colon x \in S\}$. But in this case condition (11.34) is just condition (11.35). Therefore by Theorem 11.16 the stated upper bound holds with tight rate function $\phi_{f,M}^*$. Now for $g \in F(S)$ satisfying $\sup_{x\in S} K_g(x, S) < \infty$, $\mu = P(x, \cdot)$,

$$\begin{aligned}\mathbb{E}_\mu(\exp S_n(g)) &= \int e^{g(y)} K_g^{n-1} 1(y)\mu(dy)\\ &= \int e^{g(y)} K_g^{n-1} 1(y)P(x, dy)\\ &= K_g^n 1(x).\end{aligned}$$

Therefore $\phi_M(g) = \rho(g)$, which implies $\phi_{f,M}^* = \rho_f^*$.

Remark 11.18 If E is finite-dimensional, then Theorem 11.15 and assumption 2(i) of Theorem 11.16 are not needed and assumption 2(ii) may be relaxed, replacing it with:

2(ii) There exists $r > 0$ such that $c_r < \infty$ and $d_r < \infty$.

For, $B(\rho)$ is compact for any $\rho \geq 0$ and for $\mu \in M$,

$$\begin{aligned}\mathbb{P}_\mu\left[n^{-1}S_n(f) \in (B(\rho))^c\right] &= \mathbb{P}_\mu\left[\|S_n(f)\| > n\rho\right]\\ &\leq e^{-n\rho r}\mathbb{E}_\mu\left[\exp(r\|S_n(f)\|)\right]\\ &\leq e^{-n\rho r} c_r d_r^{n-1},\end{aligned}$$

$$\overline{\lim_n}\, n^{-1} \log \sup_{\mu\in M} \mathbb{P}_\mu\left[n^{-1}S_n(f) \in (B(\rho))^c\right] \leq -\rho r + \log d_r,$$

showing that $\{\sup_{\mu\in M} \mathbb{P}_\mu[n^{-1}S_n(f) \in \cdot]\}$ is exponentially tight.

The proof that $\phi_{f,M}^*$ is tight can also be simplified. For $\|\xi\| < r$, $\mu \in M$, we have

$$\begin{aligned}\mathbb{E}_\mu\left[\exp(\langle S_n(f), \xi\rangle)\right] &\leq \mathbb{E}_\mu\left[\exp(r\|S_n(f)\|)\right]\\ &\leq c_r d_r^{n-1},\end{aligned}$$

and therefore $\phi_{f,M}(\xi) \leq \log d_r$. Assume that $u \in L_a \triangleq \{u \in E \colon \phi^*_{f,M}(u) \leq a\}$, $a \geq 0$. Then

$$\langle u, \xi\rangle \leq \phi_{f,M}(\xi) + \phi^*_{f,M}(u) \leq \log d_r + a,$$
$$r\|u\| = \sup\{\langle u, \xi\rangle \colon \|\xi\| \leq r\} \leq \log d_r + a,$$

so finally $L_a \subset B(r^{-1}(\log d_r + a))$.

11.4 The Large Deviation Principle

Let $f\colon S \to E$ be measurable, and let $\mu \in \mathcal{P}(S)$. As a consequence of Theorems 11.1 and 11.16, we have:

i. If P is irreducible, then the lower bound for $\{\mathbb{P}_\mu[n^{-1}S_n(f) \in \cdot]\}$ holds with rate function Λ^*_f.
ii. If P, f, and μ satisfy certain tightness and integrability conditions, then the upper bound for $\{\mathbb{P}_\mu[n^{-1}S_n(f) \in \cdot]\}$ holds with rate function $\phi^*_{f,\mu}$.

The following inequality holds:

$$\phi^*_{f,\mu} \leq \Lambda^*_f. \tag{11.39}$$

For, let $\xi \in E^*$. Then $\phi_{f,\mu}(\xi) = \phi_\mu(\langle f, \xi\rangle) \geq \Lambda(\langle f, \xi\rangle) = \Lambda_f(\xi)$ by de Acosta and Ney (2014, (2.8)), and therefore (11.39) holds.

We will show now that, similarly to Theorem 5.1, if $\{\mathbb{P}_\mu[n^{-1}S_n(f) \in \cdot]\}$ satisfies the large deviation principle with rate function J, then J must be bounded by $\phi^*_{f,\mu}$ and, if P is irreducible, Λ^*_f.

Theorem 11.19 *Assume that P is irreducible. Let $f\colon S \to E$ be measurable, and let $\mu \in \mathcal{P}(S)$. Suppose that $\{\mathbb{P}_\mu[n^{-1}S_n(f) \in \cdot]\}$ satisfies the large deviation principle with (lower semicontinuous) rate function J. Then*

$$\phi^*_{f,\mu} \leq J \leq \Lambda^*_f.$$

Proof.

1. We will show $\phi^*_{f,\mu} \leq J$. Let $u \in E$, and assume that $\phi^*_{f,\mu}(u) < \infty$. For $\xi \in E^*$ such that $\phi_{f,\mu}(\xi) < \infty$, $\epsilon > 0$, let

$$U_\xi = \left\{w \in E \colon \langle w, \xi\rangle - \phi_{f,\mu}(\xi) > \phi^*_{f,\mu}(u) - \epsilon\right\}.$$

Choose $\xi \in E^*$ so that $u \in U_\xi$. We have, setting $S_n(f) = S_n$,

$$\mathbb{P}_\mu[n^{-1}S_n \in U_\xi] \leq \exp\left[-n\left(\phi^*_{f,\mu}(u) - \epsilon\right)\right] \mathbb{E}_\mu\left\{\exp\left[\langle S_n, \xi\rangle - n\phi_{f,\mu}(\xi)\right]\right\}.$$

Therefore

$$\overline{\lim_n}\, n^{-1} \log \mathbb{P}_\mu[n^{-1} S_n \in U_\xi] \le -\phi^*_{f,\mu}(u) + \epsilon. \tag{11.40}$$

On the other hand, since $\{\mathbb{P}_\mu[n^{-1} S_n \in \cdot]\}$ satisfies the lower bound with rate function J,

$$\underline{\lim_n}\, n^{-1} \log \mathbb{P}_\mu[n^{-1} S_n \in U_\xi] \ge -J(u). \tag{11.41}$$

By (11.40) and (11.41), $-J(u) \le -\phi^*_{f,\mu}(u) + \epsilon$. Since ϵ is arbitrary, $\phi^*_{f,\mu}(u) \le J(u)$ is proved. The case $\phi^*_{f,\mu}(u) = \infty$ is proved by a similar argument.

2. We will show $J \le \Lambda^*_f$. Let $u \in E$, and assume that $J(u) < \infty$. Let $\epsilon > 0$,

$$U = \{w \in E\colon J(w) > J(u) - \epsilon\}.$$

By the lower semicontinuity of J, U is open. Let B be an open ball such that $u \in B$ and $\bar{B} \subset U$. Since $\{\mathbb{P}_\mu[n^{-1} S_n \in \cdot]\}$ satisfies the upper bound with rate function J,

$$\begin{aligned}\overline{\lim_n}\, n^{-1} \log \mathbb{P}_\mu \left[n^{-1} S_n \in \bar{B}\right] &\le -\inf\left\{J(w)\colon w \in \bar{B}\right\} \\ &\le -J(u) + \epsilon.\end{aligned} \tag{11.42}$$

On the other hand, by Theorem 11.1,

$$\underline{\lim_n}\, n^{-1} \log \mathbb{P}_\mu[n^{-1} S_n \in B] \ge -\Lambda^*_f(u). \tag{11.43}$$

By (11.42) and (11.43), $-\Lambda^*_f(u) \le -J(u) + \epsilon$. Since ϵ is arbitrary, $J(u) \le \Lambda^*_f(u)$ is proved. The case $J(u) = \infty$ is proved by a similar argument. □

In view of Theorems 11.1 and 11.16, we can state in Theorem 11.21 a necessary and sufficient condition for the large deviation principle for $\{\mathbb{P}_\mu[n^{-1} S_n(f) \in \cdot]\}$. We prove first the following lemma.

Lemma 11.20 *Assume:*

1. *P is irreducible and $M \subset \mathcal{P}(S)$.*
2. *For every $\xi \in E^*$,*

$$\sup_{\mu \in M} \int \exp(\langle f(y), \xi\rangle)\, \mu(dy) < \infty, \qquad \sup_{x \in S} \int \exp(\langle f(y), \xi\rangle)\, P(x, dy) < \infty.$$

3. *For every closed set $H \subset E$,*

$$\overline{\lim_n}\, n^{-1} \log \sup_{\mu \in M} \mathbb{P}_\mu \left[n^{-1} S_n(f) \in H\right] \le -\inf\left\{\Lambda^*_f(u)\colon u \in H\right\}.$$

Then $\phi_{f,M} = \Lambda_f$.

Proof.

1. $\phi_{f,M} \le \Lambda_f$: Let $\xi \in E^*$. We will apply Proposition E.3 with $\gamma_{\mu,n} = \mathbb{P}_\mu[n^{-1} S_n(f) \in \cdot]$, $J = \Lambda_f^*$, $F(u) = \langle u, \xi \rangle$, $u \in E$. By assumption 3, (E.4) in Proposition E.3 holds. We will show now that (E.5) of Proposition E.3 holds.

   Setting $S_n(f) = S_n$, since for $b > 0$ $1[F \ge b] \le e^{-nb} e^{nF}$, we have

$$\begin{aligned}\int_{[F \ge b]} \exp(nF)\, d\gamma_{\mu,n} &\le e^{-nb} \int \exp(2nF)\, d\gamma_{\mu,n} \\ &= e^{-nb} \int \exp\langle S_n, \eta \rangle\, d\mathbb{P}_\mu,\end{aligned} \tag{11.44}$$

   where $\eta = 2\xi$. But, setting $s_n = \sum_{j=0}^{n-1} f(x_j)$,

$$\begin{aligned}&\int \exp\langle S_n, \eta \rangle\, d\mathbb{P}_\mu \\ &\quad = \int \mu(dx_0) \int P(x_0, dx_1) \int \dots \int P(x_{n-2}, dx_{n-1}) e^{\langle s_n, \eta \rangle} \\ &\quad \le \left(\int e^{\langle f(x_0), \eta \rangle} \mu(dx_0) \right) \left(\sup_{x \in S} \int e^{\langle f(y), \eta \rangle} P(x, dy) \right)^{n-1}.\end{aligned} \tag{11.45}$$

   Let $c = \log \sup_{x \in S} \int e^{\langle f(y), \eta \rangle} P(x, dy)$. Then by (11.44), (11.45), and assumption 2,

$$\overline{\lim_n}\, n^{-1} \log \sup_{\mu \in M} \int_{[F \ge b]} \exp(nF)\, d\gamma_{\mu,n} \le c - b,$$

   and therefore (E.5) of Proposition E.3 does hold. By that proposition,

$$\begin{aligned}\phi_{f,M}(\xi) &= \overline{\lim_n}\, n^{-1} \log \sup_{\mu \in M} \mathbb{E}_\mu \exp\langle S_n, \xi \rangle \\ &= \overline{\lim_n}\, n^{-1} \log \sup_{\mu \in M} \int \exp(nF)\, d\gamma_{\mu,n} \\ &\le \sup \{F(u) - J(u) \colon u \in E\} \\ &= \sup \left\{ \langle u, \xi \rangle - \Lambda_f^*(u) \colon u \in E \right\} \le \Lambda_f(\xi).\end{aligned}$$

2. The inequality $\phi_{f,M} \ge \Lambda_f$ follows from de Acosta and Ney (2014, (2.8)) with $g = \langle f, \xi \rangle$. □

Extending in an obvious way the definition of uniformity set, given in Chapter 1 for $\{\mathbb{P}_\mu[L_n \in \cdot]\}$, to $\{\mathbb{P}_\mu = [n^{-1} S_n(f) \in \cdot]\}$, we have:

Theorem 11.21 *Assume:*

1. *P is irreducible and $M \subset \mathcal{P}(S)$ satisfies* (2.1).

2. *P, f, and M satisfy the assumptions of Theorem 11.16.*

Then the following conditions are equivalent:

$$\phi_{f,M} = \Lambda_f. \tag{11.46a}$$

M is a uniformity set for both the upper and lower bounds for

$$\{\mathbb{P}_\mu[n^{-1}S_n(f) \in \cdot]\} \text{ with tight rate function } \Lambda_f^*. \tag{11.46b}$$

Proof. (11.46a) $\Longrightarrow$ (11.46b): Since (11.46a) implies $\phi_{f,M}^* = \Lambda_f^*$, (11.46b) follows from Theorems 11.1 and 11.16. The second condition on M in Theorem 11.1 follows from (11.34) by Markov's inequality.

(11.46b) $\Longrightarrow$ (11.46a): Follows from Lemma 11.20. □

11.5 The Zero Set of Λ_f^*

We will now study the set $Z(\Lambda_f^*) = \{u \in E: \Lambda_f^*(u) = 0\}$. For the convex analysis notions used below, see Appendix K. For the definition of geometric ergodicity, see Appendix B, Section B.3. We will use the notation $\pi(f) = \int f \, d\pi$.

Theorem 11.22 *Assume that P is irreducible and an invariant probability measure π exists (by Lemma 5.6, π is unique). Assume also that $\int \|f\| \, d\pi < \infty$. Then:*

1. $\pi(f) \in Z(\Lambda_f^*)$.
2. *If Λ_f is finite in a neighborhood of 0 and is Gâteaux differentiable at 0, then $\nabla\Lambda_f(0) = \pi(f)$ and $Z(\Lambda_f^*) = \{\pi(f)\}$.*
3. *Conversely, if assumptions* 2(i) *and* (11.35) *of Theorem 11.16 hold, $\Lambda_f(0) = 0$, and $Z(\Lambda_f^*) = \{\pi(f)\}$, then Λ_f is Gâteaux differentiable at 0.*
4. *If* (11.35) *holds and P is geometrically ergodic, then the assumption in* 2 *above holds.*

The following lemma will be needed to prove the necessity of Gâteaux differentiability in Theorem 11.22.3. For $g \in F(S)$, let

$$b(g) = \sup_{x \in S} \int e^{g(y)} P(x, dy).$$

Lemma 11.23 *Assume that P is irreducible and assumptions* 2(i) *and* (11.35) *of Theorem 11.16 hold. Then Λ_f is w^*-sequentially continuous.*

Proof.

1. Let $g \in F(S)$ be such that for all $r > 0$, $b(r|g|) < \infty$. For $m, n \in \mathbb{N}$, $m \leq n$, let $T_{m,n}(g) = \sum_{j=m}^{n} g(X_j)$, and let

$$\Lambda^{(m)}(g) = \overline{\lim_n}\, n^{-1} \log \mathbb{E}_\nu \left[(\exp T_{m,n}(g))\, s(X_n) \right],$$

where $s = 1_C$ and $(1_C, \nu)$ is a small pair. Then

$$\Lambda^{(m)}(g) = \Lambda(g). \tag{11.47}$$

For, let $q > 1$, and let p be its conjugate exponent. Then by Hölder's inequality,

$$\begin{aligned}
&\mathbb{E}_\nu \left[(\exp T_{m,n}(g))\, s(X_n) \right] \\
&\quad = \mathbb{E}_\nu \left\{ \left[\exp\left(-T_{1,m-1}(g)\right) \left(\exp T_n(g)\right) \right] s(X_n) \right\} \\
&\quad \leq \left\{ \mathbb{E}_\nu \left[\exp\left(-T_{1,m-1}(pg)\right) \right] \right\}^{1/p} \left\{ \mathbb{E}_\nu \left[(\exp T_n(qg))\, s(X_n) \right] \right\}^{1/q},
\end{aligned}$$

and since the first factor is bounded above by $(b(p|g|))^{(m-1)/p}$, we have, by Proposition 2.7 of de Acosta and Ney (2014),

$$\Lambda^{(m)}(g) \leq q^{-1} \Lambda(qg). \tag{11.48}$$

Taking into account (C.1) and the subsequent discussion, and the fact that $\Lambda(qg)$ is bounded above by $b(q|g|)$, the convex function $q \mapsto \Lambda(qg)$ is finite, hence continuous. Therefore (11.48) implies $\Lambda^{(m)}(g) \leq \Lambda(g)$. Proceeding similarly, we have

$$\Lambda(q^{-1} g) \leq q^{-1} \Lambda^{(m)}(g),$$

which implies $\Lambda(g) \leq \Lambda^{(m)}(g)$, and then (11.47).

2. Suppose $\xi_n, \xi \in E^*$ and $w^*\text{-}\lim_n \xi_n = \xi$. Let H be a compact subset of E, $w_H(\eta) = \sup_{u \in H} |\langle u, \eta \rangle|$, $\eta \in E^*$. Then

$$\lim_n w_H(\xi_n - \xi) = 0. \tag{11.49}$$

For, let $\eta_n = \xi_n - \xi$. Then $w^*\text{-}\lim_n \eta_n = 0$ and by the uniform boundedness principle (see, e.g., Corollary III.14.4 of Conway, 1985), $c = \sup_n \|\eta_n\| < \infty$. Given $\delta > 0$, let $F \subset H$ be a finite set such that $\sup_{u \in H} d(u, F) < \delta$. Then for all $u \in H$, $z \in F$,

$$|\langle u, \eta_n \rangle| \leq |\langle u - z, \eta_n \rangle| + |\langle z, \eta_n \rangle|,$$

and therefore

$$w_H(\eta_n) \leq c\delta + \sup_{z \in F} |\langle z, \eta_n \rangle|$$

$$\overline{\lim_n}\, w_H(\eta_n) \leq c\delta.$$

Since δ is arbitrary, (11.49) is proved.

3. As in the proof of Theorem 11.16, by Theorem 11.15 and assumptions 2(i) and (11.35) of Theorem 11.16, there exist $m \in \mathbb{N}$ and a compact, convex, symmetric set $K \subset E$ such that, in the notation of Lemma 11.14,

$$\kappa = \sup_{x \in S} \int \exp\left(q_K(f(y))\right) P^m(x, dy) < \infty,$$

where q_K is the Minkowski functional of K. Let $H = (2m)K$. We have, for all $\epsilon > 0$, $u \in E$, $\xi \in E^*$,

$$|\langle u, \xi \rangle| \leq q_H(u)\,(w_H(\xi) + \epsilon)\,. \tag{11.50}$$

For, if $q_H(u) = \infty$ there is nothing to prove. If $q_H(u) = 0$ then, since $H \subset \{x \in E\colon \|x\| \leq r\}$ for some $r > 0$, it follows that $u = 0$. If $0 < q_H(u) < \infty$, then $(q_H(u))^{-1}u \in H$, and therefore $|\langle (q_H(u))^{-1}u, \xi \rangle| \leq w_H(\xi)$. (11.50) is proved.

4. Let p, q be as in part 1, and let $\xi, \eta \in E^*$ be such that $pw_H(\xi - \eta) < 1$. Then

$$\Lambda_f(\xi) \leq q^{-1}\Lambda_f(q\eta) + w_H(\xi - \eta)\log \kappa. \tag{11.51}$$

To prove (11.51), let (s, ν) be as in part 1, $n \geq m$. Set $g = \langle f, \xi \rangle$, $h = \langle f, \eta \rangle$. By Hölder's inequality,

$$\begin{aligned} &\mathbb{E}_\nu\left[(\exp T_{m,n}(g))\, s(X_n)\right] \\ &\quad = \mathbb{E}_\nu\left[(\exp T_{m,n}(h))\, s(X_n)\right](\exp T_{m,n}(g - h)) \\ &\quad \leq \left\{\mathbb{E}_\nu\left[(\exp T_{m,n}(qh))\, s(X_n)\right]\right\}^{1/q} \left\{\mathbb{E}_\nu\left[(\exp T_{m,n}(p(g - h)))\right]\right\}^{1/p}. \end{aligned} \tag{11.52}$$

For any $\epsilon > 0$ such that $pw_H(\xi - \eta) + \epsilon < 1$, by (11.50),

$$\begin{aligned} T_{m,n}(p(g - h)) &= \langle T_{m,n}(f), p(\xi - \eta) \rangle \\ &\leq q_H(T_{m,n}(f))\,(pw_H(\xi - \eta) + \epsilon)\,, \end{aligned}$$

and therefore

$$\mathbb{E}_\nu\left[\exp T_{m,n}(p(g - h))\right] \leq \left\{\mathbb{E}_\nu\left[\exp T_{m,n}(q_H(f))\right]\right\}^{pw_H(\xi-\eta)+\epsilon}. \tag{11.53}$$

By part 1, (11.52)–(11.53), and (11.26),

$$\Lambda_f(\xi) \leq q^{-1}\Lambda_f(q\eta) + \left(w_H(\xi - \eta) + p^{-1}\epsilon\right)\log \kappa.$$

Letting $\epsilon \to 0$ we obtain (11.51).

5. Assume that w^*-$\lim_n \xi_n = \eta$. By part 2, $w_H(\xi_n - \eta) \to 0$, so from (11.51) we have

$$\overline{\lim_n}\, \Lambda_f(\xi_n) \leq q^{-1}\Lambda_f(q\eta).$$

Arguing as in part 1, it follows that

$$\overline{\lim_n}\, \Lambda_f(\xi_n) \le \Lambda_f(\eta). \tag{11.54}$$

Similarly,

$$\Lambda_f(q^{-1}\eta) \le q^{-1}\Lambda_f(\xi_n) + w_H\left(q^{-1}(\eta - \xi_n)\right)\log \kappa,$$

and then

$$\Lambda_f(\eta) \le \underline{\lim_n}\, \Lambda_f(\xi_n). \tag{11.55}$$

By (11.54) and (11.55) the conclusion follows. □

Proof of Theorem 11.22.

1. Using the notation in Proposition 11.11 with $F = \|f\| + 1$, we have:
   (i) $\pi \in \mathcal{P}_F(S)$.
   (ii) $\langle f, \xi \rangle \in B_F(S)$ for all $\xi \in E^*$.

   Then

$$\begin{aligned}
\Lambda_f^*(\pi(f)) &= \sup\left\{\langle \pi(f), \xi\rangle - \Lambda_f(\xi) \colon \xi \in E^*\right\} \\
&= \sup\left\{\int \langle f, \xi\rangle \, d\pi - \Lambda(\langle f, \xi\rangle) \colon \xi \in E^*\right\} \\
&\le \sup\left\{\int h \, d\pi - \Lambda(h) \colon h \in B_F(S)\right\} \\
&= \Lambda_{(F)}^{(*)}(\pi) = I_\psi(\pi),
\end{aligned}$$

   by Proposition 11.11. But by Lemmas 5.5 and 5.7, $I_\psi(\pi) = 0$.
2. We have $Z(\Lambda_f^*) \subset \partial\Lambda_f(0)$. For, if $\Lambda_f^*(u) = 0$ then, since $\Lambda_f(0) \le 0$, for all $\xi \in E^*$,

$$\langle u, \xi\rangle \le \Lambda_f(\xi) - \Lambda_f(0),$$

   showing that $u \in \partial\Lambda_f(0)$. Since $\partial\Lambda_f(0) = \{\nabla\Lambda_f(0)\}$ and $\pi(f) \in Z(\Lambda_f^*)$ by part 1, it follows that $\nabla\Lambda_f(0) = \pi(f)$ and $Z(\Lambda_f^*) = \{\pi(f)\}$.
3. We have $Z(\Lambda_f^*) = \partial\Lambda_f(0) \cap E$. For, if $u \in E$ and $\langle u, \xi\rangle \le \Lambda_f(\xi) - \Lambda_f(0) = \Lambda_f(\xi)$ for all $\xi \in E^*$, then $\Lambda_f^*(u) = 0$. The opposite inclusion was shown in part 2; therefore $\partial\Lambda_f(0) \cap E = \{\pi(f)\}$. Since by Lemma 11.23 Λ_f is w^*-sequentially continuous, we can apply Proposition K.3 and conclude: Λ_f is Gâteaux differentiable at 0.
4. By de Acosta and Ney (2014, Prop. 7.4), if P is geometrically ergodic, $g \in F(S)$ satisfies

$$b(\delta|g|) < \infty$$

for some $\delta > 0$, and $u_g(t) = \Lambda(tg)$ for $t \in \mathbb{R}$, then u_g is finite in a neighborhood of 0, $u_g'(0)$ exists, and $u_g'(0) = \pi(g)$. Applying this result to $g = \langle f, \xi \rangle$, we have: for all $\xi \in E^*$,

$$\lim_{t \to 0} t^{-1} \left(\Lambda_f(t\xi) - \Lambda_f(0) \right) = u'_{\langle f,\xi \rangle}(0) = \pi\left(\langle f, \xi \rangle\right) = \langle \pi(f), \xi \rangle .$$

We have shown: Λ_f is Gâteaux differentiable at 0 and its Gâteaux derivative at 0 is $\pi(f)$. □

11.6 Sufficient Conditions for the Large Deviation Principle

It was established in Theorem 11.21 that if P is irreducible, $M \subset \mathcal{P}(S)$ satisfies (2.1), and P, f, and M satisfy the assumptions of Theorem 11.16, then

$$\phi_{f,M} = \Lambda_f \tag{11.56}$$

is a necessary and sufficient condition for M to be a uniformity set for both the upper and lower bounds for $\{\mathbb{P}_\nu[n^{-1} S_n(f) \in \cdot]\}$ with tight rate function Λ_f^*.

In this section we will focus on sufficient conditions for (11.56). In Theorem 11.25 we present conditions which imply (11.56) for a fixed $M \subset \mathcal{P}(S)$. In Theorem 11.27 we formulate a condition on P which implies (11.56) for a class of subsets $M \subset \mathcal{P}(S)$ and $\phi_{f,x} = \Lambda_f$ for ψ-almost all $x \in S$.

Lemma 11.24 *Assume that P is irreducible, $M \subset \mathcal{P}(S)$, $h \in F(S)$, and*

1. *For all $g \in B(S)$, $\phi_M(g) = \Lambda(g)$.*
2. *For all $r > 0$, $\sup_{\mu \in M} \int e^{r|h|} \, d\mu < \infty$ and $b(r|h|) < \infty$.*

Then $\phi_M(h) = \Lambda(h)$.

Proof.

1. For all $r > 0$,

$$\lim_{c \to \infty} b\left(r|h|\mathbf{1}[|h| > c]\right) = 1. \tag{11.57}$$

For:

$$\begin{aligned} \int \exp\left(r|h|\mathbf{1}[|h| > c]\right) \, dP(x, \cdot) &= \int_{|h| \le c} \exp\left(r|h|\mathbf{1}[|h| > c]\right) \, dP(x, \cdot) \\ &\quad + \int_{|h| > c} \exp\left(r|h|\mathbf{1}[|h| > c]\right) \, dP(x, \cdot) \\ &\le P\left(x, [|h| \le c]\right) \\ &\quad + e^{-c} \int \exp\left((r+1)|h|\right) \, dP(x, \cdot). \end{aligned}$$

Therefore

$$b(r|h|\mathbf{1}[|h| > c]) \leq 1 + e^{-c} b((r+1)|h|),$$
$$\lim_{c\to\infty} b(r|h|\mathbf{1}[|h| \leq c]) \leq 1.$$

On the other hand, $b(r|h|\mathbf{1}[|h| > c]) \geq b(0) = 1$, and (11.57) follows.

2. For all $r > 0$, $\varphi \in F(S)$ such that $|\varphi| \leq r|h|$ for some $r > 0$,

$$\phi_M(\varphi) \leq \log b(|\varphi|). \tag{11.58}$$

To provc (11.58), we first note that

$$K_\varphi^n 1(x) \leq (b(|\varphi|))^n,$$

as is easily shown, and then

$$\begin{aligned}\mathbb{E}_\mu(\exp S_n(\varphi)) &= \int \mu(dx) e^{\varphi(x)} K_\varphi^{n-1} 1(x) \\ &\leq \left(\int e^\varphi \, d\mu\right)(b(|\varphi|))^{n-1},\end{aligned}$$

$$\overline{\lim_n}\, n^{-1} \log \sup_{\mu\in M} \mathbb{E}_\mu(\exp S_n(\varphi)) \leq \log b(|\varphi|),$$

proving (11.58).

3. Let (s, ν) be a small pair. By Proposition 2.7 of de Acosta and Ney (2014), the second condition in 2 implies that

$$\Lambda(\varphi) = \overline{\lim_n}\, n^{-1} \log \nu K_\varphi^n s.$$

It easily follows that

$$\Lambda(\varphi) \leq \log b(|\varphi|). \tag{11.59}$$

4. For all $r > 0$,

$$\overline{\lim_{c\to\infty}}\, \Lambda(rh\mathbf{1}[|h| \leq c]) \leq \Lambda(rh). \tag{11.60}$$

For, let $p > 0$, $q > 0$, $p + q = 1$. Then

$$\begin{aligned}\Lambda(rh\mathbf{1}[|h| \leq c]) &= \Lambda\left(p(p^{-1}rh) + q\left(-q^{-1}rh\mathbf{1}[|h| > c]\right)\right) \\ &\leq p\Lambda(p^{-1}rh) + q\Lambda\left(-q^{-1}rh\mathbf{1}[|h| > c]\right) \\ &\leq p\Lambda(p^{-1}rh) + q\Lambda\left(q^{-1}r|h|\mathbf{1}[|h| > c]\right),\end{aligned}$$

and by (11.57) and (11.59),

$$\overline{\lim_{c\to\infty}}\, \Lambda(rh\mathbf{1}[|h| \leq c]) \leq p\Lambda(p^{-1}rh). \tag{11.61}$$

The convex function $t \mapsto \Lambda(trh)$, $t \in \mathbb{R}$, is finite, hence continuous. Letting $p \to 1$ in (11.61), (11.60) follows.

5. We claim that

$$\phi_M(h) \leq \Lambda(h). \tag{11.62}$$

For, if p, q are as in part 4, for all $c > 0$,

$$\begin{aligned}\phi_M(h) &= \phi_M\left(p\left(p^{-1}h\mathbf{1}[|h| \leq c]\right) + q\left(q^{-1}h\mathbf{1}[|h| > c]\right)\right)\\ &\leq p\phi_M\left(p^{-1}h\mathbf{1}[|h| \leq c]\right) + q\phi_M\left(q^{-1}h\mathbf{1}[|h| > c]\right)\\ &= p\Lambda\left(p^{-1}h\mathbf{1}[|h| \leq c]\right) + q\phi_M\left(q^{-1}h\mathbf{1}[|h| > c]\right)\end{aligned}$$

by assumption 1. Now by (11.57), (11.58), and (11.60), taking $\overline{\lim}_{c\to\infty}$ we obtain $\phi_M(h) \leq p\Lambda(p^{-1}h)$. Arguing as in part 4, we obtain (11.62).

Finally, reversing the roles of ϕ_M and Λ, and using now (11.57), (11.59), and (11.60) with ϕ_M instead of Λ, which obviously holds, we have $\Lambda(h) \leq \phi_M(h)$. Together with (11.62) this yields the conclusion. □

Theorem 11.25 *Assume:*

1. *P is irreducible and $M \subset \mathcal{P}(S)$ satisfies* (2.1).
2. *P, f, and M satisfy the assumptions of Theorem 11.16.*
3. (i) *ϕ_M is $\sigma(B(S), \mathcal{P}(S))$-lower semicontinuous.*
   (ii) *If $0 \leq g \in B(S)$ and $\int g\, d\psi = 0$, then $\phi_M(g) = 0$.*

Then (11.46a) *and* (11.46b) *of Theorem 11.21 hold.*

Proof. In view of Theorem 11.21, it suffices to prove (11.46a) of that theorem. By Remark 8.9, assumption 3 implies that for all $g \in B(S)$, $\phi_M(g) = \Lambda(g)$.

Assumptions (11.34) and (11.35) of Theorem 11.16 imply that for all $\xi \in E^*$, $h = \langle f, \zeta\rangle$ satisfies assumption 2 of Lemma 11.24. Applying now Lemma 11.24, we have, for all $\xi \in E^*$,

$$\phi_{f,M}(\xi) = \phi_M(\langle f, \xi\rangle) = \Lambda(\langle f, \xi\rangle) = \Lambda_f(\xi).$$ □

Remark 11.26 Recall that by Theorem 6.1, the conditions

1. ϕ_M is $B(S)$-regular,
2. (i) ϕ_M is $\sigma(B(S), \mathcal{P}(S))$-lower semicontinuous, and
   (ii) ϕ_M^* is $B(S)$-tight

are equivalent. Therefore, assumption 3 of Theorem 11.25 is weaker than assumption (8.21) of Theorem 8.6. Or, for another view: assumption 3 of Theorem 11.25 is equivalent to $\phi_M = \Lambda$ on $B(S)$, whereas assumption (8.21) of Theorem 8.6 is equivalent to the joint condition $\phi_M = \Lambda$ on $B(S)$ and I_ψ is $B(S)$-tight.

The quantity $\bar{\lambda}$ was defined in (8.37).

Theorem 11.27 *Assume that P is irreducible, $\bar{\lambda} = \infty$, and*

$$M \subset \mathcal{P}(S) \text{ satisfies (8.35);}$$

$$P \text{ and } f \text{ satisfy assumptions 2(i) and (11.35) of Theorem 11.16.}$$

Then:

1. (11.46a) *and* (11.46b) *of Theorem 11.21 hold.*
2. *There exists a ψ-null set $N_f \in \mathcal{S}$ such that if $x \notin N_f$, then $\phi_{f,x} = \Lambda_f$ and therefore $\{\mathbb{P}_x[n^{-1}S_n(f) \in \cdot]\}$ satisfies the large deviation principle with tight rate function Λ_f^*.*
3. *If P is ergodic (Section B.3) and π is its unique invariant measure, then $\int \|f\|\, d\pi < \infty$ and $Z(\Lambda_f^*) = \{\pi(f)\}$.*

Proof.

1. We will show that M satisfies (2.1) and (11.34) of Theorem 11.16. For the first assertion, let $m \in \mathbb{N}$, $h \in \mathcal{S}^+ \cap B(S)$ be as in (8.35). By Proposition B.5, there exists a sequence $\{C_k\}$ of petite sets such that $C_k \uparrow S$. Then, if $x_0 \in S$ is such that $\alpha = h(x_0) > 0$, we have:

$$\sup_{\nu \in M} \nu(C_k^c) \le \alpha^{-1} \sum_{j=1}^{m} P^j(x_0, C_k^c) \downarrow 0,$$

   and therefore $\inf\{\nu(C_k)\colon \nu \in M\} > 0$ for sufficiently large k.

   To prove the second assertion, choose x_0 as above and use (11.35).

   Next, by Proposition 8.18(1), $\phi_M(g) = \Lambda(g)$ for all $g \in B(S)$. Arguing as in the proof of Theorem 11.25, we have: $\phi_{f,M} = \Lambda_f$. This proves assertion 1.
2. We claim first that for every $\xi \in E^*$ there exists a ψ-null set $N(\xi) = N(\xi, f)$ such that $\phi_{f,x}(\xi) = \Lambda_f(\xi)$ for $x \notin N(\xi)$. For, by Proposition 8.18(2), for every $m, n \in \mathbb{N}$ there exists a ψ-null set $N(m, n)$ such that, for $x \notin N(m, n)$,

$$\phi_x\left(m\langle f, \xi\rangle \mathbf{1}\left[|\langle f, \xi\rangle| \le n\right]\right) = \Lambda\left(m\langle f, \xi\rangle \mathbf{1}\left[|\langle f, \xi\rangle \le n\right]\right).$$

   Let $N(\xi) = \bigcup\{N(m, n)\colon m, n \in \mathbb{N}\}$. Then $N(\xi)$ is ψ-null and if $x \notin N(\xi)$, the argument in part 5 of the proof of Lemma 11.24 proves the claim.

   Next, since E is separable, for any $r > 0$, $\{\xi \in E^*: \|\xi\| \le r\}$ is a compact metrizable space in the w^* topology by the Banach–Alaoglu theorem (see, e.g., pages 130, 134 of Conway, 1985) and it follows that there exists a countable set $D \subset E^*$ with the following property: for every $\xi \in E^*$, there exists a sequence $\{\eta_n\} \subset D$ such that $\xi = w^*\text{-}\lim \eta_n$.

By Lemma 11.23, Λ_f is w^*-sequentially continuous. The proof of that lemma shows that also $\phi_{f,x}$ is w^*-sequentially continuous. Let $N_f = \bigcup\{N(\eta, f): \eta \in D\}$. Then N_f is ψ-null and if $x \notin N_f$ and $\xi, \{\eta_n\}$ are as in the previous paragraph, we have

$$\phi_{f,x}(\xi) = \lim_n \phi_{f,x}(\eta_n) = \lim_n \Lambda_f(\eta_n) = \Lambda_f(\xi).$$

This proves 2.

3. By (11.35),

$$\int \|f\| \, d\pi = \int \|f\| \, d(\pi P) = \int \pi(dx) \int P(x, dy)\|f(y)\| < \infty.$$

As argued in part 2 of the proof of Proposition B.9, since P is aperiodic every petite set is small. Therefore $\bar{\lambda} = \infty$ implies that P is geometrically ergodic. By Theorem 11.22, $Z(\Lambda_f^*) = \{\pi(f)\}$. □

Remark 11.28

1. The condition $\bar{\lambda} = \infty$ is satisfied if P is uniformly recurrent. For, by Proposition B.9, in this case S is petite and therefore trivially $\tau_S \equiv 1$ and $\bar{\lambda} = \infty$. On the other hand, the condition $\bar{\lambda} = \infty$ is strictly weaker than uniform recurrence; see Example III in Chapter 10 of de Acosta and Ney (2014).
2. There is a close connection between Theorems 8.19 and 11.27. Assume that E is finite-dimensional and P is irreducible. Under assumption $\tilde{\lambda} = \infty$ of Theorem 8.19, for any bounded measurable function $f: S \to E$, statements 1 and 2 of Theorem 11.27 follow, respectively, from statements 1 and 2 of Theorem 8.19 by the contraction principle via Φ_f; see Theorem 2.17.3 and Remark 2.18. Note that, in this derivation, in 2 of Theorem 11.27 the exceptional ψ-null set N (from Theorem 8.19) does not depend on f.

   On the other hand, in Theorem 11.27 statements 1 and 2 are proved under the assumption $\bar{\lambda} = \infty$, which is weaker than $\tilde{\lambda} = \infty$; recall that by Remark 7.4, $\tilde{\lambda} \leq \bar{\lambda}$. Note that in 2 the exceptional ψ-null set N_f depends on f in general.

11.7 On the Relationship between Large Deviations for Empirical Measures and Large Deviations for Additive Functionals

In the present chapter, we have studied large deviations for additive functionals, culminating in Section 11.4. In previous chapters, we studied large deviations for empirical measures, culminating in Chapter 8. We will discuss now how these two directions are related.

Theorem 11.29 *Assume that V satisfies V.1′–V.4. Let P be irreducible, $\mu \in \mathcal{P}(S)$. Then the following conditions are equivalent:*

For every $g \in V$, $\{\mathbb{P}_\mu[n^{-1}S_n(g) \in \cdot]\}$ satisfies the large deviation principle with tight rate function Λ_g^*. (11.63a)

I_ψ is V-tight. (11.63b)

For every $d \in \mathbb{N}$, $f \in V^d$, $\{\mathbb{P}_\mu[n^{-1}S_n(f) \in \cdot]\}$ satisfies the large deviation principle with tight rate function Λ_f^*. (11.64a)

I_ψ is V-tight. (11.64b)

$\{\mathbb{P}_\mu[L_n \in \cdot]\}$ satisfies the large deviation principle in the V topology with V-tight rate function I_ψ. (11.65)

Proof. (11.64) $\Longrightarrow$ (11.63): Obvious.

(11.63) $\Longrightarrow$ (11.65): Let $g \in V$. By Theorem 11.21 with $M = \{\mu\}$, $f = g$, we have $\phi_{g,\mu} = \Lambda_g$. But $\phi_\mu(g) = \phi_{g,\mu}(1) = \Lambda_g(1) = \Lambda(g)$. Since also I_ψ is V-tight, condition (8.3) of Theorem 8.1 holds, so by that theorem (11.65) holds.

(11.65) $\Longrightarrow$ (11.64): Again, by Theorem 8.1, we have $\phi_\mu \mid V = \Lambda \mid V$. Let $\xi \in \mathbb{R}^d$. Then $\langle f, \xi \rangle = \sum_{i=1}^d \xi_i f_i \in V$, so

$$\phi_{f,\mu}(\xi) = \phi_\mu\left(\langle f, \xi \rangle\right) = \Lambda\left(\langle f, \xi \rangle\right) = \Lambda_f(\xi),$$

and (11.64) follows from Theorem 11.21. □

Remark 11.30 (11.64a) can be obtained from (11.65) directly via the contraction principle. For, let $f \in V^d$ and, as in Theorem 2.17, let $\Phi_f : \mathcal{P}(S) \to \mathbb{R}^d$ be defined by $\Phi_f(\mu) = \int f\, d\mu$. Then $n^{-1}S_n(f) = \Phi_f(L_n)$ and the fact that (11.65) implies (11.64a) follows from Theorem 2.17 and Remark 2.18.

11.8 Notes

Section 11.1: Theorem 11.1 was proved in de Acosta and Ney (1998) in the case $M = \{\mu\}$.

Section 11.2: Proposition 11.12 and Theorem 11.13.2 are proved in Liu and Wu (2009). The proofs presented here are in part different from those in Liu and Wu (2009). There are common aspects: the proofs in Liu and Wu (2009) use several results from de Acosta (1988), which we have reformulated in this book, as well as the Krein–Smulian theorem, which we also invoke in Theorem 2.13 (and which was previously used in de Acosta and Ney, 1998, Lemma 3.3).

Section 11.3: The upper bound is based on de Acosta (1985), with some improvements.

The integrability result Theorem 11.15 is proved in de Acosta (1985) and has found applications in other contexts: see, e.g., Chen (2010) and Dinwoodie and Zabell (1992).

Section 11.5: The differentiability of Λ_f at 0 and its connection to $\pi(f)$ is studied in de Acosta and Ney (2014). Lemma 11.23 is based on Lemma 9.1 of de Acosta and Ney (2014).

Section 11.6: Theorem 11.27 improves Theorem 3 of de Acosta and Ney (2014). An improved form of Theorem 1 of de Acosta and Ney (2014) also follows from Theorem 11.27, taking into account Remark 11.18 concerning the case when E is finite-dimensional. In Theorem 2 of de Acosta and Ney (2014) a local large deviation result is obtained under the weaker assumption of geometric ergodicity.

Theorem 4.1.14 of Deuschel and Stroock (1989) contains the following statement: if P satisfies (4.43) and P, f satisfy (11.35) of Theorem 11.16, then S is a uniformity set for both the upper and lower bounds for $\{\mathbb{P}_x[n^{-1}T_n(f) \in \cdot]\}$ with a tight rate function J which is generated by a subadditivity argument but is not further identified. We claim that this result follows from Theorems 11.1, 11.16, and Lemma 11.24, and in fact $J = \Lambda_f^*$. For:

1. (4.43) implies that P is irreducible, S is petite, and therefore the lower bound with rate function Λ_f^* follows from Remark 11.10. Here $f\colon S \to E$ is an arbitrary measurable function.
2. (4.43) implies assumption 2(i), Theorem 11.16. For, $\sup_{x \in S} P^m(x, \cdot) \leq \lambda$ for a certain finite measure λ and $\lambda \circ f^{-1}$ is tight by a classical result since E is Polish. Therefore the upper bound with tight rate function $\phi_{f,M}^* = \rho_f^*$, where $M = \{P(x, \cdot)\colon x \in S\}$, follows from Remark 11.17(3).
3. By the proof of Corollary 7.2, (4.43) implies that ϕ is $B(S)$-regular. By Theorem 6.1, it follows that ϕ is $\sigma(B(S), \mathcal{P}(S))$-lower semicontinuous. Since Λ is $\sigma(B(S), \mathcal{P}(S))$-lower semicontinuous by Theorem 2.13, and (4.43) implies that $\phi^* = I = I_\psi = \Lambda^*$ (the first and last equalities by Theorems 4.1

and 4.2, respectively), it follows from Proposition F.3 that $\phi = \Lambda$ on $B(S)$. But if $M = \{P(x,\cdot)\colon x \in S\}$, then it is easily seen that $\phi_M(g) = \phi(g)$ for all $g \in B(S)$. Applying now Lemma 11.24 to $h = \langle f, \xi \rangle$, we have: $\phi_{f,M}(\xi) = \phi_M(\langle f, \xi \rangle) = \Lambda(\langle f, \xi \rangle) = \Lambda_f(\xi)$, and finally $\phi^*_{f,M} = \Lambda^*_f$. This concludes the proof of the claim.

Section 11.7: Regarding the relationship between the large deviation principle for empirical measures and the large deviation principle for additive functionals:

1. Theorem 11.27 cannot be proved from a large deviation principle for empirical measures by the contraction principle via $\Phi_f\colon \mathcal{P}(S) \to E$. For, by Remark 11.28(1), Theorem 11.27 is valid for any uniformly recurrent Markov kernel P, while Section 10.3 exhibits a uniformly ergodic P for which the large deviation principle for empirical measures does not hold.
2. Proposition 6 of Bryc and Dembo (1996) gives an explicit example of the large deviation principle for $\{\mathbb{P}_\nu[n^{-1}S_n(f) \in \cdot]\}$ holding for all bounded measurable functions $f\colon S \to \mathbb{R}^d$ but the large deviation principle for $\{\mathbb{P}_\nu[L_n \in \cdot]\}$ failing to hold. If S is countable and P is matrix irreducible and uniformly ergodic, then by Theorem 11.27 and Remark 11.28(1), the large deviation principle for $\{\mathbb{P}_x[n^{-1}S_n(f) \in \cdot]\}$ holds for every $x \in S$ and every bounded functional $f\colon S \to \mathbb{R}^d$ with tight rate function Λ^*_f. Proposition 5 of Bryc and Dembo (1996) provides an example with S and P as above, in which the large deviation principle fails to hold for $\{\mathbb{P}_\pi[n^{-1}S_n(f) \in \cdot]\}$, where f is a certain indicator function and π is the invariant probability measure.

Appendix A

The Ergodic Theorem for Empirical Measures and Vector-Valued Functionals of a Markov Chain

We show that the results in the title follow from the classical ergodic theorem for real-valued functionals of Markov chains (Proposition A.2).

Proposition A.1 *Assume that P is positive Harris recurrent (Section B.3) and let π be its unique invariant probability measure.*

1. *Assume that $\mathcal{F} \subset B(S)$ is separable for the uniform norm on $B(S)$. Then for any $\nu \in \mathcal{P}(S)$, $\{L_n\}$ converges $\sigma(\mathcal{P}(S), \mathcal{F})$ to π, $\mathbb{P}_\nu$-a.s..*
2. *Let $(S, \mathcal{S})$ be a separable metric space with its Borel σ-algebra and $C_b(S)$ be the space of bounded continuous functions on S. Then for any $\nu \in \mathcal{P}(S)$, $\{L_n\}$ converges $\sigma(\mathcal{P}(S), C_b(S))$ to π, $\mathbb{P}_\nu$-a.s.*

Proposition A.2 *Let P, π be as in Proposition A.1. Then for any $\nu \in \mathcal{P}(S)$, $f \in L^1(S, \mathcal{S}, \pi)$,*

$$\mathbb{P}_\nu \left[n^{-1} S_n(f) \longrightarrow \int f \, d\pi \right] = 1.$$

See, e.g., page 140 of Revuz (1984).

Lemma A.3 *Let S be a separable metric space. Then there exists an equivalent metric d on S such that there exists a countable set $\mathcal{D}$ of bounded, d-uniformly continuous functions with the following property: if μ_n, $\mu \in \mathcal{P}(S)$, then $\{\mu_n\}$ converges $\sigma(\mathcal{P}(S), C_b(s))$ to μ if and only if for all $f \in \mathcal{D}$, $\lim_n \int f \, d\mu_n = \int f \, d\mu$.*

See, e.g., Theorem 2.6.6 of Parthasarathy (1967).

Proof of Proposition A.1.

1. For $g \in B(S)$, let $N_g = [n^{-1}S_n(g) \not\to \int g\,d\pi]$. Let $C \subset B(S)$ be countable and such that $\overline{C} \supset \mathcal{F}$. By Proposition A.2, for each $g \in B(S)$, $\mathbb{P}_\nu[N_g] = 0$, and therefore $\mathbb{P}_\nu[N] = 0$, where $N = \bigcup\{N_h : h \in C\}$. We claim for $\omega \notin N$, $g \in \mathcal{F}$,
$$\int g\,dL_n(\omega) \longrightarrow \int g\,d\pi.$$
For, given $\epsilon > 0$, let $h \in C$ be such that $\|h - g\| < \epsilon$. Then
$$\int g\,dL_n(\omega) - \int g\,d\pi = \int (g-h)\,dL_n(\omega) + \left[\int h\,dL_n(\omega) - \int h\,d\pi\right] + \int (h-g)\,d\pi,$$
and therefore, since $\int h\,dL_n = n^{-1}S_n(h)$,
$$\left|\int g\,dL_n(\omega) - \int g\,d\pi\right| \leq \left|\int h\,dL_n(\omega) - \int h\,d\pi\right| + 2\|h-g\|,$$
$$\overline{\lim_n} \left|\int g\,dL_n(\omega) - \int g\,d\pi\right| \leq 2\epsilon.$$
But ϵ is arbitrary. This proves the claim.
2. The proof is similar to that of part 1, using Lemma A.3. Let N_g be as above, and let $M = \bigcup\{N_g : g \in \mathcal{D}\}$. Then $\mathbb{P}_\nu[M] = 0$, and for $\omega \notin M$, $g \in \mathcal{D}$, we have $\int g\,dL_n(\omega) \to \int g\,d\pi$. But this implies that $\{L_n(\omega)\}$ converges $\sigma(\mathcal{P}(S), C_b(S))$ to π. □

Let E be a separable Banach space. For $\mu \in \mathcal{P}(S)$, we define
$$L^1(S, \mathcal{S}, \mu; E) = \left\{f : S \longrightarrow E : f \text{ is measurable and } \int \|f\|\,d\mu < \infty\right\}.$$

Proposition A.4 *Let P, π be as in Proposition A.1. Then the conclusion of Proposition A.2 holds for every $f \in L^1(S, \mathcal{S}, \pi; E)$.*

Proof. If f is an E-valued simple function, say $f = \sum_{i=1}^k x_i \mathbf{1}_{A_i}$, $A_i \in \mathcal{S}$, $x_i \in E$ for $i = 1, \ldots, k$, then the conclusion follows at once from Proposition A.2:
$$\begin{aligned} n^{-1}S_n(f) &= \sum_{i=1}^k x_i \left(n^{-1}S_n(\mathbf{1}_{A_i})\right) \\ &\longrightarrow \sum_{i=1}^k x_i \int \mathbf{1}_{A_i}\,d\pi = \int f\,d\pi, \quad \mathbb{P}_\nu\text{-a.s.} \end{aligned} \tag{A.1}$$

If $f \in L^1(S, \mathcal{S}, \pi; E)$, then there exists a sequence $\{f_k\}$ of E-valued simple functions such that

$$\|f_k\| \le \|f\|, \quad f_k \longrightarrow f \text{ pointwise}, \quad \text{and} \quad \int \|f - f_k\| \, d\pi < k^{-1} \tag{A.2}$$

(see pages 101–102 of Neveu, 1972). By Proposition A.2, setting $\pi(h) = \int h \, d\pi$ for $h \in L^1(S, \mathcal{S}, \pi)$ or $h \in L^1(S, \mathcal{S}, \pi; E)$,

$$n^{-1} S_n(\|f - f_k\|) \longrightarrow \pi(\|f - f_k\|), \quad \mathbb{P}_\nu\text{-a.s.} \tag{A.3}$$

Next,

$$\left\|n^{-1} S_n(f) - \pi(f)\right\| \le n^{-1} S_n(\|f - f_k\|) + \left\|n^{-1} S_n(f_k) - \pi(f_k)\right\| + \|\pi(f_k) - \pi(f)\|,$$

and for fixed k, by (A.1)–(A.3),

$$\overline{\lim_n} \left\|n^{-1} S_n(f) - \pi(f)\right\| \le 2k^{-1}, \quad \mathbb{P}_\nu\text{-a.s.}$$

Since k is arbitrary, the conclusion follows. □

Notes. Proposition A.1(2) is proved on page 102 of Stroock (1984). Proposition A.4 is proved in Theorem A.3 of de Acosta (1988).

Appendix B

Irreducible Kernels, Small Sets, and Petite Sets

B.1 Irreducible Kernels

In this section and Section B.2 we will present some background material from Nummelin (1984) and de Acosta and Ney (2014). Although in this work we are primarily interested in the kernels K_g, $g \in F(S)$, it will be useful in Appendix C to consider more general nonnegative kernels.

Definition B.1 (Nummelin, 1984, Definition 1.1) A (nonnegative) kernel K on $(S, \mathcal{S})$ is *σ-finite* if there exists a $\mathcal{S} \otimes \mathcal{S}$-measurable function $f\colon S \times S \to \mathbb{R}$ such that $f(x, y) > 0$ for all $(x, y) \in S \times S$ and for all $x \in S$,

$$Kf(x) = \int f(x, y)K(x, dy) < \infty.$$

The more common definition of the σ-finiteness of K in the literature is the condition

i. For every $x \in S$, $K(x, \cdot)$ is σ-finite.

It is easy to see that Definition B.1 implies condition i for each $x \in S$, let $E_n(x) = \{y\colon f(x, y) > n^{-1}\}$. Then, $E_n(x) \in \mathcal{S}$, $E_n(x) \uparrow S$ and

$$K(x, E_n(x)) \leq n \int f(x, y)K(x, dy) < \infty.$$

Actually it is not difficult to show that condition i is equivalent to

ii. There exists $g\colon S \times S \to \mathbb{R}$ such that $g(x, \cdot)$ is $\mathcal{S}$-measurable for all $x \in S$, $g(x, y) > 0$ for all $(x, y) \in S \times S$, and for all $x \in S$, $Kg(x) = \int g(x, y)K(x, dy) < \infty$.

Definition B.1, stipulating the joint measurability of f, is slightly stronger than condition ii. All kernels K will be assumed to be σ-finite.

Two kernels K and H on $(S, \mathcal{S})$ are *equivalent* if for each $x \in S$ the measures $K(x, \cdot)$ and $H(x, \cdot)$ are equivalent in the sense of mutual absolute continuity. If a kernel K is σ-finite, then there exists an equivalent sub-Markov kernel Q; that is, $Q(x, S) \leq 1$ for all $x \in S$. For, let f, Kf be as above and define for $(x, y) \in S \times S$, $A \in \mathcal{S}$,

$$q(x,y) = \mathbf{1}_{(0,\infty]}(K(x,S))(Kf(x))^{-1} f(x,y) + \mathbf{1}_{\{0\}}(K(x,S))$$
$$Q(x,A) = \int_A q(x,y)K(x,dy).$$

Then Q is a sub-Markov kernel and Q is equivalent to K; moreover, for all $n \in \mathbb{N}$, Q^n is equivalent to K^n. If $K(x, S) > 0$ for all $x \in S$, then Q is a Markov kernel.

K is *irreducible* if there exists a σ-finite measure φ on $(S, \mathcal{S})$ such that for all $x \in S$, $A \in \mathcal{S}$ with $\varphi(A) > 0$, there exists $n \in \mathbb{N}$ such that $K^n(x, A) > 0$. In this context, φ is called a *K-irreducibility measure*. Clearly, K is irreducible if and only if Q is irreducible and K and Q have the same class of irreducibility measures. Since obviously $K(x, S) = 0$ implies $K^n(x, S) = 0$ for all $n \in \mathbb{N}$, if K is irreducible then by the previous remark Q is a Markov kernel. A K-irreducibility measure ψ is *K-maximal* if for all K-irreducibility measures φ, we have $\varphi \ll \psi$. It is obvious that any two K-maximal irreducibility measures are equivalent; also, φ is K-maximal if and only if it is Q-maximal. A K-maximal irreducibility probability measure always exists. For, let φ be a K-irreducibility measure. By the σ-finiteness of φ, there is an equivalent probability measure which, abusing notation, we will also denote φ. Let ψ be the probability measure defined by

$$\psi = \sum_{n=1}^{\infty} 2^{-n} \varphi Q^n.$$

Then ψ is a Q-maximal (K-maximal) irreducibility probability measure. For a proof, see de Acosta and Ney (2014), Lemma 2.1.

We will prove now some direct consequences of the definitions of an irreducible kernel K and a maximal irreducibility probability measure ψ.

A set $F \in \mathcal{S}$ is K-closed if $F \neq \emptyset$ and $K(x, F^c) = 0$ for all $x \in F$.

Proposition B.2

1. $\psi K \ll \psi$.
2. *If F is K-closed, then $\psi(F^c) = 0$.*
3. *Let $A \in \mathcal{S}$ and assume $\psi(A) = 0$. Let*

$$A^0 = \{x \in S : G\mathbf{1}_A(x) = 0\},$$

where $G = \sum_{n=0}^{\infty} K^n$ is the potential kernel of K. Then A^0 is K-closed and $A \subset (A^0)^c$.

Proof.

1. Suppose that $A \in \mathcal{S}$ and $\int (Q\mathbf{1}_A)\, d\psi = (\psi Q)(A) > 0$. By the irreducibility of Q, for all $x \in S$,
$$\sum_{n=1}^{\infty} Q^{n+1}(x, A) = \sum_{n=1}^{\infty} Q^n(Q\mathbf{1}_A)(x) > 0.$$
This shows that the probability measure ψQ is a Q-irreducibility measure. Therefore $\psi Q \ll \psi$. Since $\psi K \equiv \psi Q$, also $\psi K \ll \psi$.
2. It is easily seen that for all $x \in F$, $n \in \mathbb{N}$, $K^n(x, F^c) = 0$. Therefore, by the irreducibility of K, $\psi(F^c) > 0$ is impossible.
3. It follows from statement 1 that $\psi G \ll \psi$. Therefore
$$\int (G\mathbf{1}_A)\, d\psi = \psi G(A) = 0,$$
which implies $\psi((A^0)^c) = 0$, so $A^0 \neq \emptyset$. A^0 is K-closed: for, if $x \in A^0$, then
$$\int K(x, dy) G\mathbf{1}_A(y) = KG\mathbf{1}_A(x) \leq G\mathbf{1}_A(x) = 0,$$
and therefore $K(x, (A^0)^c) = 0$. Finally, $A \subset (A^0)^c$, since $A^0 \subset A^c$. □

Given an irreducible kernel K, in the next section and in Appendix C the symbol ψ denotes a K-maximal irreducibility probability measure. As stated in Chapter 1, in the remainder of the text ψ denotes a P-maximal irreducibility probability measure.

B.2 Small Functions, Sets, and Measures

Let K be an irreducible kernel. The class of sets $A \in \mathcal{S}$ such that $\psi(A) > 0$ will be denoted $\mathcal{S}^+$. Let $F(S)$ be the space of real-valued measurable functions defined on S. Abusing notation, we write $h \in \mathcal{S}^+$ if $h \in F(S)$, $h \geq 0$, and $\int h\, d\psi > 0$. $\mathcal{M}^+ = \mathcal{M}^+(S)$ will denote the space of finite nonzero measures on $(S, \mathcal{S})$.

Definition B.3 A function $t \in B(S)$, $t \geq 0$, is K-*small* if there exist $m \in \mathbb{N}$, $\alpha > 0$, $\lambda \in \mathcal{M}^+(S)$ such that
$$K^m \geq \alpha(t \otimes \lambda), \tag{B.1}$$
that is,
$$K^m(x, A) \geq \alpha t(x)\lambda(A), \qquad x \in S,\ A \in \mathcal{S}.$$
A set $C \in \mathcal{S}$ is K-small if $\mathbf{1}_C$ is K-small.

A measure $\lambda \in \mathcal{M}^+(S)$ is K-small if there exist $m \in \mathbb{N}$, $\alpha > 0$, $t \in \mathcal{S}^+ \cap B(S)$ such that (B.1) holds. A pair (t, λ), where $t \in \mathcal{S}^+ \cap B(S)$ and $\lambda \in \mathcal{M}^+(S)$ is K-small if (B.1) holds for some $m \in \mathbb{N}$, $\alpha > 0$. The constant α will sometimes be absorbed into t or into λ.

If $t \in \mathcal{S}^+ \cap B(S)$ is K-small and $\lambda \in \mathcal{M}^+(S)$ is K-small, then (t, λ) is K-small (de Acosta and Ney, 2014, Lemma 2.2). If $K = P$, we will simply write "small" in all cases. If $g \in B(S)$, then (t, λ) is K_g-small if and only if it is small (see de Acosta and Ney, 2014, (2.2)).

If K is irreducible and $\lambda \in \mathcal{M}^+(S)$ is a K-small measure, then it is a K-irreducibility measure and therefore $\lambda \ll \psi$. For, given $x \in S$, since $t \in \mathcal{S}^+$ there exists $p \in \mathbb{N}$ such that $K^p t(x) > 0$. Then, if $A \in \mathcal{S}$ and $\lambda(A) > 0$, we have $K^{p+m}(x, A) \geq \alpha K^p t(x) \lambda(A) > 0$.

A fundamental result on an irreducible kernel K is the minorization theorem, which asserts the existence of a K-small function $t \in \mathcal{S}^+ \cap B(S)$ satisfying (B.1). It is proved in Nummelin (1984, Thm. 2.1) under the assumption that all powers K^n are σ-finite. This condition is satisfied by a Markov kernel P and, more generally, by K_g if $g \in B(S)$. It may not be satisfied for $g \in F(S)$, but it does hold for the kernel $K_{g\wedge 1}$. If (t, λ) is a $K_{g\wedge 1}$-small pair, then it is also a K_g-small pair: for some $m \in \mathbb{N}$, $\alpha > 0$,

$$K_g^m \geq K_{g\wedge 1}^m \geq \alpha(t \otimes \lambda).$$

Therefore for all $g \in F(S)$ there exists a minorization of the form (B.1) with $t \in \mathcal{S}^+ \cap B(S)$ for K_g if it is irreducible.

Let K be irreducible and assume that all powers K^n are σ-finite. Then if in (B.1) $t \in \mathcal{S}^+$ it may be assumed without loss of generality that $\int t \, d\lambda > 0$. For, let $m \in \mathbb{N}$, $\alpha > 0$, $t \in \mathcal{S}^+ \cap B(S)$, $\lambda \in \mathcal{M}^+$ be such that (B.1) holds. By the irreducibility of K, $\sum_{n=1}^{\infty} K^n t(x) > 0$ for all $x \in S$, and it follows that

$$\sum_{n=1}^{\infty} \lambda K^n t = \int \left(\sum_{n=1}^{\infty} K^n t \right) d\lambda > 0.$$

Then for some $n \in \mathbb{N}$, we have

$$K^{m+n} \geq \alpha(t \otimes \lambda K^n),$$

with $\int t \, d(\lambda K^n) > 0$. However, λK^n need not be finite. Let $x_0 \in S$ be such that $t(x_0) > 0$. Since

$$K^{m+n}(x_0, \cdot) \geq \alpha t(x_0) \lambda K^n$$

and $K^{m+n}(x_0, \cdot)$ is σ-finite, it follows that λK^n is σ-finite. Therefore there exists $\lambda' \in \mathcal{M}^+(S)$ such that $\int t\, d\lambda' > 0$ and

$$K^{m+n} \geq \alpha(t \otimes \lambda').$$

Taking into account the previous discussion on the minorization of K_g for $g \in F(S)$, we summarize: if K is irreducible and either K^n is σ-finite for all $n \in \mathbb{N}$ or $K = K_g$ for some $g \in F(S)$, there exists a minorization of the form (B.1) with $t \in \mathcal{S}^+ \cap B(S)$, $\lambda \in \mathcal{M}^+(S)$, and $\int t\, d\lambda > 0$.

B.3 Petite Functions, Sets, and Measures

The notion of K-small function (set, measure) may be extended as follows.

Definition B.4 Assume that P is irreducible. A function $t \in B(S)$, $t \geq 0$, is *K-petite* if there exist $m \in \mathbb{N}$, $\alpha > 0$, $\lambda \in \mathcal{M}^+(S)$ such that

$$\sum_{j=1}^{m} K^j \geq \alpha(t \otimes \lambda). \tag{B.2}$$

A set $C \in \mathcal{S}$ is K-petite if 1_C is K-petite. A measure $\lambda \in \mathcal{M}^+(S)$ is K-petite if there exist $m \in \mathbb{N}$, $\alpha > 0$, $t \in \mathcal{S}^+ \cap B(S)$ such that (B.2) holds. A pair (t, λ), where $t \in \mathcal{S}^+ \cap B(S)$ and $\lambda \in \mathcal{M}^+(S)$, is K-petite if (B.2) holds for some $m \in \mathbb{N}$, $\alpha > 0$.

If $t \in \mathcal{S}^+ \cap B(S)$ is K-petite and $\lambda \in \mathcal{M}^+(S)$ is K-petite, then (t, λ) is a K-petite pair; this is a simple extension of Lemma 2.2 of de Acosta and Ney (2014). If $K = P$, we will simply write "petite." Of course, a K-small function (set, measure, pair) is K-petite. Similarly to the case of K-small measures, if $\lambda \in \mathcal{M}^+(S)$ is K-petite, then $\lambda \ll \psi$.

Proposition B.5 *Assume that P is irreducible.*

1. *If C_1 and C_2 are petite, then so is $C_1 \cup C_2$.*
2. *There exists an increasing sequence $\{C_n\}_{n \in \mathbb{N}}$ of petite sets such that $C_n \in \mathcal{S}^+$ and $C_n \uparrow S$.*

Proof.

1. Let C be petite and $A \in \mathcal{S}^+$. Then for some $n \in \mathbb{N}$, $\beta > 0$,

$$\sum_{i=1}^{n} P^i 1_A \geq \beta 1_C. \tag{B.3}$$

For, let

$$\sum_{i=1}^{m} P^i \geq \gamma(\mathbf{1}_C \otimes \lambda)$$

for some $m \in \mathbb{N}$, $\gamma > 0$, $\lambda \in \mathcal{M}^+$. Let $q \in \mathbb{N}$ be such that $\lambda P^q \mathbf{1}_A > 0$. Then

$$\left(\sum_{i=1}^{m} P^i\right) P^q \geq \gamma\,(\mathbf{1}_C \otimes \lambda P^q)\,,$$

$$\left(\sum_{i=1}^{m} P^i\right) P^q \mathbf{1}_A \geq \gamma\,(\lambda P^q \mathbf{1}_A)\,\mathbf{1}_C,$$

and therefore (B.3) holds for $n = m + q$, some $\beta > 0$.

Next, suppose $A \in \mathcal{S}^+$ is small: for some $p \in \mathbb{N}$, $\alpha > 0$, $\mu \in \mathcal{M}^+$,

$$P^p \geq \alpha(\mathbf{1}_A \otimes \mu).$$

Then for any petite set C there exist $r \in \mathbb{N}$, $\delta > 0$ such that

$$\sum_{k=1}^{r} P^k \geq \delta(\mathbf{1}_C \otimes \mu). \tag{B.4}$$

For, by (B.3),

$$\left(\sum_{i=1}^{n} P^i\right) P^p \geq \alpha\left(\sum_{i=1}^{n} P^i \mathbf{1}_A\right) \otimes \mu \geq \alpha\beta(\mathbf{1}_C \otimes \mu),$$

and (B.4) follows.

Now let C_1, C_2 be petite sets. By (B.4), there exist $r \in \mathbb{N}$, $\delta > 0$ such that for $i = 1, 2$,

$$\sum_{k=1}^{r} P^k \geq \delta(\mathbf{1}_{C_i} \otimes \mu),$$

and therefore

$$\begin{aligned}\sum_{k=1}^{r} P^k &\geq \frac{\delta}{2}\,[(\mathbf{1}_{C_1} + \mathbf{1}_{C_2}) \otimes \mu] \\ &\geq \frac{\delta}{2}(\mathbf{1}_{C_1 \cup C_2} \otimes \mu).\end{aligned}$$

This proves statement 1.

2. Let $s \leq 1$, $s \in \mathcal{S}^+$ be a small function, and for $n \in \mathbb{N}$, $f_n = \sum_{j=1}^{n} 2^{-j} P^j s$. Let $m \in \mathbb{N}$, $\alpha > 0$, $\nu \in \mathcal{P}(S)$ be such that $P^m \geq \alpha(s \otimes \nu)$. Then

$$\sum_{j=1}^{m+n} P^j \geq \left(\sum_{j=1}^{n} 2^{-j} P^j\right) P^m \geq \alpha(f_n \otimes \nu).$$

Next, let $f = \sum_{j=1}^{\infty} 2^{-j} P^j s$. By irreducibility, $f(x) > 0$ for all $x \in S$. Let $C_k = [f \geq k^{-1}]$; then $C_k \uparrow S$ and $\psi(C_k) > 0$ for all sufficiently large k. It remains to show that C_k is petite.

Given $k \in \mathbb{N}$, choose $N \in \mathbb{N}$ such that

$$2^{-N} < \frac{1}{2k}.$$

Then $C_k \subset [f_N \geq 1/2k]$, and therefore

$$(2k)^{-1}\mathbf{1}_{C_k} \leq (2k)^{-1}\mathbf{1}\left[f_N \geq (2k)^{-1}\right] \leq f_N,$$

$$\alpha(2k)^{-1}\mathbf{1}_{C_k} \otimes \nu \leq \alpha(f_N \otimes \nu) \leq \sum_{j=1}^{m+N} P^j. \qquad \square$$

Proposition B.6 *Assume that P is irreducible. Suppose $M \subset \mathcal{P}(S)$ and there exists a petite set $B \in \mathcal{S}^+$ such that*

$$\lim_n \sup_{\nu \in M} \mathbb{P}_\nu[\tau_B > n] = 0.$$

Then:

1. *There exists a petite set $C \in \mathcal{S}^+$ such that*

$$\inf \{\nu(C) \colon \nu \in M\} > 0.$$

2. *In particular, if $M \subset S$, then M is petite.*

Proof.

1. For $p \in \mathbb{N}$, $0 < \epsilon < 1/2$, let

$$C(p, \epsilon) = \left\{ x \in S : \sum_{j=1}^{p} P^j(x, B) > \epsilon \right\}.$$

By irreducibility, there exist p, ϵ such that

$$\psi\,(C(p, \epsilon)) > 0. \tag{B.5}$$

Next, let $q > p$ be such that

$$\sup_{\nu \in M} \mathbb{P}_\nu \left\{ \bigcap_{j=1}^{q} [X_j \in B^c] \right\} = \sup_{\nu \in M} \mathbb{P}_\nu[\tau_B > q] < 1 - 2\epsilon.$$

Then for all $\nu \in M$,

$$\nu \sum_{j=1}^{q} P^j \mathbf{1}_B = \sum_{j=1}^{q} \mathbb{P}_\nu[X_j \in B] \geq \mathbb{P}_\nu \left\{ \bigcup_{j=1}^{q} [X_j \in B] \right\} \geq 2\epsilon.$$

Therefore, setting $C = C(q, \epsilon)$,

$$\begin{aligned} 2\epsilon &\leq \int_C \nu(dx) \sum_{j=1}^{q} P^j(x, B) + \int_{C^c} \nu(dx) \sum_{j=1}^{q} P^j(x, B) \\ &\leq \int_C \nu(dx) \sum_{j=1}^{q} P^j(x, B) + \epsilon, \end{aligned}$$

and it follows that

$$\inf_{\nu \in M} \nu(C) \geq \epsilon q^{-1}.$$

Since B is petite, for some $k \in \mathbb{N}$, $\alpha > 0$, $\lambda \in \mathcal{P}(S)$, we have

$$\sum_{l=1}^{k} P^l \geq \alpha(\mathbf{1}_B \otimes \lambda).$$

Then

$$\begin{aligned} \left(\sum_{j=1}^{q} P^j \right) \left(\sum_{l=1}^{k} P^l \right) &\geq \sum_{j=1}^{q} P^j \left(\alpha(\mathbf{1}_B \otimes \lambda) \right) \\ &= \alpha \left(\sum_{j=1}^{q} P^j \mathbf{1}_B \right) \otimes \lambda \geq \alpha\epsilon(\mathbf{1}_C \otimes \lambda). \end{aligned} \tag{B.6}$$

(B.6) shows that C is petite. Also, by (B.5), $\psi(C) > 0$.

2. If $M \subset S$, then we have shown: $0 < \inf\{\delta_x(C) : x \in M\}$. Therefore $M \subset C$, so M is petite. □

Definition B.7 Assume that P is irreducible. A set $M \subset \mathcal{P}(S)$ is *uniform* if for all $A \in \mathcal{S}^+$,

$$\limsup_{n} \sup_{\nu \in M} \mathbb{P}_\nu[\tau_A > n] = 0.$$

Of course, this definition makes sense for subsets of S.

Corollary B.8 *Suppose $M \subset \mathcal{P}(S)$ is uniform. Then the conclusions of Proposition B.6(1)–B.6(2) hold.*

We recall some classical definitions. P is *uniformly* recurrent if it is irreducible and S is uniform; P is *uniformly ergodic* if it is uniformly recurrent and aperiodic. We have:

Proposition B.9 *Assume that P is irreducible. Then:*

1. *P is uniformly recurrent if and only if S is petite.*
2. *P is uniformly ergodic if and only if S is small.*

Proof.

1. By Corollary B.8, if P is uniformly recurrent then S is petite. In the converse direction, if S is petite there exist $\alpha > 0$, $m \in \mathbb{N}$, $\nu \in \mathcal{P}(S)$ such that

$$\sum_{j=1}^{m} P^j \geq \alpha(\mathbf{1}_S \otimes \nu). \tag{B.7}$$

Let $B \in \mathcal{S}^+$. By irreducibility, there exists $p \in \mathbb{N}$ such that $\nu P^p(B) > 0$, so by (B.7),

$$\inf_{x \in S} \sum_{j=1}^{q} P^j(x, B) \geq \beta, \tag{B.8}$$

where $q = m + p$ and $\beta = \alpha \nu P^p(B)$. We claim:

$$\inf_{x \in S} \mathbb{P}_x \left\{ \bigcup_{j=1}^{q} [X_j \in B] \right\} \geq \beta q^{-1}. \tag{B.9}$$

For, suppose not. Then there exists $x_0 \in S$ such that

$$\mathbb{P}_{x_0} \left\{ \bigcup_{j=1}^{q} [X_j \in B] \right\} < \beta q^{-1},$$

and therefore

$$\sum_{j=1}^{q} P^j(x_0, B) = \sum_{j=1}^{q} \mathbb{P}_{x_0}[X_j \in B] < q(\beta q^{-1}) = \beta,$$

contradicting (B.8). It follows from (B.9) that

$$\sup_{x \in S} \mathbb{P}_x[\tau_B > q] = \sup_{x \in S} \mathbb{P}_x \left\{ \bigcap_{j=1}^{q} [X_j \in B^c] \right\} \leq \gamma,$$

where $\gamma = 1 - \beta q^{-1}$. Finally, by repeated applications of the Markov property, for all $k \in \mathbb{N}$,

$$\sup_{x \in S} \mathbb{P}_x[\tau_B > kq] \leq \gamma^k.$$

This proves that S is uniform.

2. We show first if S is small, then P is aperiodic. The aperiodicity of P is equivalent to the following condition: for all $x \in S$, $A \in \mathcal{S}^+$, there exists $n_0 \in \mathbb{N}$ such that for all $n \geq n_0$, $P^n(x, A) > 0$ (see, e.g., de Acosta and Ney, 2014, Proposition 2.10).

   Let $m \in \mathbb{N}$, $\alpha > 0$, $\nu \in \mathcal{P}(S)$ be such that $P^m \geq \alpha(\mathbf{1}_S \otimes \nu)$. Given $A \in \mathcal{S}^+$, let $n \in \mathbb{N}$ be such that $\beta = \nu P^n(A) > 0$. Then for $k \geq 1$,

$$P^{m+n+k}\mathbf{1}_A \geq P^k(\alpha\mathbf{1}_S \otimes \nu)P^n\mathbf{1}_A \geq \alpha\beta,$$

   proving the condition stated above. Together with statement 1, this shows that P is uniformly ergodic.

   In the converse direction, assume that $A \in \mathcal{S}$ is petite: for some $\alpha > 0$, $m \in \mathbb{N}$, $\nu \in \mathcal{P}(S)$,

$$\sum_{j=1}^{m} P^j \geq \alpha(\mathbf{1}_A \otimes \nu).$$

   Let (t, λ) be a P-small pair such that $\int t\, d\lambda > 0$. By the proof of Proposition 2.10 of de Acosta and Ney (2014), there exists $p \in \mathbb{N}$ such that for $n \geq p$,

$$P^n \geq \beta_n(t \otimes \lambda) \quad \text{for some } \beta_n > 0.$$

   By irreducibility, there exists $q \in \mathbb{N}$ such that $\nu P^q t > 0$. Let $\beta > 0$ be such that for $p \leq n \leq p + m + q$,

$$P^n \geq \beta(t \otimes \lambda).$$

   We have: for $1 \leq n \leq m + q$,

$$P^{p+m+q} = P^n P^{(p+m+q)-n} \geq P^n\beta(t \otimes \lambda).$$

   Therefore

$$\begin{aligned}(m+q)P^{p+m+q} &\geq \left(\sum_{n=1}^{m+q} P^n\right)\beta(t \otimes \lambda) \geq \left(\sum_{n=1}^{m} P^n\right)P^q\beta(t \otimes \lambda)\\ &\geq \alpha(\mathbf{1}_A \otimes \nu)P^q\beta(t \otimes \lambda)\\ &= \alpha\beta(\nu P^q t)\mathbf{1}_A \otimes \lambda,\end{aligned}$$

$$P^{p+m+q} \geq \gamma(\mathbf{1}_A \otimes \lambda),$$

   with $\gamma = (m+q)^{-1}\alpha\beta(\nu P^q t) > 0$. This shows that A is small. But by statement 1 we can take $A = S$. □

We close this section with some definitions, stated for easy reference (see, e.g., Nummelin, 1984). P is *Harris recurrent* if it is irreducible and for all $x \in S$, $A \in \mathcal{S}^+$, $\mathbb{P}_x[V_A = \infty] = 1$, where $V_A = \sum_{j=1}^{\infty} \mathbf{1}_A(X_j)$; it is *positive Harris recurrent* if it is Harris recurrent and has an invariant probability measure (necessarily unique, by Lemma 5.6). P is *ergodic* if it is positive Harris recurrent and aperiodic. P is *geometrically ergodic* if it is ergodic and for some small set C,

$$\lambda(C) = \sup\left\{\lambda \geq 0 : \sup_{x \in C} \mathbb{E}_x e^{\lambda\tau} < \infty\right\} > 0,$$

where $\tau = \tau_C$.

B.4 Feller Kernels on a Polish Space and Petite Sets

Let S be a Polish space. We recall that a Markov kernel Q on $(S, \mathcal{S})$ is *Feller* (respectively, *strong Feller*) if $Qf \in C_b(S)$ for all $f \in C_b(S)$ (respectively, $f \in B(S)$).

Proposition B.10 *Assume that S is Polish and P is irreducible.*

1. *Suppose that there exists a petite set $C \in \mathcal{S}^+$ such that for all $n \in \mathbb{N}$, $P^n \mathbf{1}_C$ is lower-semicontinuous. Then every compact subset of S is petite.*
2. *Suppose that P is Feller and there exists an open petite set $C \in \mathcal{S}^+$. Then the conclusion of* 1 *holds.*
3. *Suppose that P is strong Feller. Then the conclusion of* 1 *holds.*

Proof.

1. Let $f = \sum_{n=1}^{\infty} 2^{-n} P^n \mathbf{1}_C$. Then f is lower-semicontinuous. Also, $f(x) > 0$ for all $x \in S$ by irreducibility. Therefore if $K \subset S$ is compact, we have

$$\alpha = \inf\{f(x) : x \in K\} > 0.$$

Let $k \in \mathbb{N}$ be such that $2^{-k} < \alpha/2$. Then

$$\sum_{n=1}^{k} 2^{-n} P^n \mathbf{1}_C \geq f - 2^{-k} \geq \left(\frac{\alpha}{2}\right) \mathbf{1}_K.$$

Let $m \in \mathbb{N}$, $\beta > 0$, $\lambda \in \mathcal{P}(S)$ be such that

$$\sum_{j=1}^{m} P^j \geq \beta(\mathbf{1}_C \otimes \lambda).$$

Then

$$\left(\sum_{n=1}^{k} P^n\right)\left(\sum_{j=1}^{m} P^j\right) \geq \beta\left(\sum_{n=1}^{k} P^n \mathbf{1}_C\right) \otimes \lambda \geq \beta\left(\frac{\alpha}{2}\right)(\mathbf{1}_K \otimes \lambda),$$

showing that K is petite.

2. If P is Feller and C is open, then $P^n \mathbf{1}_C$ is lower-semicontinuous for all $n \in \mathbb{N}$.
3. If P is strong Feller, then for every $C \in \mathcal{S}$, $P^n \mathbf{1}_C$ is continuous for all $n \in \mathbb{N}$. □

Proposition B.11 *Assume that S is Polish, and P is irreducible and strong Feller. Then there exists a small open set $U \in \mathcal{S}^+$.*

Proof. Let $C \in \mathcal{S}^+$ be a small set. Then by irreducibility there exists $k \in \mathbb{N}$ such that

$$\psi(A_k) > 0, \quad \text{where } A_k = \left\{x \colon P^k(x, C) > 0\right\}.$$

Let $b > 0$ be such that

$$U = \left\{x \colon P^k \mathbf{1}_C(x) > b\right\} \in \mathcal{S}^+.$$

Since C is small, there exist $m \in \mathbb{N}$, $\alpha > 0$, $\lambda \in \mathcal{P}(S)$ such that

$$P^m \geq \alpha(\mathbf{1}_C \otimes \lambda).$$

Then

$$P^{m+k} \geq \alpha(P^k \mathbf{1}_C) \otimes \lambda \geq \alpha b(\mathbf{1}_U \otimes \lambda).$$

This shows that U is small. U is open by the strong Feller property. □

Notes. The definition of K-small function (set) given here differs from that given in Nummelin (1984, Chap. 2) and de Acosta and Ney (2014, Chap. 2), in that we do not assume $t \in \mathcal{S}^+$; this condition is introduced here when relevant.

In de Acosta and Ney (2014), K-petite measures are called K-weakly small measures.

The definition of petite set given here is equivalent to that given in Meyn and Tweedie (1993, Chap. 5). Proposition B.5 and the fact that if P is aperiodic and A is petite then A is small (proof of Proposition B.9) are proved in Meyn and Tweedie (1993). The proofs presented here are phrased somewhat differently.

Proposition B.9(2) is proved in Nummelin (1984, Chap. 6).

The relationship between compact sets and petite sets is discussed in Meyn and Tweedie (1993, Chap. 6); this is the basis for Propositions B.10 and B.11.

Appendix C

The Convergence Parameter

In this appendix we define the convergence parameter of an irreducible kernel K and discuss its properties, on the basis of the framework presented in Section B.1–Section B.3 of Appendix B. It will be assumed that either K^n is σ-finite for all $n \in \mathbb{N}$ or $K = K_g$ for some $g \in F(S)$. By the discussion in Section B.2, it then follows that there exists a minorization of the form (B.1) with $t \in \mathcal{S}^+ \cap B(S)$, $\lambda \in \mathcal{M}^+(S)$, and $\int t \, d\lambda > 0$.

For an irreducible kernel K and a K-petite pair (t, λ), let

$$R(K; (t, \lambda)) = \sup \left\{ r \geq 0 \colon \sum_{n=0}^{\infty} r^n (\lambda K^n t) < \infty \right\},$$

the radius of convergence of the power series with coefficients $\{\lambda K^n t\}$. It turns out that $R(K; (t, \lambda))$ does not depend on (t, λ).

Lemma C.1 *Let K be irreducible and let (t_i, λ_i) $(i = 1, 2)$ be K-petite pairs. Then*

$$\sum_{n=0}^{\infty} r^n (\lambda_1 K^n t_1) < \infty \text{ if and only if } \sum_{n=0}^{\infty} r^n (\lambda_2 K^n t_2) < \infty.$$

In particular, $R(K; (t, \lambda))$ does not depend on (t, λ).

Proof. Let $m \in \mathbb{N}$, $\alpha > 0$ be such that

$$H = \sum_{j=1}^{m} K^j \geq \alpha (t_1 \otimes \lambda_1).$$

By the irreducibility of K, and hence of H, there exists $p \in \mathbb{N}$ such that $\beta = \lambda_2 H^p t_1 > 0$. Also, there exist $q \in \mathbb{N}$ and $\gamma > 0$ such that $H^{q+1} t_2 \geq \gamma t_1$.

For, again by irreducibility, $\lambda_1 H^q t_2 > 0$ for some $q \in \mathbb{N}$ and therefore $HH^q t_2 \geq \alpha t_1(\lambda_1 H^q t_2)$. Now let $\gamma = \alpha(\lambda_1 H^q t_2)$. We have, for $n \in \mathbb{N}$,

$$\begin{aligned}\lambda_2 H^{p+1} K^n H^{q+1} t_2 &\geq \lambda_2 H^p \left(\alpha(t_1 \otimes \lambda_1)\right) K^n \gamma t_1 \\ &= \alpha\beta\gamma(\lambda_1 K^n t_1).\end{aligned}$$

Now $H^{p+1}H^{q+1} \leq c\sum_{i=1}^{k} K^i$ for some constant c, where $k = p + q + 2$, and therefore

$$\begin{aligned}\sum_{n=0}^{\infty} r^n(\lambda_2 H^{p+1} K^n H^{q+1} t_2) &\leq c \sum_{n=0}^{\infty} r^n \left[\lambda_2 K^n \left(\sum_{i=1}^{k} K^i\right) t_2\right] \\ &= c \sum_{i=1}^{k} \sum_{n=0}^{\infty} r^n(\lambda_2 K^{n+i} t_2).\end{aligned}$$

It follows that $\sum_{n=0}^{\infty} r^n(\lambda_2 K^n t_2) < \infty$ implies $\sum_{n=0}^{\infty} r^n(\lambda_1 K^n t_1) < \infty$. Reversing the roles of (t_1, λ_1) and (t_2, λ_2), the converse implication is obtained. □

For an irreducible kernel K, the common value of $R(K; (t, \lambda))$ for (t, λ) a K-petite pair is called the *convergence parameter of K* and is denoted $R(K)$.

By an elementary fact about power series,

$$R(K) = \left[\overline{\lim_n}\, (\lambda K^n t)^{1/n}\right]^{-1}. \tag{C.1}$$

We have: $0 \leq R(K) < \infty$. For, let (t, λ) be a K-small pair with $\int t\, d\lambda > 0$, say $K^m \geq t \otimes \lambda$. Then

$$\lambda K^{mn} t \geq \lambda(t \otimes \lambda)^n t = \left(\int t\, d\lambda\right)^{n+1}$$

and therefore by (C.1),

$$R(K) \leq \left[\overline{\lim_n}\, (\lambda K^{mn} t)^{1/mn}\right]^{-1} \leq \left(\int t\, d\lambda\right)^{-1/m}.$$

Proposition C.3 gives a refinement of (C.1).

For the proof of Proposition C.3 we must first discuss briefly the notions of period and cycle associated with an irreducible kernel K.

Let (t, λ) be a K-small pair such that $\int t\, d\lambda > 0$ and

$$\begin{aligned}D(K; (t, \lambda)) &= \{n \geq 1 : K^n \geq \alpha_n(t \otimes \lambda) \text{ for some } \alpha_n > 0\}, \\ d(K; (t, \lambda)) &= \text{greatest common divisor } D(K; (t, \lambda)).\end{aligned}$$

The number $d(K; (t, \lambda))$ does not depend on the particular choice of (t, λ) satisfying $\int t\, d\lambda > 0$; it is called the *period of K* and it is denoted $d(K)$ (see, e.g., Section 2.4 of de Acosta and Ney, 2014).

For $m \in \mathbb{N}$, a finite sequence $\{E_0, \dots, E_{m-1}\}$ of disjoint nonempty sets in $\mathcal{S}$ is called an *m-cycle* (associated with K) if for $0 \le i < m-1$, $x \in E_i$, we have $K(x, E_{i+1}^c) = 0$, and $K(x, E_0^c) = 0$ for $x \in E_{m-1}$. It is proved in Nummelin (1984, Th. 2.2) that a d-cycle exists and is unique in a suitable sense, where $d = d(K)$. Let

$$F = \bigcup_{i=0}^{d-1} E_i.$$

Then F is K-closed: if $x \in K$, then $K(x, F^c) = 0$. It follows that $\psi(F^c) = 0$. It is easily seen that for all $n \in \mathbb{N}$, $i = 0, \dots, d-1$, $x \in E_i$,

$$K^n(x, E_j^c) = 0 \qquad \text{if } j \equiv i + n \pmod{d}. \tag{C.2}$$

The following lemma contains the facts related to the period d and d-cycle associated with K which are needed in Proposition C.3. Let $d\mathbb{N} = \{hd\colon h \in \mathbb{N}\}$.

Lemma C.2 *Assume that $K^{hd} \ge \alpha(t \otimes \lambda)$ for some $h \in \mathbb{N}$, $\alpha > 0$, and K-small pair (t, λ) such that $\int t\, d\lambda > 0$. Then:*

$$\textit{For all } n \notin d\mathbb{N},\ \lambda K^n t = 0. \tag{C.3}$$

$$\textit{There exist } p \in \mathbb{N}, c > 0, \textit{ such that for } j = 1, \dots, h-1,$$
$$K^{jd} t \le c K^{phd} t. \tag{C.4}$$

Proof. We first prove that there exists $i \in \{0, \dots, d-1\}$ such that

$$[t > 0] \subset E_i \cup N, \quad \text{where } N = F^c; \tag{C.5}$$
$$\lambda(E_i^c) = 0. \tag{C.6}$$

We have $[t > 0] \not\subset N$. For, if $[t > 0] \subset N$, then since $\psi(N) = 0$ it follows that $\int t\, d\psi = 0$, which is impossible given that $\lambda \ll \psi$ and $\int t\, d\lambda > 0$. Therefore there exists $i \in \{0, \dots, d-1\}$ such that $[t > 0] \cap E_i \ne \emptyset$. If $x \in [t > 0] \cap E_i$, then by (C.2),

$$0 = K^{hd}(x, E_i^c) \ge \alpha t(x) \lambda(E_i^c),$$

which implies $\lambda(E_i^c) = 0$. i is unique: for, if $[t > 0] \cap E_j \ne \emptyset$ for some $j \ne i$, the previous argument yields $\lambda(E_j^c) = 0$, which is impossible. This proves (C.5) and (C.6).

Next, by (C.5), $t \le a(\mathbf{1}_{E_i} + \mathbf{1}_N)$, where $a = \sup t$, and for $n \in \mathbb{N}$,

$$K^n t \le a K^n \mathbf{1}_{E_i} + a K^n \mathbf{1}_N. \tag{C.7}$$

Since λK^n is a K-small measure, hence $\lambda \ll \psi$, we have for any $n \in \mathbb{N}$,

$$\lambda K^n \mathbf{1}_N = 0. \tag{C.8}$$

If $n \notin d\mathbb{N}$, say $n = kd + \ell$ with $\ell \in \{1, \ldots, d-1\}$, then by (C.2) for $x \in E_i$ we have $K^n(x, E_j^c) = 0$ if $j \equiv i + (kd + \ell) \pmod d$, or $j \equiv i + \ell \pmod d$. Since $j \neq i$, we have $E_i \subset E_j^c$ and therefore for $x \in E_i$,

$$K^n 1_{E_i}(x) = 0. \tag{C.9}$$

By (C.6), $\lambda(E_i^c) = 0$. Now (C.3) follows from (C.7)–(C.9).

For $p \in \mathbb{N}$, writing $(p-1)h = j + ((p-1)h - j)$, we have

$$\begin{aligned} K^{phd} &= K^{hd} K^{(p-1)hd} \\ &\geq \alpha\left(K^{jd} t \otimes \lambda K^{((p-1)h-j)d}\right), \\ K^{phd} t &\geq \alpha c(p, j) K^{jd} t, \end{aligned}$$

where $c(p, j) = \lambda K^{((p-1)h-j)d} t$.

Since the set $D(K; (t, \lambda))$ is closed under addition, by a well-known number-theoretic result (Seneta, 1981, p. 248), there exists $\ell_0 \in \mathbb{N}$ such that

$$D(K; (t, \lambda)) \supset \{\ell d \colon \ell \geq \ell_0, \ell \in \mathbb{N}\}.$$

If $\ell \geq \ell_0$, then $K^{\ell d} \geq \beta(t \otimes \lambda)$ for some $\beta > 0$ and therefore

$$\lambda K^{\ell d} t \geq \beta \left(\int t \, d\lambda \right)^2 > 0. \tag{C.10}$$

Choosing p such that $(p-1)h - (h-1) \geq \ell_0$, by (C.10) we have $c(p, j) > 0$ for $j = 1, \ldots, h-1$, and (C.4) follows. □

The key feature of the following result is the expression of $(-\log R(K))$ as a certain supremum. By our assumptions on the class of irreducible kernels we consider, the assumption in Proposition C.3 is automatically satisfied.

Proposition C.3 *Let K be an irreducible kernel on $(S, \mathcal{S})$. Let $m \in \mathbb{N}$, $t \in \mathcal{S}^+ \cap B(S)$, and $\lambda \in \mathcal{M}^+(S)$ be such that $\int t \, d\lambda > 0$ and*

$$K^m \geq t \otimes \lambda.$$

Then

$$\begin{aligned} L &= \lim_n (nm)^{-1} \log \lambda K^{(n-1)m} t \text{ exists and} \\ L &= \sup_n (nm)^{-1} \log \lambda K^{(n-1)m} t. \end{aligned} \tag{C.11}$$

$$L = -\log R(K). \tag{C.12}$$

Proof. Let $b_n = \lambda K^{(n-1)m} t$. Then $\{b_n\}_{n \geq 1}$ is supermultiplicative. For,

$$\begin{aligned} b_{n+k} = \lambda K^{(n+k-1)m} t &= \lambda K^{(n-1)m} K^m K^{(k-1)m} \\ &\geq \lambda K^{(n-1)m} (t \otimes \lambda) K^{(k-1)m} \\ &= \left(\lambda K^{(n-1)m} t\right)\left(\lambda K^{(k-1)m} t\right) = b_n b_k. \end{aligned}$$

Also, since

$$K^{nm} \geq \left(\int t \, d\lambda\right)^{n-1} (t \otimes \lambda),$$

we have for $n \geq 1$,

$$b_n \geq \left(\int t \, d\lambda\right)^{n+1} > 0.$$

Let $a_n = -\log b_n$. Then $\{a_n\}_{n \geq 1}$ is subadditive and for $n \geq 1$, $a_n < \infty$. Therefore by Fekete's lemma (Rassoul-Agha and Seppäläinen, 2015, p. 63), $\ell = \lim_n (n^{-1} a_n)$ exists and

$$\ell = \inf_n (n^{-1} a_n). \tag{C.13}$$

Therefore $L = \lim_n (nm)^{-1} \log b_n$ exists and equals $(-m^{-1}\ell)$. Taking into account (C.13), assertion (C.11) follows.

By (C.3) of Lemma C.2,

$$R(K) = \sup\left\{r \geq 0 \colon \sum_{n=0}^{\infty} r^{nd} \lambda K^{nd} t < \infty\right\}.$$

Let

$$\begin{aligned} R' &= \sup\left\{\rho \geq 0 \colon \sum_{n=0}^{\infty} \rho^{nm} \lambda K^{nm} t < \infty\right\} \\ &= \sup\left\{\rho \geq 0 \colon \sum_{n=0}^{\infty} \rho^{nm} \lambda K^{(n-1)m} t < \infty\right\}. \end{aligned}$$

We claim that $R(K) = R'$. Since $m \in D(K; (t, \lambda))$, it follows that $m = hd$ for some $h \in \mathbb{N}$. Therefore $R(K) \leq R'$. On the other hand, writing $n = kh + j$ for some $k \in \mathbb{N}$ and $j \in \{0, \ldots, h-1\}$, we have

$$\sum_{n=0}^{\infty} r^{nd} \lambda K^{nd} t = \sum_{j=0}^{h-1} \sum_{k=0}^{\infty} r^{(kh+j)d} \lambda K^{(kh+j)d} t. \tag{C.14}$$

By (C.4) of Lemma C.2, there exist $p \in \mathbb{N}$, $c > 0$ such that

$$\begin{aligned} K^{(kh+j)d} t &= K^{khd} K^{jd} t \\ &\leq c K^{(k+p)hd} t, \end{aligned}$$

for $j = 0, \ldots, h-1$. Therefore (C.14) continues as

$$\sum_{n=0}^{\infty} r^{nd} \lambda K^{nd} t = \cdots \leq c \sum_{j=0}^{h-1} r^{jd-phd} \sum_{k=0}^{\infty} r^{(k+p)hd} \lambda K^{(k+p)hd} t,$$

and it follows that $R(K) \geq R'$, proving the claim.

Now $(R')^m$ is the radius of convergence of the power series with coefficients $\{\lambda K^{(n-1)m} t\}$, and therefore by (C.1),

$$(R')^m = \left[\overline{\lim_n} \left(\lambda K^{(n-1)m} t \right)^{1/n} \right]^{-1}.$$

Finally, by (C.11) and the claim,

$$\begin{aligned} -\log R(K) &= -\log R' \\ &= -m^{-1} \log\left[(R')^m\right] = L. \end{aligned}$$

□

The main application of Proposition C.3 is to the proof of Theorem 2.13. Here we present several other consequences.

Corollary C.4 *Let K_j, $j \in \mathbb{N}$, K be irreducible kernels on $(S, \mathcal{S})$. Assume: for all $x \in \mathcal{S}$, $A \in \mathcal{S}$, $K_j(x, A) \uparrow K(x, A)$. Then $R(K_j) \downarrow R(K)$.*

Proof. By the irreducibility of K_1, there exist $m \in \mathbb{N}$ and a K_1-small pair (t, λ) such that $\int t \, d\lambda > 0$ and $K_1^m \geq t \otimes \lambda$, and therefore $K_j^m \geq t \otimes \lambda$ for $j \geq 1$ and $K^m \geq t \otimes \lambda$.

By a simple extension of the monotone convergence theorem (Royden, 1988, pp. 268–270), for all $p \in \mathbb{N}$,

$$\lambda K_j^p t \uparrow \lambda K^p t. \tag{C.15}$$

Let $\Lambda(K_j) = -\log R(K_j)$, $\Lambda(K) = -\log R(K)$. Then by Proposition C.3 and (C.15),

$$\begin{aligned} \lim_j \Lambda(K_j) = \sup_j \Lambda(K_j) &= \sup_j \sup_n (nm)^{-1} \log \lambda K_j^{(n-1)m} t \\ &= \sup_n \sup_j (nm)^{-1} \log \lambda K_j^{(n-1)m} t \\ &= \sup_n (nm)^{-1} \log \lambda K^{(n-1)m} t = \Lambda(K), \end{aligned}$$

and the conclusion follows. □

Let P be irreducible. Then K_g is irreducible for all $g \in F(S)$ and $\Lambda\colon F(S) \to (-\infty, \infty]$ is defined by

$$\Lambda(g) = -\log R(K_g).$$

Corollary C.5 *Let P be irreducible. Let g_j, $j \in \mathbb{N}$, $g \in F(S)$, and assume that $g_j \uparrow g$ pointwise. Then $\Lambda(g_j) \uparrow \Lambda(g)$.*

Proof. By monotone convergence, $K_{g_j}(x, A) \uparrow K_g(x, A)$ for all $x \in S$, $A \in \mathcal{S}$. Apply now Corollary C.4. □

Corollary C.6 *Assume that K and K^p are irreducible, where $p \in \mathbb{N}$. Then $R(K^p) = (R(K))^p$.*

Proof. By the irreducibility of K^p, there exist $m \in \mathbb{N}$ and a K^p-small pair (t, λ) such that $\int t\, d\lambda > 0$ and $(K^p)^m \geq t \otimes \lambda$. By Proposition C.3, applied to K^p,

$$- \log R(K^p) = \sup_n (nm)^{-1} \log \lambda (K^p)^{(n-1)m} t. \tag{C.16}$$

Since (t, λ) is a K-small pair, applying Proposition C.3 to K we have

$$- \log R(K) = \sup_n (npm)^{-1} \log \lambda K^{(n-1)pm} t. \tag{C.17}$$

Equalities (C.16) and (C.17) imply the conclusion. □

Notes. The definition of convergence parameter given here is equivalent to that given in Nummelin (1984); see Definition 3.2 and Proposition 3.4 there.

The observation that a subadditivity property is present in the definition of Λ was made in de Acosta and Ney (2014, Prop. 2.13).

Corollary C.4 was proved in de Acosta (1988, Th. 2.1) by a different approach, using the renewal characterization of the convergence parameter.

Corollary C.6 is implicit in Nummelin (1984, Prop. 3.5) and explicitly stated and proved for P aperiodic in de Acosta and Ney (2014, Cor. 2.14).

Appendix D

Approximation of P by P_t

Let P be an irreducible Markov kernel on $(S, \mathcal{S})$. We wish to approximate P in a suitable sense by another Markov kernel satisfying the minorization condition (2.3).

For $0 < t < 1$, let P_t be the Markov kernel

$$P_t = (1 - t)P \sum_{n=0}^{\infty} t^n P^n.$$

By the basic minorization theorem for irreducible kernels (see Appendix B), there exist $m \in \mathbb{N}$, $C \in \mathcal{S}^+$, $\nu \in \mathcal{P}(S)$, $\beta > 0$ such that

$$P^m \geq \beta(\mathbf{1}_C \otimes \nu). \tag{D.1}$$

As pointed out in Section B.2 of Appendix B, we may assume without loss of generality that $\nu(C) > 0$. From (D.1) it follows that

$$\begin{aligned} P_t &\geq (1 - t)t^{m-1} P^m \\ &\geq \alpha_t(\mathbf{1}_C \otimes \nu), \end{aligned} \tag{D.2}$$

with $\alpha_t = \beta(1 - t)t^{m-1}$, which is the desired minorization condition.

Let $\mathbb{P}_\mu^{(t)}$ be the Markovian probability measure on $(S^{\mathbb{N}_0}, \mathcal{S}^{\mathbb{N}_0})$ with transition kernel P_t and initial distribution μ, and let $\mathbb{E}_\mu^{(t)}$ be the associated expectation functional.

Let $\sigma_t(n)$ ($n \in \mathbb{N}$) be the nth partial sum of an i.i.d. geometric $(1-t)$ sequence of r.v.'s, and let $\sigma_t(0) = 0$; that is, $\{\sigma_t(n) - \sigma_t(n-1) \colon n \geq 1\}$ is an i.i.d. sequence with

$$\mathbb{P}\left[\sigma_t(1) = j\right] = (1 - t)t^{j-1}, \qquad j \in \mathbb{N},$$

defined on some probability space $(\Omega', \mathcal{A}, \mathbb{P})$, with associated expectation functional $\mathbb{E}$. Let $\Phi_t \colon \Omega' \times S^{\mathbb{N}_0} \to S^{\mathbb{N}_0}$ be defined by

$$\Phi_t\Big(\omega, (x_j)_{j\geq 0}\Big) = \left(x_{\sigma_t(n)(\omega)}\right)_{n\geq 0}.$$

$\mathbb{E}_\mu$ and $\mathbb{E}_\mu^{(t)}$ are related as follows:

Lemma D.1 *For every bounded measurable function* $F\colon S^{\mathbb{N}_0} \to \mathbb{R}^+$, $\mu \in \mathcal{P}(S)$,

$$\mathbb{E}\mathbb{E}_\mu(F \circ \Phi_t) = \mathbb{E}_\mu^{(t)} F.$$

Proof. By well-known arguments, it suffices to prove: for every $n \in \mathbb{N}$, every bounded measurable function $f_j\colon S \to \mathbb{R}^+$, $j = 0, \ldots, n$,

$$\mathbb{E}\mathbb{E}_\mu\left[\prod_{j=0}^{n} f_j\left(X_{\sigma_t(j)}\right)\right] = \mathbb{E}_\mu^{(t)}\left[\prod_{j=0}^{n} f_j(X_j)\right]. \tag{D.3}$$

We prove (D.3) by induction. For $n = 0$, both sides of (D.3) equal $\int f_0\, d\mu$. Assume that (D.3) holds for n. Setting $\tau_{n+1} = \sigma_t(n+1) - \sigma_t(n)$, we have

$$\begin{aligned}\mathbb{E}\mathbb{E}_\mu\left[\prod_{j=0}^{n+1} f_j\left(X_{\sigma_t(j)}\right)\right] &= \mathbb{E}\mathbb{E}_\mu\left[\prod_{j=0}^{n} f_j\left(X_{\sigma_t(j)}\right) f_{n+1}\left(X_{\sigma_t(n)+\tau_{n+1}}\right)\right]\\ &= \sum_{i=1}^{\infty}(1-t)t^{i-1}\mathbb{E}\mathbb{E}_\mu\left[\prod_{j=0}^{n} f_j\left(X_{\sigma_t(j)}\right) f_{n+1}\left(X_{\sigma_t(n)+i}\right)\right].\end{aligned} \tag{D.4}$$

Now

$$\mathbb{E}_\mu\left[\prod_{j=0}^{n} f_j\left(X_{\sigma_t(j)}\right) f_{n+1}\left(X_{\sigma_t(n)+i}\right)\right] = \mathbb{E}_\mu\left[\prod_{j=0}^{n} f_j\left(X_{\sigma_t(j)}\right) P^i f_{n+1}\left(X_{\sigma_t(n)}\right)\right].$$

From this equality, (D.4), and the inductive hypothesis,

$$\begin{aligned}\mathbb{E}\mathbb{E}_\mu\left[\prod_{j=0}^{n+1} f_j\left(X_{\sigma_t(j)}\right)\right] &= \sum_{i=1}^{\infty}(1-t)t^{i-1}\mathbb{E}_\mu^{(t)}\left[\prod_{j=0}^{n} f_j(X_j)P^i f_{n+1}(X_n)\right]\\ &= \mathbb{E}_\mu^{(t)}\left\{\prod_{j=0}^{n} f_j(X_j)\left[\sum_{i=1}^{\infty}(1-t)t^{i-1}P^i\right] f_{n+1}(X_n)\right\}\\ &= \mathbb{E}_\mu^{(t)}\left[\prod_{j=0}^{n} f_j(X_j)P_t f_{n+1}(X_n)\right] = \mathbb{E}_\mu^{(t)}\left[\prod_{j=0}^{n+1} f_j(X_j)\right],\end{aligned}$$

proving (D.3). □

Lemma D.1 is the basis for Lemma 2.7, which is one aspect of the approximation of P by P_t needed in the proof of Theorem 2.1. Another aspect is Proposition D.2. Actually, only claim (D.5) of that result is used in Theorem 2.1, but the more comprehensive statement given below appears to be worth proving.

For $g \in B(S)$, $0 < t < 1$, let

$$K_g^{(t)}(x, A) = \int e^{g(y)} \mathbf{1}_A(y) P_t(x, dy), \qquad x \in S,\ A \in \mathcal{S},$$
$$\Lambda_t(g) = -\log R\left(K_g^{(t)}\right),$$

where, as in Appendix C, $R(K_g^{(t)})$ is the convergence parameter of the irreducible kernel $K_g^{(t)}$.

Proposition D.2 *Assume that P is irreducible and let $g \in B(S)$.*

$$\textit{For } 0 < t < 1, \quad \Lambda_t(g) \geq \Lambda(g) + \log(1-t). \tag{D.5}$$

$$\textit{If } g \geq 0,\ 0 < t < e^{-\Lambda(g)}, \quad \Lambda_t(g) \leq \Lambda(g) + \log\left\{(1-t)\big/\left(1 - te^{\Lambda(g)}\right)\right\}. \tag{D.6}$$

$$\lim_{t \to 0^+} \Lambda_t(g) = \Lambda(g). \tag{D.7}$$

Proof. Since $P_t \geq (1-t)P$, we have

$$R\left(K_g^{(t)}\right) \leq (1-t)^{-1} R(K_g),$$
$$\Lambda_t(g) \geq \Lambda(g) + \log(1-t).$$

Let (s, ν) be a small pair. Then it is also a K_g-small pair and a $K_g^{(t)}$-small pair, and we have

$$\Lambda(g) = \overline{\lim_n}\, n^{-1} \log \nu K_g^n s,$$
$$\Lambda_t(g) = \overline{\lim_n}\, n^{-1} \log \nu \left(K_g^{(t)}\right)^n s.$$

Given $\epsilon > 0$, let $n_0 \in \mathbb{N}$ be such that for $n \geq n_0$,

$$n^{-1} \log \nu K_g^n s \leq \Lambda(g) + \epsilon,$$

or

$$\mathbb{E}_\nu \left(\exp \sum_{j=1}^n g(X_j)\right) s(X_n) \leq \exp\left[n\left(\Lambda(g) + \epsilon\right)\right]. \tag{D.8}$$

By Lemma D.1 with

$$F\left((x_j)_{j \geq 0}\right) = \left(\exp \sum_{j=1}^n g(x_j)\right) s(x_n),$$

and since $\sigma_t(n) \geq n$ for all $n \in \mathbb{N}$, $0 < t < 1$, from (D.8) we have, for $n \geq n_0$,

$$\begin{aligned} \nu\left(K_g^{(t)}\right)^n s = \mathbb{E}_\nu^{(t)} F &= \mathbb{E}\mathbb{E}_\nu \left[\exp \sum_{j=1}^{n} g\left(X_{\sigma_t(j)}\right)\right] s\left(X_{\sigma_t(n)}\right) \\ &\leq \mathbb{E}\mathbb{E}_\nu \left[\exp \sum_{j=1}^{\sigma_t(n)} g(X_j)\right] s\left(X_{\sigma_t(n)}\right) \\ &\leq \mathbb{E} \exp\left[\sigma_t(n)\left(\Lambda(g) + \epsilon\right)\right]. \end{aligned} \tag{D.9}$$

By an elementary computation, for $0 < \lambda < -\log t$,

$$\mathbb{E} \exp\left(\lambda \sigma_t(n)\right) = \left[\frac{(1-t)e^\lambda}{1 - te^\lambda}\right]^n, \tag{D.10}$$

and (D.9) continues as

$$\nu\left(K_g^{(t)}\right)^n s = \cdots \leq \left[\frac{(1-t)e^{\Lambda(g)+\epsilon}}{1 - te^{\Lambda(g)+\epsilon}}\right]^n$$

for $0 < t < e^{-(\Lambda(g)+\epsilon)}$. Therefore

$$\begin{aligned} \Lambda_t(g) &= \overline{\lim_n}\, n^{-1} \log \nu\left(K_g^{(t)}\right)^n s \\ &\leq \Lambda(g) + \epsilon + \log\left\{(1-t)\big/\left(1 - te^{\Lambda(g)+\epsilon}\right)\right\}. \end{aligned}$$

Since ϵ is arbitrary, assertion (D.6) follows.

By (D.5),

$$\underline{\lim}_{t \to 0^+} \Lambda_t(g) \geq \Lambda(g).$$

Let $c = \inf g$. Letting $h = g - c$, we have

$$\begin{aligned} \Lambda_t(g) &= \Lambda_t(h) + c, \\ \Lambda(g) &= \Lambda(h) + c, \end{aligned}$$

so now by (D.6),

$$\begin{aligned} \overline{\lim}_{t \to 0^+} \Lambda_t(g) &= \overline{\lim}_{t \to 0^+} \Lambda_t(h) + c \\ &\leq \Lambda(h) + c = \Lambda(g). \end{aligned}$$

□

Notes. The Markov kernel P_t and a form of Lemma 2.7 appear in the proof of the lower bound in Donsker and Varadhan (1976). Lemma D.1 is implicit in de Acosta (1988).

Appendix E

On Varadhan's Theorem

Proposition E.1 *Let E be a topological space, $\mathcal{E}$ a σ-algebra of subsets of E. Let $\{\gamma_n : n \in \mathbb{N}\}$ be a sequence of subprobability measures on $\mathcal{E}$.*

Let $J : E \to \overline{\mathbb{R}^+}$ be lower semicontinuous. Assume: for every set $B \in \mathcal{E}$,

$$\overline{\lim_n}\, n^{-1} \log \gamma_n(B) \leq -\inf\left\{J(x) : x \in \bar{B}\right\}, \tag{E.1}$$

$$\underline{\lim_n}\, n^{-1} \log \gamma_n(B) \geq -\inf\left\{J(x) : x \in B^0\right\}. \tag{E.2}$$

Then for every $F : E \to \mathbb{R}$ which is bounded above, measurable, and continuous,

$$\lim_n n^{-1} \log \int \exp(nF)\, d\gamma_n = \sup\{F(x) - J(x) : x \in E\}. \tag{E.3}$$

Proof.

1. Let $m = \sup\{F(x) : x \in E\}$. Let $a > 0$, $\epsilon > 0$ and $\{I_j : 1 \leq j \leq k\}$ be a family of closed intervals, with length $(I_j) < \epsilon$ for all j, such that

$$[-a, m] = \bigcup_{j=1}^{k} I_j.$$

Let $C_j = F^{-1}(I_j)$, $1 \leq j \leq k$, $C_0 = F^{-1}(-\infty, -a]$. Then

$$\begin{aligned}\int \exp(nF)\, d\gamma_n &\leq \sum_{j=0}^{k} \int_{C_j} \exp(nF)\, d\gamma_n \\ &\leq \sum_{j=1}^{k} \exp\left\{n\left[\inf(F \mid C_j) + \epsilon\right]\right\} \gamma_n(C_j) + e^{-na},\end{aligned}$$

and therefore

$$\begin{aligned}
&\overline{\lim_n}\, n^{-1} \log \int \exp(nF)\, d\gamma_n \\
&\quad \le \max\left\{ \max_{1\le j\le k} \left[\inf(F \mid C_j) + \epsilon + \overline{\lim_n}\, n^{-1} \log \gamma_n(C_j)\right], -a\right\} \\
&\quad \le \max\left\{ \max_{1\le j\le k} \left[\inf(F \mid C_j) - \inf(J \mid C_j) + \epsilon\right], -a\right\} \\
&\quad \le \max\left\{ \max_{1\le j\le k} \left[\sup\left\{F(x) - J(x) \colon x \in C_j\right\} + \epsilon\right], -a\right\} \\
&\quad = \max\left[\sup\{F(x) - J(x) \colon x \in E\} + \epsilon, -a\right].
\end{aligned}$$

But ϵ and a are arbitrary. Therefore

$$\overline{\lim_n}\, n^{-1} \log \int \exp(nF)\, d\gamma_n \le \sup\{F(x) - J(x) \colon x \in E\}.$$

2. Assume first that

$$L = \sup\{F(x) - J(x) \colon x \in E\} < \infty.$$

Given $\epsilon > 0$, let $x_0 \in E$ be such that

$$F(x_0) - J(x_0) > L - (\epsilon/2),$$

and

$$B = \{x \in E \colon F(x) > F(x_0) - (\epsilon/2)\}.$$

Then, since B is open,

$$\begin{aligned}
\int \exp(nF)\, d\gamma_n &\ge \int_B \exp(nF)\, d\gamma_n \\
&\ge \exp\{n\,[F(x_0) - (\epsilon/2)]\}\, \gamma_n(B),
\end{aligned}$$

$$\begin{aligned}
\underline{\lim_n}\, n^{-1} \log \int \exp(nF)\, d\gamma_n &\ge F(x_0) - (\epsilon/2) - \inf\{J(x) \colon x \in B\} \\
&\ge F(x_0) - J(x_0) - (\epsilon/2) > L - \epsilon.
\end{aligned}$$

But ϵ is arbitrary. The proof is similar if $L = \infty$. □

Remark E.2 It follows from the proof of Proposition E.1 that if F is bounded, measurable, and continuous, and (1) (E.1) holds for every set B which is the inverse image of a closed interval of $\mathbb{R}$ by F, (2) (E.2) holds, then the conclusion of Proposition E.1 holds.

The following result extends the upper bound in Proposition E.1 to a family of subprobability measures and certain unbounded functions.

Proposition E.3 *Let E, $\mathcal{E}$, and J be as in Proposition E.1. Let $\{\gamma_{\alpha,n}: \alpha \in A, n \in \mathbb{N}\}$ be a family of subprobability measures on $\mathcal{E}$, where A is an index set, and assume for every set $B \in \mathcal{E}$,*

$$\overline{\lim_n}\, n^{-1} \log \sup_{\alpha \in A} \gamma_{\alpha,n}(B) \leq -\inf\left\{J(x):\ x \in \bar{B}\right\}. \tag{E.4}$$

Let $F: E \to \mathbb{R}$ be measurable and continuous and assume:

$$\lim_{b\to\infty} \overline{\lim_n}\, n^{-1} \log \sup_{\alpha \in A} \int_{[F \geq b]} \exp(nF)\, d\gamma_{\alpha,n} = -\infty. \tag{E.5}$$

Then

$$\overline{\lim_n}\, n^{-1} \log \sup_{\alpha \in A} \int \exp(nF)\, d\gamma_{\alpha,n} \leq \sup\{F(x) - J(x):\ x \in E\}. \tag{E.6}$$

Proof. By the same method as step 1 in the proof of Proposition E.1, by assumption (E.4) and for any $b > 0$,

$$\overline{\lim_n}\, n^{-1} \log \sup_{\alpha \in A} \int \exp\left(n(F \wedge b)\right)\, d\gamma_{\alpha,n} \leq \sup\{F(x) \wedge b - J(x):\ x \in E\}.$$

Let

$$\ell(b) = \overline{\lim_n}\, n^{-1} \log \sup_{\alpha \in A} \int_{[F \geq b]} \exp(nF)\, d\gamma_{\alpha,n}.$$

Then

$$\int \exp(nF)\, d\gamma_{\alpha,n} \leq \int \exp\left(n(F \wedge b)\right)\, d\gamma_{\alpha,n} + \int_{[F \geq b]} \exp(nF)\, d\gamma_{\alpha,n},$$

$$\begin{aligned}\overline{\lim_n}\, n^{-1} \log \sup_{\alpha \in A} \int \exp(nF)\, d\gamma_{\alpha,n} &\leq \max\left\{\sup\{F(x) \wedge b - J(x):\ x \in E\}, \ell(b)\right\}\\ &\leq \max\left\{\sup\{F(x) - J(x):\ x \in E\}, \ell(b)\right\}.\end{aligned}$$

Letting $b \to \infty$, by assumption (E.5), the result follows. □

One of the main applications of Proposition E.3 in this work is as follows. Let $E = \mathcal{P}(S)$, endowed with the V topology, where V is a vector subspace of $B(S)$; let $\mathcal{E} = \mathcal{B}(\mathcal{P}(S), B(S))$; and let $\Phi_g: \mathcal{P}(S) \to \mathbb{R}$ be defined by $\Phi_g(\mu) = \int g\, d\mu$, where $g \in V$.

Corollary E.4 *Let* $\{\gamma_{\alpha,n}\colon \alpha \in A, n \in \mathbb{N}\}$ *be a family of subprobability measures on* $\mathcal{B}(\mathcal{P}(S), B(S))$. *Let* $J\colon \mathcal{P}(S) \to \overline{\mathbb{R}^+}$ *be V-lower semicontinuous and assume for every measurable set* $B \subset \mathcal{P}(S)$,

$$\overline{\lim_n}\, n^{-1} \log \sup_{\alpha\in A} \gamma_{\alpha,n}(B) \le -\inf\{J(\mu)\colon \mu \in \mathrm{cl}_V(B)\}. \tag{E.7}$$

Let

$$\Gamma_A(g) = \overline{\lim_n}\, n^{-1} \log \sup_{\alpha\in A} \int \exp(n\Phi_g)\, d\gamma_{\alpha,n}. \tag{E.8}$$

Then for all $g \in V$,

$$\Gamma_A(g) \le \sup\left\{\int g\, d\mu - J(\mu)\colon \mu \in \mathcal{P}(S)\right\}. \tag{E.9}$$

Notes. Proposition E.1 is due to S.R.S. Varadhan. This result, as well as Proposition E.3, is proved e.g., in Rassoul-Agha and Seppäläinen (2015, Chap. 3).

Appendix F

The Duality Theorem for Convex Functions

Let (X, Y) be a pair of real vector spaces in duality. That is, there is a bilinear form $\langle\cdot,\cdot\rangle: X \times Y \to \mathbb{R}$, which we assume to be separating: for each nonzero $x \in X$ (respectively, $y \in Y$) there exists $z \in Y$ (respectively, $w \in X$) such that $\langle x, z\rangle \neq 0$ (respectively, $\langle w, y\rangle \neq 0$). The weak topology $\sigma(X, Y)$ is the smallest topology on X such that for each $y \in Y$, the function $x \mapsto \langle x, y\rangle$ is continuous; the topology $\sigma(Y, X)$ on Y is defined similarly.

For $f: X \to [-\infty, \infty]$, its *convex conjugate* $f^*: Y \to [-\infty, \infty]$ is defined by

$$f^*(y) = \sup\{\langle x, y\rangle - f(x): x \in X\}.$$

The convex conjugate of a function $g: Y \to [-\infty, \infty]$ is defined similarly.

We state now the classical duality theorem (see, e.g., Rassoul-Agha and Seppäläinen, 2015, p. 53). $f: X \to \overline{\mathbb{R}}$ is *proper* if $-\infty \notin f(X)$ and f is not identically equal to ∞.

Proposition F.1 *Suppose $f: X \to \overline{\mathbb{R}}$ is proper. Then the following conditions are equivalent:*

$$f \text{ is convex and } \sigma(X, Y)\text{-lower semicontinuous.} \tag{F.1}$$

$$\text{For all } x \in X,\ f(x) = \sup\{\langle x, y\rangle - f^*(y): y \in Y\}. \tag{F.2}$$

We will apply Proposition F.1 to the case when $X = \mathcal{M}(S)$, the vector space of finite signed measures on $(S, \mathcal{S})$ and $Y = V$, where V satisfies V.1–V.3 as in Chapter 3. For $\mu \in \mathcal{M}(S)$, $g \in V$, $\langle \mu, g\rangle = \int g\, d\mu$. The bilinear form $\langle\cdot,\cdot\rangle$ is separating. For, if $\mu, \nu \in \mathcal{M}(S)$ are such that $\int g\, d\mu = \int g\, d\nu$ for all $g \in V$, then $\mu = \nu$ by Lemma G.2; the other separating property is obvious.

Corollary F.2 *Assume that $J: \mathcal{P}(S) \to (-\infty, \infty]$ is not identically equal to ∞, convex, and $\sigma(\mathcal{P}(S), V)$-lower semicontinuous. Then for all $\mu \in \mathcal{P}(S)$,*

$$J(\mu) = \sup\left\{\int g\, d\mu - J'(g) : g \in V\right\},$$

where

$$J'(g) = \sup\left\{\int g\, d\mu - J(\mu) : \mu \in \mathcal{P}(S)\right\}.$$

Proof. Define $\tilde{J}: \mathcal{M}(S) \to (-\infty, \infty]$ by

$$\tilde{J}(\mu) = \begin{cases} J(\mu) & \text{if } \mu \in \mathcal{P}(S) \\ \infty & \text{otherwise.} \end{cases}$$

Clearly $\tilde{J}$ is $\sigma(\mathcal{M}(S), V)$-lower semicontinuous, convex, and proper. By Proposition F.1, for $\mu \in \mathcal{P}(S)$,

$$J(\mu) = \tilde{J}(\mu) = \sup\left\{\int g\, d\mu - \tilde{J}^*(g) : g \in V\right\}.$$

But

$$\begin{aligned} \tilde{J}^*(g) &= \sup\left\{\int g\, d\mu - \tilde{J}(\mu) : \mu \in \mathcal{M}(S)\right\} \\ &= \sup\left\{\int g\, d\mu - J(\mu) : \mu \in \mathcal{P}(S)\right\} = J'(g). \end{aligned}$$

□

Proposition F.3 *Let $\Gamma: V \to \mathbb{R}$ satisfy:*

1. *Γ is convex.*
2. *For all $c \in \mathbb{R}$, $\Gamma(c) = \Gamma(0) + c$.*
3. *If $f, g \in V$ and $f \le g$, then $\Gamma(f) \le \Gamma(g)$.*

Then:

If $\Gamma^*: \mathcal{M}(S) \to \overline{\mathbb{R}}$ is defined

$$\Gamma^*(\mu) = \sup\left\{\int g\, d\mu - \Gamma(g) : g \in V\right\}, \tag{F.3}$$

then $\operatorname{dom} \Gamma^* \subset \mathcal{P}(S)$.

If Γ is $\sigma(V, \mathcal{P}(S))$-lower semicontinuous, then for all $g \in V$,

$$\Gamma(g) = \sup\left\{\int g\, d\mu - \Gamma^*(\mu) : \mu \in \mathcal{P}(S)\right\}. \tag{F.4}$$

Proof.

1. Assume that $\mu \in \mathcal{M}(S)$ and $\Gamma^*(\mu) < \infty$. Then if $\ell_\mu(g) = \int g\, d\mu$, we have
   (i) $g \in B(S)$, $g \geq 0$ imply $\ell_\mu(g) \geq 0$.
   (ii) $\ell_\mu(1) = 1$.

   To prove 1(i), we first prove it for $g \in V$, $g \geq 0$. For, assume $\ell_\mu(g) < 0$. We have: for $t < 0$, $\Gamma(tg) \leq \Gamma(0)$,

$$\begin{aligned} t\ell_\mu(g) = \ell_\mu(tg) &\leq \Gamma(tg) + c \\ &\leq \Gamma(0) + c, \end{aligned}$$

   where $c = \Gamma^*(\mu)$. Letting $t \to -\infty$, we obtain a contradiction. This proves 1(i) for $g \in V$, $g \geq 0$. Next, let

$$\mathcal{H} = \left\{ g \in B(S) : \ell_\mu(g^+) \geq 0 \right\}.$$

   Then $\mathcal{H}$ is closed under monotone pointwise limits and contains V (recall that if $g \in V$, then $g^+ \in V$). By Proposition I.1, $\mathcal{H} = B(S)$. This proves 1(i).

   To prove 1(ii): for all $t \in \mathbb{R}$,

$$\begin{aligned} t\ell_\mu(1) = \ell_\mu(t) &\leq \Gamma(t) + c \\ &= \Gamma(0) + t + c, \\ t\left(\ell_\mu(1) - 1\right) &\leq \Gamma(0) + c, \end{aligned}$$

   which implies $\ell_\mu(1) = 1$. 1(i) and 1(ii) prove (F.3).
2. It is easily seen that Γ is $\sigma(V, \mathcal{P}(S))$-lower semicontinuous if and only if it is $\sigma(V, \mathcal{M}(S))$-lower semicontinuous. Therefore by Proposition F.1, we have for all $g \in V$,

$$\Gamma(g) = \sup \left\{ \int g\, d\mu - \Gamma^*(\mu) : \mu \in \mathcal{M}(S) \right\},$$

   and by (F.3),

$$\Gamma(g) = \sup \left\{ \int g\, d\mu - \Gamma^*(\mu) : \mu \in \mathcal{P}(S) \right\}.$$

□

Appendix G

Daniell's Theorem

We state the following classical result in a form suitable for our purposes.

Proposition G.1 *Let $(S, \mathcal{S})$ be a measurable space, and let V be a vector space contained in $B(S)$, satisfying V.1–V.3 (Chapter 3). Let $\ell: V \to \mathbb{R}$ be linear and such that*

$$g \in V,\ g \geq 0 \text{ imply } \ell(g) \geq 0; \tag{G.1}$$

$$\ell(1) = 1; \tag{G.2}$$

$$0 \leq g_k \in V,\ g_k \downarrow 0 \text{ pointwise imply } \ell(g_k) \longrightarrow 0. \tag{G.3}$$

Then there exists a unique probability measure μ on $\mathcal{S}$ such that for all $g \in V$,

$$\ell(g) = \int g\, d\mu. \tag{G.4}$$

The following lemma is an extension of the uniqueness assertion (G.4) of Proposition G.1.

Lemma G.2 *Let $\mu, \nu \in \mathcal{M}(S)$ and assume that for all $g \in V$, $\int g\, d\mu = \int g\, d\nu$. Then $\mu = \nu$.*

Proof. Let $\mathcal{H} = \{g \in B(S)\} : \int g\, d\mu = \int g\, d\nu\}$. Then $\mathcal{H}$ is closed under pointwise monotone limits by dominated convergence. Since $\mathcal{H} \supset V$, by Proposition I.1 we conclude that $\mathcal{H} = B(S)$. □

Notes. Proposition G.1 is due to P.J. Daniell; see, e.g., Neveu (1964, Chap. 2) and Dudley (2002, Chap. 4).

Appendix H

Relative Compactness in the V Topology

Let $(S, \mathcal{S})$ be a measurable space, and assume that $V \subset B(S)$ satisfies V.1–V.3 (Chapter 3). Let $M \subset \mathcal{P}(S)$. The following result is a necessary and sufficient condition for the relative compactness of M in the V topology.

Proposition H.1 *Let $M \subset \mathcal{P}(S)$. The following conditions are equivalent:*

$$M \text{ is } V\text{-relatively compact.} \tag{H.1}$$

$$0 \le f_k \in V,\ f_k \downarrow 0 \text{ pointwise imply} \quad \lim_k \sup_{\mu \in M} \int f_k \, d\mu = 0. \tag{H.2}$$

Proof. (H.1) $\Longrightarrow$ (H.2): Assume that $0 \le f_k \in V$, $f_k \downarrow 0$ pointwise. Let $F_k : \mathcal{P}(S) \to \mathbb{R}^+$ be defined by $F_k(\mu) = \int f_k \, d\mu$. Then F_k is V-continuous, $\{F_k\}$ is decreasing, and $F_k \downarrow 0$ pointwise on $\mathcal{P}(S)$ by dominated convergence. Since M is V-relatively compact, by Dini's theorem,

$$\sup_{\mu \in M} F_k(\mu) \longrightarrow 0.$$

(H.2) $\Longrightarrow$ (H.1): Let U be the unit ball of V^*, and for $\mu \in \mathcal{P}(S)$, let $\ell_\mu \in U$ be defined as in Chapter 3. Let $\{\mu_\alpha : \alpha \in D\}$ be a net in M, where D is a directed set. Since $\{\ell_{\mu_\alpha} : \alpha \in D\} \subset U$, by the Banach–Alaoglu theorem (Conway, 1985) there exist a subnet $\{\ell_{\mu_\beta} : \beta \in D'\}$, D' a certain directed set, and $\ell \in U$ such that $\{\ell_{\mu_\beta} : \beta \in D'\}$ converges w^* to ℓ; that is, for all $f \in V$,

$$\lim_\beta \ell_{\mu_\beta}(f) = \ell(f).$$

Clearly ℓ satisfies:

1. $0 \le f \in V$ implies $\ell(f) \ge 0$.
2. $\ell(1) = 1$.

3. $0 \le f_k \in V$, $f_k \downarrow 0$ pointwise imply $\ell(f_k) \to 0$.

For,

$$\ell(f_k) = \lim_{\beta} \ell_{\mu_\beta}(f_k) \le \sup_{\mu \in M} \int f_k \, d\mu \longrightarrow 0$$

by (H.2). By Proposition G.1, there exists a probability measure μ on $\mathcal{S}$ such that for all $f \in V$,

$$\ell(f) = \int f \, d\mu.$$

Therefore $\{\mu_\beta : \beta \in D'\}$ converges to μ in the V topology. This proves that M is V-relatively compact. □

Remark H.2 The implication (H.1) $\Longrightarrow$ (H.2) is valid for any V topology; conditions V.1–V.3 are not used in its proof.

The implication (H.1) $\Longrightarrow$ (H.2) in Proposition H.1 has the following useful variant.

Proposition H.3 *Let $M \subset \mathcal{P}(S)$, $\lambda \in \mathcal{P}(S)$. Assume:*

$$M \text{ is } V\text{-compact.} \tag{H.3}$$

$$M \subset \mathcal{P}(S, \lambda). \tag{H.4}$$

Then $0 \le f_k \in V$, $\{f_k\}$ is decreasing, $\int f_k \, d\lambda \to 0$ imply

$$\lim_k \sup_{\mu \in M} \int f_k \, d\mu = 0. \tag{H.5}$$

Proof. Let F_k be as in the proof of Proposition H.1. If $0 \le f_k \in V$, $\{f_k\}$ is decreasing, and $\int f_k \, d\lambda \to 0$, then by (H.4) $F_k(\mu) \to 0$ for every $\mu \in M$. As in the proof of Proposition H.1, we have

$$\lim_k \sup_{\mu \in M} F_k(\mu) = 0.$$

□

Appendix I

A Monotone Class Theorem

We will prove a monotone class theorem for functions suitable for applications in the present work.

As in Chapter 3, V will denote a vector space contained in $B(S)$, satisfying V.1–V.3.

Proposition I.1 *Let $(S, \mathcal{S})$ be a measurable space. Let $\mathcal{H}$ be a class of real-valued bounded measurable functions such that:*

1. *$\mathcal{H}$ is closed under pointwise monotone limits: $g_n \in \mathcal{H}$, $g \in B(S)$, $g_n \uparrow g$ (respectively, $g_n \downarrow g$) pointwise imply $g \in \mathcal{H}$.*
2. *$\mathcal{H} \supset V$.*

Then $\mathcal{H} = B(S)$.

Let us recall that a class $\mathcal{M}$ of subsets of S is *monotone* if $A_n \in \mathcal{M}$, $A_n \uparrow A$ (respectively, $A_n \downarrow A$) imply $A \in \mathcal{M}$. For a class $\mathcal{F}$ of subsets of S, $\sigma(\mathcal{F})$ is the σ-algebra generated by $\mathcal{F}$.

Lemma I.2 *Let $\mathcal{F}$ be a class of subsets of S such that:*

1. *$A, B \in \mathcal{F}$ imply $A \cap B \in \mathcal{F}$.*
2. *$A \in \mathcal{F}$ implies $A^c = \bigcap_{n=1}^{\infty} B_n$ for a certain decreasing sequence $\{B_n\} \subset \mathcal{F}$.*

Let $\mathcal{M}$ be a monotone class such that $\mathcal{F} \subset \mathcal{M}$. Then $\sigma(\mathcal{F}) \subset \mathcal{M}$.

Proof. Let $m(\mathcal{F})$ be the smallest monotone class containing $\mathcal{F}$. We will show: $m(\mathcal{F})$ is an algebra. Since a monotone algebra is a σ-algebra, it will follow that $\sigma(\mathcal{F}) \subset m(\mathcal{F}) \subset \mathcal{M}$.

Consider $\mathcal{G} = \{A : A^c \in m(\mathcal{F})\}$. Since $m(\mathcal{F})$ is monotone, so is $\mathcal{G}$. Let $A \in \mathcal{F}$. Then there exists a decreasing sequence $\{B_n\} \subset \mathcal{F}$ such that $A^c = \bigcap_{n=1}^{\infty} B_n$. Therefore $A^c \in m(\mathcal{F})$. It follows that $\mathcal{F} \subset \mathcal{G}$, which implies that $m(\mathcal{F}) \subset \mathcal{G}$ and $m(\mathcal{F})$ is closed under complementation.

Let $\mathcal{G}_1 = \{A\colon A \cap B \in m(\mathcal{F}) \text{ for all } B \in \mathcal{F}\}$. Then $\mathcal{G}_1$ is a monotone class and $\mathcal{F} \subset \mathcal{G}_1$. Therefore $m(\mathcal{F}) \subset \mathcal{G}_1$.

Let $\mathcal{G}_2 = \{B\colon A \cap B \in m(\mathcal{F}) \text{ for all } A \in m(\mathcal{F})\}$. Then $\mathcal{G}_2$ is a monotone class. Since $m(\mathcal{F}) \subset \mathcal{G}_1$, it follows that $\mathcal{F} \subset \mathcal{G}_2$. Therefore $m(\mathcal{F}) \subset \mathcal{G}_2$ and $m(\mathcal{F})$ is closed under intersection, hence $m(\mathcal{F})$ is an algebra. □

Lemma I.3 *Let $\mathcal{F} = \{f^{-1}((1,\infty))\colon f \in V\}$. Then:*

1. *$A, B \in \mathcal{F}$ imply $A \cap B \in \mathcal{F}$.*
2. *$A \in \mathcal{F}$ implies $A^c = \bigcap_{n=1}^{\infty} B_n$ for a certain decreasing sequence $\{B_n\} \subset \mathcal{F}$.*
3. *$A \in \mathcal{F}$ implies $\mathbf{1}_A = \lim_n g_n$ for a certain nonnegative increasing sequence $\{g_n\} \subset V$.*

Proof.

1. Let $A = f^{-1}((1,\infty))$, $B = g^{-1}((1,\infty))$ with $f, g \in V$. Then $f \wedge g \in V$ and $A \cap B = (f \wedge g)^{-1}((1,\infty)) \in \mathcal{F}$.
2. Let $A \in \mathcal{F}$, $A = f^{-1}((1,\infty))$ with $f \in V$. For $a > 1$, let $g_a = 2 - a^{-1} f$. Then $g_a \in V$ and

$$\{x \in S\colon g_a(x) > 1\} = \{x \in S\colon f(x) < a\}.$$

Let $a_n = 1 + n^{-1}$,

$$B_n = \{x \in S\colon f(x) < a_n\} = \{x \in S\colon g_{a_n}(x) > 1\} \in \mathcal{F}.$$

Then $B_n \downarrow \{x \in S\colon f(x) \le 1\} = A^c$.

3. For A as above, let

$$g_n = n(f - f \wedge 1) \wedge 1 \in V.$$

Then $g_n \uparrow \mathbf{1}_A$. □

Lemma I.4 *Let $\mathcal{H}$, V be as in Proposition I.1. Let $g \in B(S)$, and assume for all $f \in V$, $g + f \in \mathcal{H}$. Then for all $\alpha \in \mathbb{R}$, $A \in \mathcal{S}$, $f \in V$,*

$$g + \alpha \mathbf{1}_A + f \in \mathcal{H}.$$

Proof. Let

$$\mathcal{M} = \{A \subset S : g + \alpha \mathbf{1}_A + f \in \mathcal{H} \text{ for all } \alpha \in \mathbb{R}, f \in V\}.$$

Then, by property 1 of $\mathcal{H}$, $\mathcal{M}$ is a monotone class.

We claim that $\mathcal{F} \subset \mathcal{M}$, where $\mathcal{F}$ is as in Lemma I.3. For, if $A \in \mathcal{F}$, then by Lemma I.3(3) there exists a nonnegative increasing sequence $\{g_n\} \subset V$ such that $g_n \uparrow \mathbf{1}_A$. Since, by assumption, for $\alpha \in \mathbb{R}$ and $f \in V$ we have $g + \alpha g_n + f \in \mathcal{H}$, it follows that $g + \alpha \mathbf{1}_A + f$, being a monotone limit of functions in $\mathcal{H}$, belongs to $\mathcal{H}$. Hence $A \in \mathcal{M}$.

By Lemmas I.2 and I.3, $\sigma(\mathcal{F}) \subset \mathcal{M}$. But taking into account property V.1 of V and the definition of $\mathcal{F}$, it is clear that $\sigma(\mathcal{F}) = \mathcal{S}$. □

Proof of Proposition I.1

1. We will show that every simple $\mathcal{S}$-measurable function belongs to $\mathcal{H}$. Let

$$E = \left\{ k \in \mathbb{N}\colon \text{ for all } \alpha_i \in \mathbb{R}, A_i \in \mathcal{S}, i = 1, \dots, k, f \in V, \sum_{i=1}^{k} \alpha_i \mathbf{1}_{A_i} + f \in \mathcal{H} \right\}.$$

Since $V \subset \mathcal{H}$, Lemma I.4 with $g = 0$ shows that $1 \in E$. Assume that $k \in E$. Let $\alpha_i \in \mathbb{R}$, $A_i \in \mathcal{S}$, $i = 1, \dots, k+1$. By Lemma I.4 with $g = \sum_{i=1}^{k} \alpha_i \mathbf{1}_{A_i}$, we have

$$\sum_{i=1}^{k+1} \alpha_i \mathbf{1}_{A_i} + f = \sum_{i=1}^{k} \alpha_i \mathbf{1}_{A_i} + \alpha_{k+1} \mathbf{1}_{A_{k+1}} + f \in \mathcal{H}$$

for all $f \in V$. Therefore $k + 1 \in E$, and consequently $E = \mathbb{N}$. Now taking $f = 0$, (1) is proved.
2. $B(S) \subset \mathcal{H}$. For, given $f \in B(S)$, let $c = \inf f$, $g = f - c$. Then there exists an increasing sequence $\{s_n\}$ of nonnegative simple functions such that $s_n \uparrow g$, and therefore $s_n + c \uparrow f$, pointwise. By (1) above and property 1 of $\mathcal{H}$, we have $f \in \mathcal{H}$. □

Notes Lemma I.4 is based on Lemma A.12 of Rassoul-Agha and Seppäläinen (2015).

Proposition I.1 improves Lemma A.11 of Rassoul-Agha and Seppäläinen (2015) and is of a form different from the commonly quoted monotone class theorems for classes of functions on a measurable space (e.g., Revuz, 1984, Thm. 3.3, Chp. 0).

Appendix J

On the Axioms V.1–V.3 and V.1′–V.4

We prove several properties of a vector space $V \subset B(S)$ satisfying V.1–V.3 (Proposition J.1) and V.1′–V.4 (Proposition J.3).

In the following result we prove an approximation property of a function in $B(S)$ by functions in V and an absolute continuity criterion in terms of V.

Let $C(V) = \{g \in B(S)\colon$ there exists a decreasing sequence $\{f_k\} \subset V$ such that $g = \inf_k f_k\}$. Note that if S is Polish and $V = C_b(S)$, then $C(V)$ contains the class of indicators of closed sets.

Proposition J.1 *Assume that V satisfies V.1–V.3.*

1. *Let $\mu \in \mathcal{P}(S)$. Then for every $0 \leq g \in B(S)$, $\epsilon > 0$, there exists $f \in C(V)$ such that*

$$0 \leq f \leq g \quad \textit{and} \quad \int (g - f)\, d\mu \leq \epsilon.$$

2. *Let $\lambda \in \mathcal{P}(S)$ and assume that $f \geq 0$, $f \in C(V)$, $\int f\, d\lambda = 0$ imply $\int f\, d\mu = 0$. Then $\mu \ll \lambda$.*

Lemma J.2 *Assume that V satisfies V.1–V.3. Suppose that for every $n \in \mathbb{N}$, $h_n \in C(V)$. Then $h = \inf_n h_n \in C(V)$.*

Proof. For each $n \in \mathbb{N}$, there exists a decreasing sequence $\{h_{n,k}\colon k \in \mathbb{N}\} \subset V$ such that $h_n = \inf_k h_{n,k}$. For $k \in \mathbb{N}$, let

$$f_k = \inf\left\{h_{i,j}\colon i \leq k, j \leq k\right\}.$$

Clearly $f_k \in V$ and $\{f_k\colon k \in \mathbb{N}\}$ is decreasing. Claim: $h = \inf_k f_k$. In fact, for $k \in \mathbb{N}$,

$$f_k = \inf_{i \leq k} \inf_{j \leq k} h_{i,j} \geq \inf_{i \leq k} h_i \geq h,$$

and therefore $\inf_k f_k \geq h$.

On the other hand, for each $n \in \mathbb{N}$, $k \geq n$,

$$f_k \leq \inf_{i \leq k} h_{i,k} \leq h_{n,k}.$$

Therefore, for all $n \in \mathbb{N}$,

$$\inf_k f_k \leq \inf_k h_{n,k} = h_n,$$

and it follows that

$$\inf_k f_k \leq h. \qquad \square$$

Proof of Proposition J.1

1. For $\mu \in \mathcal{P}(S)$, let

$$\mathcal{H}_\mu = \Big\{g \in B(S): \text{ for every } \epsilon > 0, \text{ there exists } f \in C(V) \text{ such that } 0 \leq f \leq g^+, \ \int (g^+ - f)\, d\mu \leq \epsilon\Big\}.$$

It suffices to prove that $\mathcal{H}_\mu = B(S)$. Obviously, $\mathcal{H}_\mu \supset V$ (recall that $f \in V$ implies $f^+ \in V$). We will prove that $\mathcal{H}_\mu$ is closed under monotone limits. It will then follow from Proposition I.1 that $\mathcal{H}_\mu = B(S)$.

Let $g_n \in \mathcal{H}_\mu$, $g \in B(S)$, $g_n \downarrow g$. For each $n \in \mathbb{N}$, there exists $h_n \in C(V)$ such that

$$0 \leq h_n \leq g_n^+ \quad \text{and} \quad \int (g_n^+ - h_n)\, d\mu \leq \epsilon 2^{-n}.$$

Let $h = \inf_n h_n$. Then $0 \leq h \leq g^+$ and $h \in C(V)$ by Lemma J.2. Also

$$g^+ - h = \inf_n g_n^+ - \inf_n h_n \leq \sup_n (g_n^+ - h_n) \leq \sum_{n=1}^{\infty} (g_n^+ - h_n),$$

$$\int (g^+ - h)\, d\mu \leq \int \sum_{n=1}^{\infty} (g_n^+ - h_n)\, d\mu = \sum_{n=1}^{\infty} \int (g_n^+ - h_n)\, d\mu \leq \sum_{n=1}^{\infty} \epsilon 2^{-n} = \epsilon.$$

This shows that $g \in \mathcal{H}_\mu$.

Next, let $g_n \in \mathcal{H}_\mu$, $g \in B(S)$, $g_n \uparrow g$. Let $n_0 \in \mathbb{N}$ be such that $\int (g^+ - g_{n_0}^+)\, d\mu \leq \epsilon/2$. There exists $h \in V$ such that $0 \leq h \leq g_{n_0}^+$ and $\int (g_{n_0}^+ - h)\, d\mu \leq \epsilon/2$. Then $h \leq g^+$ and $\int (g^+ - h)\, d\mu \leq \epsilon$, showing that $g \in \mathcal{H}_\mu$.

2. Given $\epsilon > 0$, $A \in \mathcal{S}$, let $f \in C(V)$ be such that $0 \leq f \leq \mathbf{1}_A$ and $\int(\mathbf{1}_A - f)\, d\mu \leq \epsilon$. Assume that $\lambda(A) = 0$. Then $\int f\, d\lambda = 0$ and therefore $\int f\, d\mu = 0$ and $\mu(A) \leq \epsilon$. But ϵ is arbitrary. □

In the next proposition we prove that if V satisfies V.1′–V.4, then it satisfies V.1–V.3. We also show that V is closed under several different operations.

Proposition J.3 *Assume that V satisfies V.1′–V.4. Then*

1. *V satisfies conditions V.1–V.3.*
2. *If $f \in V$, then $e^f \in V$.*
3. *If $f \in V$ and* $\inf f > 0$, *then* $\log f \in V$.
4. *For $f_i \in V$, $i = 1, \ldots, k$, let*

$$g(x) = \mathbb{E}_x\left[f_1(X_1), \ldots, f_k(X_k)\right], \ x \in S.$$

Then $g \in V$.

Proof.

1. We only have to prove: $f, g \in V$ imply $f \vee g, f \wedge g \in V$.

   To prove this, we follow Folland (1999, pp. 139–140). We claim that if $f \in V$, then $|f| \in V$. For, given $\epsilon > 0$, there exists a polynomial p on $\mathbb{R}$ such that $p(0) = 0$ and $||x| - p(x)| \leq \epsilon$ for $x \in [-1, 1]$; this is a special case of the classical Weierstrass approximation theorem. Given $f \in V$, $f \neq 0$, let $h = f/\|f\|$. Then $h(S) \subset [-1, 1]$, so $\||h| - p \circ h\| \leq \epsilon$. But $p \circ h \in V$ and since V is $\|\cdot\|$-closed, it follows that $|h| \in V$, hence $|f| = \|f\||h| \in V$. Next, for $f, g \in V$,

$$f \vee g = \frac{1}{2}(f + g + |f - g|) \in V,$$
$$f \wedge g = \frac{1}{2}(f + g - |f - g|) \in V.$$

2. For $t \in \mathbb{R}$, $n \in \mathbb{N}$, let

$$e_n(t) = \sum_{k=0}^{n} \frac{t^k}{k!}.$$

   Then $e_n(t)$ converges to e^t uniformly over bounded intervals. For $f \in V$, let $f_n = e_n \circ f$. Since f is bounded, we have

$$\|f_n - e^f\| \longrightarrow 0.$$

   But $f_n \in V$, and therefore $e^f \in V$.

3. For $t \in \mathbb{R}$, let

$$\ell_n(t) = \sum_{k=0}^{n} \frac{(-1)^k t^{k+1}}{k+1}.$$

Then for any $r \in (0, 1)$, $\ell_n(t)$ converges to $\log(1 + t)$ uniformly over $[-r, r]$. For $f \in V$ with $a = \inf f > 0$, let $b = \sup f$, $c > b$. Then

$$-1 < \frac{a}{c} - 1 \leq \frac{f}{c} - 1 \leq \frac{b}{c} - 1 < 0.$$

Let $h_n = \ell_n \circ (f/c - 1)$. Then

$$\left\| \log\left(\frac{f}{c}\right) - h_n \right\| = \left\| \log\left[1 + \left(\frac{f}{c} - 1\right)\right] - h_n \right\| \longrightarrow 0.$$

But $h_n \in V$ and therefore $\log(f/c) \in V$ and

$$\log f = \log c + \log\left(\frac{f}{c}\right) \in V.$$

4. This is proved by induction. For $k = 1$, if $f_1 \in V$ then

$$g = \mathbb{E}_\cdot [f_1(X_1)] = Pf_1 \in V.$$

Inductive step: assume that for all $f_i \in V$, $i = 1, \ldots, k$,

$$h = \mathbb{E}_\cdot [f_1(X_1) \ldots f_k(X_k)] \in V.$$

Let $f_{k+1} \in V$. Then for $x \in S$,

$$\begin{aligned} g(x) &= \mathbb{E}_x [f_1(X_1) \ldots f_{k+1}(X_{k+1})] \\ &= \mathbb{E}_x \mathbb{E}_x [f_1(X_1) \ldots f_{k+1}(X_{k+1}) \mid \mathcal{F}_k] \\ &= \mathbb{E}_x [f_1(X_1) \ldots f_k(X_k) \mathbb{E}_{X_k} f_{k+1}(X_1)] \\ &= \mathbb{E}_x [f_1(X_1) \ldots f_k(X_k) P f_{k+1}(X_k)]. \end{aligned}$$

But $Pf_{k+1} \in V$ and $f_k P f_{k+1} \in V$. Therefore by the inductive hypothesis $g \in V$. □

Appendix K

On Gâteaux Differentiability

Let E be a separable Banach space. We will consider differentiability properties of convex functions defined on E^*. We will be particularly interested in the situation when the subgradients or Gâteaux derivatives are elements of E, rather than just elements of the larger space E^{**} (Proposition K.3).

Let U be an open convex subset of E^*, and let $h \colon U \to \mathbb{R}$ be convex and lower semicontinuous. Let $\xi \in U$, $\eta \in E^*$. The *directional derivative of h at ξ along η* is denoted $h'(\xi;\eta)$ and is defined

$$h'(\xi;\eta) = \lim_{t \to 0^+} \frac{h(\xi + t\eta) - h(\xi)}{t}.$$

The function $0 < t \mapsto t^{-1}(h(\xi + t\eta) - h(\xi))$ is increasing and therefore $h'(\xi;\eta)$ exists and

$$h'(\xi;\eta) = \inf_{t>0} t^{-1}\left(h(\xi + t\eta) - h(\xi)\right).$$

Under the assumption on h, $h'(\xi;\cdot)$ is convex, positively homogeneous, and continuous on E^* (see, e.g., Lucchetti, 2006, pp. 32–34).

A linear function $\ell \in E^{**}$ is a *subgradient of h at ξ* if for all $\eta \in E^*$,

$$h(\eta) \geq h(\xi) + \ell(\eta - \xi).$$

The *subdifferential of h at ξ*, denoted $\partial h(\xi)$, is the set of all subgradients of h at ξ.

The connection between $h'(\xi;\cdot)$ and $\partial h(\xi)$ is as follows (Lucchetti, 2006, p. 38):

$$\partial h(\xi) = \left\{\ell \in E^{**} : \ell(\eta) \leq h'(\xi;\eta) \text{ for all } \eta \in E^*\right\}.$$

The function h is *Gâteaux differentiable at ξ* if there exists $\ell \in E^{**}$ such that

$$h'(\xi;\eta) = \ell(\eta) \qquad \text{for all } \eta \in E^*;$$

ℓ is the *Gâteaux derivative of h at ξ* and ℓ is denoted $\ell = \nabla h(\xi)$. If h is Gâteaux differentiable at ξ, then $\partial h(\xi) = \{\nabla h(\xi)\}$ (Lucchetti, 2006, p. 41).

There is a converse to the last statement: if h is continuous at ξ and $\partial h(\xi) = \{\ell\}$ for some $\ell \in E^{**}$, then h is Gâteaux differentiable at ξ and $\nabla h(\xi) = \ell$ (see, e.g., Lucchetti, 2006, Proposition 3.3.7). We will prove below a variant of this result: under a certain condition on h, if $\partial h(\xi) \cap E = \{x_0\}$ for some $x_0 \in E$, then h is Gâteaux differentiable at ξ and $\nabla h(\xi) = x_0$.

We will need the following result. Recall that the $\sigma(E^*, E)$ (or w^*) topology on E^* is the weakest topology such that each map $\xi \mapsto \langle x, \xi \rangle$, $x \in E$, is continuous.

Proposition K.1 *Let E be a separable Banach space, and assume that $L\colon E^* \to \mathbb{R}$ is linear. Then the following conditions are equivalent:*

1. *L is $\sigma(E^*, E)$-sequentially continuous.*
2. *L is $\sigma(E^*, E)$-continuous.*
3. *There exists $x \in E$ such that $L(\xi) = \langle x, \xi \rangle$ for all $\xi \in E^*$.*

The equivalence of Conditions 1 and 2 is proved in Conway (1985), Corollary V.12.8. The implication Condition 2 $\Longrightarrow$ 3 is proved in Conway (1985), Theorem V.1.3.

Lemma K.2 *Let U be an open convex subset of E^*, and let $h\colon U \to \mathbb{R}$ be convex. Assume:*

1. *h is $\sigma(E^*, E)$-sequentially continuous.*
2. *$L \in \partial h(\xi)$ for some $\xi \in U$.*

Then $L \in E$: there exists $x_0 \in E$ such that $L(\eta) = \langle x_0, \eta \rangle$ for all $\eta \in E^*$.

Proof. Clearly assumption 1 implies that h is (norm) continuous. Let $p(\eta) = h'(\xi; \eta)$, $q(t, \eta) = t^{-1}(h(\xi + t\eta) - h(\xi))$ for $\eta \in E^*$, $t > 0$. Then $q(t, \cdot)$ is $\sigma(E^*, E)$-sequentially continuous and therefore

$$p = \inf_{t>0} q(t, \cdot)$$

is $\sigma(E^*, E)$-sequentially upper semicontinuous.

L is $\sigma(E^*, E)$-sequentially continuous. Clearly, it suffices to show that this property holds at $0 \in E^*$. If $\{\eta_n\} \subset E^*$ and $\sigma(E^*, E) - \lim_n \eta_n = 0$, then

$$\overline{\lim_n}\, L(\eta_n) \le \overline{\lim_n}\, p(\eta_n) \le p(0) = 0,$$

and

$$\underline{\lim}_n L(\eta_n) = \underline{\lim}_n \,(-L(-\eta_n))$$
$$= -\overline{\lim}_n L(-\eta_n)$$
$$\geq -\overline{\lim}_n p(-\eta) \geq p(0) = 0,$$

showing that $\lim_n L(\eta_n) = 0 = L(0)$. Applying now Proposition K.1, we have: there exists $x_0 \in E$ such that for all $\eta \in E^*$, $L(\eta) = \langle x_0, \eta\rangle$. □

Proposition K.3 *Let E be a separable Banach space. Let U be an open convex subset of E^* and assume that $h\colon U \to \mathbb{R}$ is convex. Suppose:*

1. *h is $\sigma(E^*, E)$-sequentially continuous.*
2. *For a certain $\xi_0 \in E^*$ and $x_0 \in E$, $\partial h(\xi_0) \cap E = \{x_0\}$.*

Then, h is Gâteaux differentiable at ξ_0 and $\nabla h(\xi_0) = x_0$.

Proof. For $\xi \in E^*$, let $p(\xi) = h'(\xi_0; \xi)$. Now fix $\xi \in E^*$, $\xi \neq 0$. We define the linear function ℓ_ξ: span $\{\xi\} \to \mathbb{R}$ by

$$\ell_\xi(\eta) = ap(\xi) \qquad \text{for } \eta = a\xi.$$

Then

$$\ell_\xi(\eta) \leq p(\eta) \qquad \text{for } \eta \in \text{span}\,\{\xi\}. \tag{K.1}$$

For: if $\eta = a\xi$ with $a \geq 0$, then

$$\ell_\xi(\eta) = ap(\xi) = p(a\xi) = p(\eta).$$

By the convexity of p and $p(0) = 0$, we have $-p(\xi) \leq p(-\xi)$. Therefore if $\eta = (-a)\xi$ with $a > 0$,

$$\ell_\xi(\eta) = (-a)p(\xi) \leq ap(-\xi)$$
$$= p\,(a(-\xi)) = p(\eta),$$

proving (K.1).

Since $p\colon E^* \to \mathbb{R}$ is sublinear (subadditive and positively homogeneous), by the Hahn–Banach theorem (see, e.g., Conway, 1985, Theorem III.6.2), there exists $L_\xi\colon E^* \to \mathbb{R}$ linear and such that $L_\xi(\eta) = \ell_\xi(\eta)$ for $\eta \in$ span $\{\xi\}$ and

$$L_\xi(\eta) \leq p(\eta) \qquad \text{for } \eta \in E^*. \tag{K.2}$$

By assumption 1 and Lemma K.2, there exists $y_\xi \in E$ such that for all $\eta \in E^*$, $L_\xi(\eta) = \langle y_\xi, \eta\rangle$. But then by (K.2) and assumption 2, we have $y_\xi = x_0$.

Therefore for all $\xi \in E^*$,

$$p(\xi) = \ell_\xi(\xi) = L_\xi(\xi) = \langle x_0, \xi \rangle.$$

This says that h is Gâteaux differentiable at ξ_0 and $\nabla h(\xi_0) = x_0$. □

Note. The Hahn–Banach argument in the proof of Proposition K.3 is found, e.g., in Lucchetti (2006), Proposition 3.3.7.

References

Bahadur, Raghu R., and Zabell, Sandy L. 1979. Large deviations of the sample mean in general vector spaces. *Ann. Probab.*, **7**(4), 587–621.

Baxter, John R., Jain, Naresh C., and Varadhan, Srinivasa R. S. 1991. Some familiar examples for which the large deviation principle does not hold. *Comm. Pure Appl. Math.*, **44**(8–9), 911–23.

Bolthausen, Erwin. 1987. Markov process large deviations in τ-topology. *Stochastic Process. Appl.*, **25**(1), 95–108.

Bryc, Włodzimierz, and Dembo, Amir. 1996. Large deviations and strong mixing. *Ann. Inst. H. Poincaré Probab. Statist.*, **32**(4), 549–69.

Chen, Xia. 2010. Random Walk Intersections: Large Deviations and Related Topics. Mathematical Surveys and Monographs, vol. 157. American Mathematical Society, Providence, RI.

Conway, John B. 1985. A Course in Functional Analysis. Graduate Texts in Mathematics, vol. 96. Springer-Verlag, New York.

Dawson, Donald A., and Gärtner, Jürgen. 1987. Large deviations from the McKean–Vlasov limit for weakly interacting diffusions. *Stochastics*, **20**(4), 247–308.

de Acosta, Alejandro. 1985. Upper bounds for large deviations of dependent random vectors. *Z. Wahrsch. Verw. Gebiete*, **69**(4), 551–65.

de Acosta, Alejandro. 1988. Large deviations for vector-valued functionals of a Markov chain: Lower bounds. *Ann. Probab.*, **16**(3), 925–60.

de Acosta, Alejandro. 1990. Large deviations for empirical measures of Markov chains. *J. Theoret. Probab.*, **3**(3), 395–431.

de Acosta, Alejandro. 1994a. On large deviations of empirical measures in the τ topology. *J. Appl. Probab.*, **31A**, 41–7.

de Acosta, Alejandro. 1994b. Projective systems in large deviation theory II. Some applications. Pages 241–50 of: Hoffmann-Jørgensen, Jørgen, Kuelbs, James, and Marcus, Michael B. (eds), *Probability in Banach Spaces 9 (Sandjberg, 1993). Progr. Probab.*, vol. 35. Birkhäuser, Boston, MA.

de Acosta, Alejandro, and Ney, Peter. 1998. Large deviation lower bounds for arbitrary additive functionals of a Markov chain. *Ann. Probab.*, **26**(4), 1660–82.

de Acosta, Alejandro, and Ney, Peter. 2014. Large deviations for additive functionals of Markov chains. *Mem. Amer. Math. Soc.*, **228**(1070), vi+108.

Dembo, Amir, and Zeitouni, Ofer. 1998. Large Deviations Techniques and Applications. Second edn. Applications of Mathematics, vol. 38. Springer-Verlag, New York.

den Hollander, Frank. 2000. Large Deviations. Fields Institute Monographs, vol. 14. American Mathematical Society, Providence, RI.

Deuschel, Jean-Dominique, and Stroock, Daniel W. 1989. Large Deviations. Pure and Applied Mathematics, vol. 137. Academic Press, Boston, MA.

Dinwoodie, Ian H. 1993. Identifying a large deviation rate function. *Ann. Probab.*, **21**(1), 216–31.

Dinwoodie, Ian H., and Ney, Peter. 1995. Occupation measures for Markov chains. *J. Theoret. Probab.*, **8**(3), 679–91.

Dinwoodie, Ian H., and Zabell, Sandy L. 1992. Large deviations for exchangeable random vectors. *Ann. Probab.*, **20**(3), 1147–66.

Donsker, Monroe D., and Varadhan, Srinivasa R. S. 1975. Asymptotic evaluation of certain Markov process expectations for large time I. *Comm. Pure Appl. Math.*, **28**, 1–47.

Donsker, Monroe D., and Varadhan, Srinivasa R. S. 1976. Asymptotic evaluation of certain Markov process expectations for large time III. *Comm. Pure Appl. Math.*, **29**(4), 389–461.

Donsker, Monroe D., and Varadhan, Srinivasa R. S. 1983. Asymptotic evaluation of certain Markov process expectations for large time IV. *Comm. Pure Appl. Math.*, **36**(2), 183–212.

Dudley, Richard M. 2002. Real Analysis and Probability. Cambridge Studies in Advanced Mathematics, vol. 74. Cambridge University Press, Cambridge. Revised reprint of the 1989 original.

Dunford, Nelson, and Schwartz, Jacob T. 1958. Linear Operators Part I: General Theory. Pure and Applied Mathematics, vol. 7. Interscience Publishers (John Wiley & Sons), New York.

Dupuis, Paul, and Ellis, Richard S. 1997. A Weak Convergence Approach to the Theory of Large Deviations. Wiley Series in Probability and Statistics. John Wiley & Sons, New York.

Ellis, Richard S. 1988. Large deviations for the empirical measure of a Markov chain with an application to the multivariate empirical measure. *Ann. Probab.*, **16**(4), 1496–508.

Feng, Jin, and Kurtz, Thomas G. 2006. Large Deviations for Stochastic Processes. Mathematical Surveys and Monographs, vol. 131. American Mathematical Society, Providence, RI.

Folland, Gerald B. 1999. Real Analysis. Second edn. Pure and Applied Mathematics. John Wiley & Sons, New York.

Gao, Fuqing, and Wang, Qinghua. 2003. Upper bound estimates of the Cramér functionals for Markov processes. *Potential Anal.*, **19**(4), 383–98.

Garling, D. J. H. 2018. Analysis on Polish Spaces and an Introduction to Optimal Transportation. London Mathematical Society Student Texts, vol. 89. Cambridge University Press, Cambridge.

Gärtner, Jürgen. 1977. On large deviations from the invariant measure. *Theor. Probab. Appl.*, **22**(1), 24–39.

Hernández-Lerma, Onésimo, and Lasserre, Jean Bernard. 2003. *Markov Chains and Invariant Probabilities*. Progress in Mathematics, vol. 211. Birkhäuser Verlag, Basel.

Iscoe, Ian, Ney, Peter, and Nummelin, Esa. 1985. Large deviations of uniformly recurrent Markov additive processes. *Adv. Appl. Math.*, **6**(4), 373–412.

Jain, Naresh C. 1990. Large deviation lower bounds for additive functionals of Markov processes. *Ann. Probab.*, **18**(3), 1071–98.

Jiang, Yi Wen, and Wu, Li Ming. 2005. Large deviations for empirical measures of not necessarily irreducible countable Markov chains with arbitrary initial measures. *Acta Math. Sin. (Engl. Ser.)*, **21**(6), 1377–90.

Léonard, Christian. 1992. Large deviations in the dual of a normed space. Preprint.

Liu, Wei, and Wu, Liming. 2009. Identification of the rate function for large deviations of an irreducible Markov chain. *Electron. Commun. Probab.*, **14**, 540–51.

Lucchetti, Roberto. 2006. Convexity and Well-Posed Problems. CMS Books in Mathematics/Ouvrages de Mathématiques de la SMC, vol. 22. Springer, New York.

Meyer, Paul-André. 1966. Probability and Potentials. Blaisdell Publishing Co., Waltham.

Meyn, Sean P., and Tweedie, Richard L. 1993. Markov Chains and Stochastic Stability. Communications and Control Engineering Series. Springer-Verlag London, Ltd., London.

Neveu, Jacques. 1964. Bases Mathématiques du Calcul des Probabilités. Masson et Cie, Éditeurs, Paris.

Neveu, Jacques. 1972. Martingales à Temps Discret. Masson et Cie, Éditeurs, Paris.

Ney, Peter, and Nummelin, Esa. 1987. Markov additive processes II: Large deviations. *Ann. Probab.*, **15**(2), 593–609.

Nummelin, Esa. 1984. General Irreducible Markov Chains and Nonnegative Operators. Cambridge Tracts in Mathematics, vol. 83. Cambridge University Press, Cambridge.

Parthasarathy, K. R. 1967. Probability Measures on Metric Spaces. Probability and Mathematical Statistics, Academic Press, New York-London.

Rassoul-Agha, Firas, and Seppäläinen, Timo. 2015. A Course on Large Deviations with an Introduction to Gibbs Measures. Graduate Studies in Mathematics, vol. 162. American Mathematical Society, Providence, RI.

Revuz, Daniel. 1984. Markov Chains. Second edn. North-Holland Mathematical Library, vol. 11. North-Holland Publishing Co., Amsterdam.

Royden, Halsey L. 1988. *Real Analysis*. Third edn. Macmillan Publishing Company, New York.

Schaefer, Helmut H. 1966. Topological Vector Spaces. The Macmillan Co., New York.

Seneta, Eugene. 1981. Nonnegative Matrices and Markov Chains. Second edn. Springer Series in Statistics. Springer-Verlag, New York.

Stroock, Daniel W. 1984. An Introduction to the Theory of Large Deviations. Universitext. Springer-Verlag, New York.

Wu, Liming. 2000a. Some notes on large deviations of Markov processes. *Acta Math. Sin. (Engl. Ser.)*, **16**(3), 369–94.

Wu, Liming. 2000b. Uniformly integrable operators and large deviations for Markov processes. *J. Funct. Anal.*, **172**(2), 301–76.

Author Index

Subject Index

Printed in the United States
by Baker & Taylor Publisher Services